THE CHANGING NATURE OF
PHYSICAL GEOGRAPHY

THE CHANGING NATURE
OF PHYSICAL GEOGRAPHY

K. J. GREGORY

Emeritus Professor of Geography
University of London

A member of the Hodder Headline Group
LONDON
Co-published in the USA by
Oxford University Press Inc., New York

First published in Great Britain in 2000 by
Arnold, a member of the Hodder Headline Group,
338 Euston Road, London NW1 3BH

http://www.arnoldpublishers.com

Co-published in the United States of America by
Oxford University Press Inc.,
198 Madison Avenue, New York, NY10016

British Library Cataloguing in Publication Data
A catalogue record for this book is available from the British Library

Library of Congress Cataloging-in-Publication Data
A catalog record for this book is available from the Library of Congress

ISBN 0 340 74118 X (hb)
ISBN 0 340 74119 8 (pb)

1 2 3 4 5 6 7 8 9 10

Production Editor: Anke Ueberberg
Production Controller: Fiona Byrne
Cover Design: Patrick Knowles

Typeset in 10/12pt Sabon by Phoenix Photosetting, Chatham, Kent
Printed and bound in Great Britain by MPG Books Ltd, Bodmin, Cornwall

What do you think about this book? Or any other Arnold title?
Please send your comments to feedback.arnold@hodder.co.uk

Contents

Part V   CONCLUSION

Preface

Liphook has provided an excellent environment in which to complete much of this book, in the locality where Gilbert White wrote his very influential *The Illustrated Natural History of Selborne* (White, 1789), although I cannot aspire to his level of achievement. An invitation from the publishers to consider writing a new version of *The Nature of Physical Geography* came at the time when I had decided that I could retire early from my post as Warden of Goldsmiths College University of London, but before my intention was widely known. It was therefore a propitious moment when I could anticipate not only infinite amounts of available time but also a relatively easy first project to accomplish. Neither has proved to be the case. As many others have found, so-called retirement leaves less time than one at first expects and, what is more, it is even more difficult to survey the whole of physical geography than it was in 1985. In particular I am extremely grateful for the award of an Emeritus Leverhulme Fellowship, to help accomplish the work that I retired to achieve, because this has meant that I have been able to return to my interest in river channels from the management and quality issues that were necessarily prominent during my time at Goldsmiths. It remains for the reader, and the reviewers, to judge whether, as a result of these circumstances, it has been advisable to write a new book on *The Changing Nature of Physical Geography*.

Inevitably some of the present volume reflects, and relates to, the contents of the 1985 book on *The Nature of Physical Geography*. However, so much has changed since writing the 1985 book, and continues to do so, that it seemed appropriate to modify the title to *The Changing Nature of Physical Geography*. Periods as Dean of Science, as Deputy Vice Chancellor, as Warden, as Chair of Research Assessment Exercise Panels in 1992 and 1996, and as UK geographer on the Royal Society International Unions Committee (1995–1999) have inevitably influenced and coloured what is now written. I have been fortunate to have benefited from the contacts with many colleagues in these various activities. In addition to recognizing all the individuals mentioned in the preface to the 1985 book, I wish to gratefully acknowledge all of the professional interaction and friendship with academic colleagues and research students over many years, all of whom have contributed very significantly to making my career enjoyable and durable.

A number of these people have been good enough to comment upon the book outline, the synopsis, particular chapters, or to discuss particular aspects, and I am grateful to Professor Tony Brown, Professor Mike Clark, Professor Paul Curran, Dr Peter Downs, Professor Vince Gardiner, Professor Will Graf, Professor Angela Gurnell, Professor Janet Hooke, Professor John Lewin, Professor Geoff Petts, Professor Bruce Rhoads, Dr David Sear, Professor Leszek Starkel, Professor Colin Thorn and Professor Des Walling. I am also grateful to the University of Southampton, particularly to Professor Paul Curran, for the facilities that were provided to me as a Visiting Professor, and to the University of Birmingham where I am an Honorary Professor. The resources of the University of London library have been much appreciated on frequent occasions. Help from my son, Dr J.K. Gregory, is much appreciated as he established the PC that has been vital to the creation of this text and he responded with good humour to my queries as I retrained – all a part of life-long learning. Other members of the family have been very supportive: Caroline, Sarah, Raj, Ginnie and Tim (for his diagram scanning expertise), and also our wonderful grandchildren. It is customary to thank one's wife and this I do, but not in any routine way; Chris has supported all the years leading up to this volume, has provided exceptionally high quality sustenance during its preparation, has carefully read and corrected the drafts, and then helped with the finalization of the text. With support like this the book ought to be good!

This book is written in the knowledge that its execution is an extremely difficult, if not impossible, one. I appreciate many of the book's limitations, but I would be delighted to hear from readers of their reactions. Possibly all that can now be realistically provided, in view of the amount of information and literature available so readily and rapidly to the physical geographer, is a guide to some of the literature from a particular time-dependent perspective. It is perhaps impossible to improve upon the sentiments of Gilbert White at Selborne on 1 January 1788:

> If the writer should . . . by any means, through his researches, have lent an helping hand towards the enlargement of the boundaries of historical and topographical knowledge . . . his purpose will be fully answered.

Ken Gregory
Liphook, Hampshire
1 October 1999

Acknowledgements

The publishers would like to thank the following for permission to include figures which are copyright and are used by permission:

Addison Wesley Longman Limited (Fig. 4.4); *American Journal of Science* (Fig. 6.3; Fig. 7.1C); Philip Allan, publishers of *Geography Review* (Fig. 7.3); Professor N. Arnell and Institute of Hydrology (Fig. 9.3); Edward Arnold (Publishers) Ltd. (Fig. 3.2, Fig. 9.1); Blackwell Publishers (Fig. 4.2); Cambridge University Press (Fig. 6.1); Professor G.M. Foody and Professor P.A. Curran (Fig. 9.2); Kluwer Academic Publishers (Table 8.3); Oxford University Press and Methuen Publishers (Fig. 6.5); Prentice Hall Inc. (Fig. 4.1; Fig. 6.2); Routledge Publishers (Fig. 3.4, Fig. 5.1A); Routledge Publishers and Professor I.G. Simmons (Fig. 3.3B); Royal Geographical Society (with the Institute of British Geographers) (Fig. 4.3B); Royal Scottish Geographical Society (Fig. 9.4); Springer-Verlag and Dr R.J. Huggett (Fig. 3.3A); Professor D.E. Walling (Fig. 5.2); John Wiley & Sons Ltd (Fig. 4.3A, Fig. 5.1B and C; Fig. 7.1, Fig. 7.2, Fig. 7.4).

Full citations may be found by consulting the captions and the References.

PART

I

INTRODUCTION

PART

I

INTRODUCTION

|1|

Prologue

This chapter suggests how this book complements others recently produced (1.1), outlines background considerations (1.2), and concludes with an approach to physical geography expressed as a physical geography equation (1.3), and an explanation of the book structure (1.4).

1.1 Introduction to the literature

Presumption was included in the title of the first chapter in *The Nature of Physical Geography* (1985) because it was admitted from the beginning that it is indeed presumptuous for one physical geographer to attempt to survey the discipline with insight into its major subdivisions. However, we expect our students to cope with the breadth of the discipline and so this book attempts to do so. The original book was conceived after a suggestion from the publishers that there should be a book to do for physical geography what *Geography and Geographers: Anglo-American Human Geography Since 1945* (Johnston, 1979, 1997) has done for human geography. In 1985 it was not easy to provide a survey of physical geography, which has not been affected so obviously by several paradigms and in which most philosophical debates have focused on its component parts, possibly reflecting the fact that unlike human geographers, the physical breed have not until recently seen themselves in an integral way and have generally been much less concerned with the philosophy of their discipline.

Haines-Young and Petch (1986) focused on the nature and methods of physical geography, providing a volume that was, with its numerous illustrations, entirely complementary to *The Nature of Physical Geography* (Gregory, 1985). Clark, Gregory and Gurnell (1987) felt that it was an appropriate time to review *Horizons in Physical Geography* to complement a similar volume produced for human geography (Gregory and Walford,

1989) because it was then more than 20 years since Chorley and Haggett had laid the foundation for the 'new geography' with *Frontiers in Geographical Teaching* (1965) and *Models in Geography* (1967). In the *Horizons* volumes it was argued that a hallmark of geography in the late 1980s was its diversity, and so in both volumes topics and themes were identified such that particular authors could approach them from their own perspective. In the concluding chapter, Clark, Gregory and Gurnell (1987) suggested that three recurrent themes featured in the priorities for physical geography. Firstly, physical geography had become, and should remain, an effective natural science with a strong reliance on the development and application of accurate monitoring, analytical and modelling techniques, with the focus on process remaining as strong as that on form, although the emphasis placed upon understanding of material properties appeared overdue for greater emphasis. Secondly, a diagnostic characteristic of physical geography in 1987 was identified as a revival of interest in the concepts that motivate investigation and explanation, where the elucidation of temporal behaviour with important new models for temporal change were more evident than spatial concepts. A third dominating theme was suggested to be the massive swing towards management-consciousness and the associated interest in socioeconomic attributes. Few 'natural systems' are independent of socioeconomic influence, so the broadening of interest had rendered physical geography much better equipped to comment on real as opposed to ideal landscape or atmospheric systems. Global environmental change (Chapter 9) has now received considerable attention, and Slaymaker and Spencer (1998) proposed that physical geography needed to rediscover its original mandate to focus on the study of the interlinkages between society and environment and between spheres of environment (see Chapter 9, p. 232). They suggested (Slaymaker and Spencer, 1998, Fig. 1.3) the existence of a sparsely inhabited niche between the earth system science approach to global environmental change and the prevailing approach to physical geography (see Chapters 4 and 11, p. 100, 286).

These publications are amongst the comparatively few books that have focused since 1985 on the nature of physical geography. In the branches of the subject, especially geomorphology, the project instigated by R.J. Chorley to provide the *History of the Study of Landforms* has realised three impressive volumes (Chorley *et al.*, 1964, 1973; Beckinsale and Chorley, 1991), and more recently Rhoads and Thorn (1996) edited an impressive volume on *The Scientific Nature of Geomorphology*. Earlier books dealing with geomorphology included *The Earth in Decay: A History of British Geomorphology 1578–1878* (Davies, 1968), focusing on early developments in Britain; a book by Pitty (1982) began its life as early chapters in a geomorphology textbook which were subsequently extracted and updated to provide a survey of the nature of geomorphology; and Tinkler (1985) published a short history of geomorphology and then a history of geomorphology from Hutton to Hack (Tinkler, 1989). Since the first international geomorphological conference in 1985, a number of publications have been stimulated by the International Association of Geomorphologists as exemplified by the national surveys collected by Walker and Grabau (1993). Annual Binghamton conferences stimulated a series of volumes, many of which provide benchmark surveys of parts of geomorphology (e.g. Vitek and Giardino, 1992).

Other volumes published since 1985 have focused on branches of physical geography, on subdivisions of the branches, or have been stimulated by individuals, such volumes providing insight into developments and trends. One example is the volume edited by Costa *et al.* (1995) as a tribute to 'one dimension of a multidimensional and sagacious man', M. Gordon Wolman. Also in 1995, Professor Des Walling and I were invited to a dinner at the University of Birmingham each expecting a book to be presented as a surprise to the other! In fact, two edited volumes had been compiled in great secrecy, with contributions covering sediment and water quality in river catchments (Foster *et al.*, 1995) presented to Des, and changing river channels (Gurnell and Petts, 1995) to me. These volumes implicitly give much insight into the development of hydrogeomorphology up to 1995, but the relatively small number of such volumes indicates that the majority of physical geographers still do not wish to venture too far into philosophical issues or matters concerned with physical geography as a whole. One consequence is that students, for example when preparing for examinations, may have to be reminded that where physical geography appears in a question, it is not synonymous with geomorphology.

There has been more discussion of the nature of human geography and of geography as a whole. *Geography and Geographers: Anglo-American Human Geography Since 1945* had by 1997 reached its fifth edition, and at the end of Chapter 2 on geography in its 'modern period', three approaches are identified, namely exploration, environmental determinism and possibilism, and the region and regional geography, and in this last named there is some mention of physical geography (Johnston, 1997). *Explanation in Geography* (Harvey, 1969) was of very considerable significance, achieving a substantial impact, and subsequently in *Geography, Ideology and Social Concern*, Stoddart (1981) distinguished the history of a subject as a chronology of events from the history of geographical ideas. A number of books have focused on change and challenge, especially at global scales, including some related to branches of physical geography (e.g. Bennett and Estall, 1991; Johnston, 1993; Johnston *et al.*, 1995). In a seminal volume on *The Geographer's Art*, Haggett (1990) considered the purpose and practice of geography, identifying three essential geographic characteristics which make up a geographic trinity, namely emphasis on location, on land and people relations, and on regional synthesis. He contended that this threefold character of geography is now widely accepted, although individual geographers or University departments will lay emphasis on one aspect of the trinity at any one time. Haggett (1983, 1990) showed that geography is particularly dependent on the flow of concepts and techniques from other sciences, including

- earth sciences group – including geomorphology, climatology, Quaternary studies;
- ecology group – including biogeography, natural resources, conservation, resource planning;
- regional science group;
- area studies group;
- urban studies group.

Environmental studies link the earth sciences and ecology groups. Other sciences listed are geology, meteorology, biology, planning, engineering, mathematics and computer science, as well as links to human geography, econometrics, sociology, economics, anthropology, history and linguistics (see also Section 1.2 below). Bird (1989) included reference to physical geography, viewed the discipline of geography as a whole, advocated a form of critical rationalism within which he expressed PAME, which was the acronym for Pragmatic Analytical (combined) Methodology Epistemology, and did not agree with Johnston (1983c) that the links between physical and human geography were so tenuous. A Rediscovering Geography Committee established by the National Research Council in the USA published *Rediscovering Geography: New Relevance for Science and Society* (National Research Council, 1997) as the first comprehensive assessment of geography in the US for almost 30 years, and as a reflection of the renaissance in the discipline of geography in the previous decade. Whereas geography in the US had not previously embraced all physical geography, in 1997 it was rediscovered that 'Physical geography has evolved into a number of overlapping subfields although the three major subdivisions are biogeography, climatology, and geomorphology' (National Research Council, 1997), although stimulating ideas had been presented eight years (Gaile and Willmott, 1989) and four years previously (Abler *et al.*, 1992). Stoddart (1987) attempted to suggest how to reclaim the high ground with a vision of '. . . a real geography – a reinvented unified geography . . . and at the same time a committed geography. It is a geography which reaches out to the future . . .' (Stoddart, 1987, p. 333). Thus Stoddart urged that the integrity of geography as a discipline should be maintained, and continued to stress the need for comprehensive awareness. Debates in the literature can often illustrate the polarities which exist (e.g. Richards and Wrigley, 1996; Stoddart, 1996).

In addition to geographers writing about their discipline, it is necessary to be cognizant of developments in other disciplines, these being of three kinds. Firstly there are developments in related disciplines including those in geology, environmental sciences, biology, meteorology and engineering, for example (see Section 1.2). Secondly, there are developments in science and in methodological thinking which relate to all disciplines and, for example, the advent of paradigms and models of scientific thought (Section 1.2, developed in Chapter 3). Thirdly, there are more popular, general developments catalysed by public interest in the environment. Disciplinary boundaries are erected for convenience, and the study of physical geography or of any other discipline should not be dominated by any particular viewpoint, approach or methodology to the exclusion of all others. In his contribution to a volume of essays collected in honour of Professor R.J. Chorley, Slaymaker (1997) advocated a pluralist problem-focused geomorphology, and concluded that:

> Geomorphology in the 1990s has encouraged little philosophical debate because normal science is assumed to be adequate to provide explanation in geomorphology. There is no recognisable central concept in geomorphology and we have no problem focus.

He applauded the vision of Chorley and we would all now agree about the need for pluralism and for philosophical debate (e.g. St Onge, 1981).

As in the 1985 book on *The Nature of Physical Geography*, it is necessary to avoid spreading the net too widely and suggesting that physical geography embraces work traditionally and properly undertaken by other disciplines. Some people have proseletyzed geography, including physical geography, as an integrating discipline embracing a number of other disciplines which may not wish to be embraced, or indeed have sufficient respect for the pretender doing the embracing. I believe with many others since the salutory paper by Schaefer (1953) that geography is a discipline which is not exceptionalist (Johnston, 1997, p. 55) in the sense that it incorporates several others. The view that will emerge in this volume is that there is a distinctive core area of study appropriate for physical geography, and that physical geographers are able to utilize one of several approaches which are the themes for chapters 4 to 8. Students of the subject should find this reassuring because there is no imperative for them to be superhuman: they are not expected to study, be adept in, and to integrate a whole range of disciplines, but they do need to be aware of the way in which physical geography relates to human geography and to other disciplines.

1.2 Background considerations and context

Of the publications noted above in physical geography, in its branches, and in geography as a whole, particularly those produced since 1985, no single volume has focused on physical geography and the way in which it has developed. *The Changing Nature of Physical Geography* was chosen as the title for a new book because it maintains a sufficient link with the original (Gregory, 1985) but also introduces a focus on change which was particularly apparent in the last two decades of the twentieth century and will certainly continue to affect the discipline in the future. This book surveys not only how the present position of physical geography has arisen and how particular approaches have developed, but also attempts to provide a basis to appreciate how further changes might take place.

Seven topics can give some background and context for past developments and the present character of physical geography; topical issues for geography as a whole are referred to in Haggett (1996). *Definitions* suggest how the subject is delimited by its practitioners, and four given in Table 1.1 indicate that physical geography is concerned with the surface of the Earth, that the subject has been increasingly subdivided into separate branches so that there is a danger of the whole being lost sight of, and that there are recent developments or new disciplines rediscovering physical geography. A recent reorientation of the focus of physical geography (e.g. Slaymaker and Spencer, 1998) has seen a global, energy-based approach and a focus on sustainable development becoming increasingly apparent. One critical issue is that physical geography could become separated from human geography and isolated from the discipline of geography as a whole. This was a con-

Table 1.1 Some definitions of physical geography (compare with p. 9 and p. 288)

Publication	Definition
Gvozdetskiy *et al.* (1971)	Physical geography in the USSR has evolved as a separate discipline with its own object of study and its tasks and internal structure; it is ... a discipline concerned with the geographical shell (the landscape sphere) of the Earth and with its constituent natural geosystem (the geographical, or natural, territorial complexes).
Brown (1975)	... physical geography is at the present time internally unbalanced, and centrifugal in character – to put it dogmatically, geomorphology plays too dominant a role in the subject ... Physical geography as an integrated subject has been rediscovered by non-geographers under the guise of environmental science, and ... physical geographers have tended to hold aloof from this development.
Goudie *et al.* (1985)	... the whole field of physical geography including biogeography, climatology, ecology, geomorphology, hydrology, pedology and the Quaternary.
National Research Council (1997)	Physical geography has evolved into a number of overlapping subfields, although the three major subdivisions are biogeography, climatology, and geomorphology.

cern of Stoddart (1987) when he disagreed with the view of Johnston that the links between physical and human geography were tenuous, and insisted that it was necessary to reclaim the high ground to ensure that certain major world problems were not ignored by geographers in general and by physical geographers in particular. Stoddart (1987) saw a great danger that the discipline would disintegrate into its component specialisms and be absorbed by neighbouring disciplines. However, by 1992 it was suggested (Gregory, 1992) that change taking place in physical geography needed to be viewed in relation to changes in the physical environment, to the need to study such changes, and that this was actually strengthening the links with human geography.

An audit at that time indicated that activity in different branches of physical geography was still uneven, that further energetic approaches were required, and that physical geography could become more involved in environmental design. The issues that have been raised about the definition of physical geography are therefore:

- its extent – how much of hydrology does it include, for example, and should it extend to other planets?
- its balance – although equal effort should be accorded to all aspects of the environment, some branches of physical geography have been more actively pursued than others;
- the relationship with human geography;
- relationships with other disciplines.

In the light of the above discussion, a working definition of physical geography is therefore suggested as:

Physical geography focuses upon the character of, and processes shaping, the land-surface of the Earth and its envelope, emphasizes the spatial variations that occur and the temporal changes necessary to understand the contemporary environments of the Earth, and its purpose is to understand how the Earth's physical environment is the basis for, and is affected by, human activity. Physical geography is conventionally subdivided into geomorphology, climatology and biogeography, its study requires expertise in mathematical and statistical modelling and in remote sensing, and it benefits from collaborative links with many other disciplines such as geology, biology and engineering. In many countries, physical geography is studied and researched in association with human geography.

The definition is considered again in Chapter 11 (p. 288).

The three-fold *subdivision of physical geography* (National Research Council, 1997) has become less clear (Table 1.1). Thus *Progress in Physical Geography* 'publishes reviews of current research and theoretical developments in any aspect of geomorphology, climatology, biogeography, and human–environment interaction'. The *Encyclopaedic Dictionary of Physical Geography* (Goudie *et al.*, 1985) goes further to include hydrology

and the Quaternary, and subdivides biogeography into ecology and pedology. Brown (1975) concluded (Table 1.1) that in the 1970s physical geography was internally unbalanced, with a major dominance exerted by
geomorphology. The internal imbalance still persists and is reflected in the
subjects of the progress reports included in *Progress in Physical Geography*
over the period 1977 to 1997. Over 20 years, progress reports have varied
in number from 10–12 each year to as many as 21 in 1986. Taking the 17
themes that have occurred in at least five years, geomorphology accounts for
40%, climatology for 24%, and soils, biogeography, oceanography and
remote sensing for about 6% each, with environmental issues representing
some 12%. Another source of the breadth and balance within contemporary physical geography is *Physical Geography Abstracts*, published in
monthly issues. The 12 issues in 1995 provided 15,582 abstracts, and in
1996 there were 16,100 abstracts – a substantial growth since the 1960s
when there were fewer than 2000 articles abstracted each year. Whereas
some 16,000 abstracts per year are provided for physical geographers, the
original publications are not all written by physical geographers. In 1995
the broad categories of abstract in *Physical Geography Abstracts* indicated
that:

- synoptic geography accounted for 5.7% of the abstracts;
- meteorology and climatology for 23.1%;
- hydrology for 16.9%;
- remote sensing for 15.9%;
- soils for 13.4%;
- sedimentology for 12%;
- the Quaternary for 7.4%;
- landforms for 5.6%.

This breakdown is rather different from the subject emphasis of progress
reports in *Progress in Physical Geography*, and probably very different from
the balance of physical geography familiar to many students, certainly those
in British universities. These abstracts employ a classification that not all
would choose, and landforms includes some geomorphology but other
aspects may be included under the Quaternary, sedimentology or hydrology
categories. The abstracts are edited from the School of Environmental
Sciences, University of East Anglia, where there is no department of geography, and biogeography is included in *Ecological Abstracts*.

Whereas the contents of *Physical Geography Abstracts* give an idea of
the published material that could be of relevance to physical geographers,
we can gain some further idea of the relative emphasis upon different
branches of research within the subject by a content analysis of recent journals, or by looking at the specialisms of staff in the different branches of
physical geography in university departments. In the period 1988–1997,
papers published in *Progress in Physical Geography* included 40% on
broadly geomorphology and hydrology, 14% on climate and climate

change, 13% on biogeography and 3% on soil geography, but 30% do not fit into the original branches because 8% are remote sensing and other techniques, and 17% cross-branches. In *Australian Geographer* for the same period, geomorphology papers are dominant (36%) but there are also a number (40%) of papers that focus on the Quaternary, management, global change, and cross the traditional branches. In the subdivisions of physical geography it is apparent that emphasis has been placed upon smaller subdivisions of the subject – the reductionist or fissiparist approach; that physical geography constitutes less than half of the discipline of geography as a whole; and that there is certainly an uneven degree of emphasis across the discipline. This must lead us to question whether the emphasis is wrong, whether major global problems might have been ignored, and how the composition of physical geography varies from one country to another.

There are *variations between countries* in the way that physical geography is studied. Although it is still a major component of university geography departments in Britain, Canada, Australia and New Zealand, it has had varying fortunes in university departments in the US. In the 1950s, US geography departments tended to develop primarily as departments of human geography, and the physical geography included was overtly concerned with environmental description, including what was described as landform geography (Zakrzewska, 1967) – study of environmental evolution was the domain of researchers who were usually found in US geology departments. Subsequently in the 1970s, with the advent of studies of landscape processes, there was a revival of physical geography in university geography departments of North America. However, it is still the case that some of the best known contributors to the physical geography literature have had their professional affiliation in other disciplines. Thus L.B. Leopold was a Professor of Geology and Landscape Architecture at the University of California, Berkeley; S.A. Schumm was a Professor in the School of Forestry and Natural Resources at Colorado State University, Boulder; M.G. Wolman was Professor of Environmental Engineering at Johns Hopkins University, Baltimore; and A.N. Strahler was Professor of Geology at Columbia University, New York. The way in which geomorphology emerged as a science in the US in the late nineteenth century, when geology and geography were closely related disciplines, was summarized by Costa and Graf (1984). They showed how geography was initially dominated by physical geographers and geomorphologists who had been trained as geologists, but by the late 1930s geomorphology had declined in importance and the majority of geomorphologists in the US were no longer located in geography departments. In the 1980s geomorphology was no longer required of undergraduate geology majors in many US universities, and this decline coincided with the increasing importance of physical geography in geography departments, embracing not only geomorphology but also natural hazards and natural resource management. In the US physical geography is much less evident in many university geography departments than it is in departments in the UK, and the position of the subject as a whole is less strong because prior to higher education, geography is not a subject which large numbers of students have already studied. Students in the US have to be attracted to geography, whereas in the UK the subject is one of the five most popular subjects for students in the 16–18 age group (p. 285).

Variation is found from country to country in the extent to which geography is a single university department. In English-speaking countries there has been a long tradition of geography as a single discipline, with teaching and research concentrated in the university setting. However, since the 1970s, there has been a trend to establish new departments of environmental sciences which include many of the subdisciplines that have traditionally been studied as part of geography. A further variation which was originally very significant in eastern Europe and Russia was for university departments to undertake geographical teaching and learning, whereas much of the prime research was located in the National Academy. Thus in Poland the Polish Academy of Sciences was the organization that undertook the best-funded research, and the separation of Academy and University was a distinction which persisted until the early 1990s. Although it is necessary to appreciate the different ways in which physical geography is organized, there is strength in diversity, and so it is beneficial for the subject to be studied in particular ways in specific countries. Environment itself has had an influence, of course. In the UK, with a very rich and detailed stratigraphic record, geology developed as a discipline by largely ignoring the record of the Quaternary and some of the Tertiary – geologists had enough to contend with elsewhere. However, in the US the stratigraphic record was less complete and Quaternary deposits, as well as the results of Quaternary history, were much more extensive, so geologists naturally turned their research interests in that direction.

In the early twentieth century it was usual for *the training of physical geographers* to occur in other disciplines, simply because geography degrees including physical geography were not available. Thus the climatologist A.A. Miller was initially a graduate in mathematics, and S.W. Wooldridge took his first degree in geology, it being on the basis of his geological research that Wooldridge was elected FRS in 1969. L.B. Leopold took his first degree in civil engineering and his doctorate in geology. More recently it is not unknown for physical geographers to take their first degrees in subjects other than geography, M.J. Kirkby being, for example, a graduate in mathematics. Although the majority of physical geographers in the English-speaking world are now usually trained in university geography departments, variations between countries mean that some training, especially in aspects of information technology and remote sensing, may occur in other university departments, especially at postgraduate level. The number of Masters degree courses has increased, and the boundaries between disciplines have become more permeable so that it is no longer unknown for biologists or mathematicians to migrate to train and undertake research in physical geography. Postgraduate training to doctoral level may be achieved exclusively within a university department of geography, but often there are collaborative supervision arrangements with other university departments such as geology, biology or archaeology, or there may be collaborative arrangements with external organizations. Thus, in the UK, research students trained in a university geography department may be supported by a studentship which involves research supervision from staff of national environment organizations, such as the Environment Agency or the Meteorological Office, or consultancy firms. In the US and elsewhere it is not unusual for the training of physical geographers to be associated with consultancy projects. This is to be commended in view of the increasing emphasis in physical geography towards applied problems.

Data available to, and used by, physical geographers can be of several kinds. First and most obvious is empirical data, derived from observation or from experiment, and including information derived from fieldwork, or from laboratory analyses that succeed fieldwork, in the investigation of soils, for example. Data may be collected by national agencies, and meteorological statistics, climatic data or hydrological data are instances of this kind. The use of remote sensing has opened up a very large and exciting new source of data for physical geography (Chapter 3, p. 68), and this provides an information resource which has revolutionized research throughout physical geography to varying degrees. A second major category of data is provided by modelling, furnished by mathematical or statistical models or by physical or hardware models. An atmospheric mathematical model offers an example of the first, whereas physical hardware models have often been constructed for coastal environments and can provide data relating to strictly controlled situations. Such modelling approaches depend to some extent upon knowledge of an empirical kind, and if data have been collected for several decades it is often necessary to see how those data can be used in a mathematical or statistical model to provide information for a much longer period of time than the observational record (see Chapter 5, p. 131).

Data, however generated, cannot be used in research without a conceptual approach and a scientific framework (Chapter 3, p. 56). It is therefore necessary to be aware of the approaches to science over recent decades; there has been a series of philosophical scientific approaches which have necessarily had an impact in physical geography. Kuhn (1962) introduced the idea of paradigms as an integral part of his book on *The Structure of Scientific Revolutions,* using the term paradigm to express a generally accepted set of assumptions and procedures which are useful in both the subjects and methods of scientific enquiry. Certain paradigms or particular approaches may dominate at any one stage of development of the discipline, so that paradigmatic shifts may be detected from one approach to another. This means that when reading the work of physical geographers it is important to remember that the time when the research was done or the script written can have a major influence on the approach that was taken; equally the particular standpoint of the author can be influential. Dominance of approaches, or paradigms, enabled the idea of what is fashionable to be explored in a stimulating paper by Sherman (1996), by adopting the concept of fashion change proposed by Sperber (1990). Sperber contended that changes in the goals, subjects, methods and philosophies of science can often be attributed to the emergence of an opinion or what he sees as a new fashion leader. The fashion process relies upon fashion dudes to advance their disciplines (and Sherman also includes their careers!), and in relation to geomorphology Sherman (1996) instanced Davis, Gilbert, Strahler and Chorley as fashion dudes. The reader could doubtless think of other fashions and of other individuals who have initiated and exploited fashions in physical geography.

Related to the progress of physical geography it is also important to think of the available literature for reference. The substantial growth in the literature is indicated by an increase from some 2000 abstracts each year in the 1960s to 16,000 per year in the late 1990s. Chorley (1987) argued that perhaps the single most important fact regarding hydrology in general, and geographical interest in the hydrosphere in particular, is the rapidity in its growth. He noted that the *Journal of Hydrology* in 1963 was 22 mm thick, but in 1983 it was 126 mm thick. The process has not ended, however, because in 1997 the same journal was 183 mm thick. The literature of physical geography can also be regarded as a data source in itself. In addition to books and edited volumes, there are four types of scientific research journal possibly of particular interest to physical geographers. Firstly there are the journals directed to the field of geography as a whole, which include physical geography contributions. Secondly there are journals that are essentially physical geography serials. Thirdly there are journals that deal with a sub-branch of physical geography and may be largely the outlet for, and edited by, physical geographers, or they may involve other disciplines. Such journals are numerous and continue to increase in number, with significant additions such as *Regulated Rivers* (1987–) and *Hydrological Processes* (1986–). Fourthly, there are journals primarily associated with other disciplines in which physical geographers may publish the results of research, ranging from general periodicals such as *Nature, Science*, or the *Philosophical Transactions of the Royal Society*, to those which are discipline-associated, such as *Weather* or *Journal of Soil Science*, and some which embrace more than one discipline such as *Environmental Management, Catena* or *Earth Science Reviews*.

A number of *influences upon physical geography* affect all disciplines. Firstly there are strands associated with what may be described as the intellectual climate of the times. Physical geography must be seen as in no way divorced from mainstream scientific philosophical thought (Chapter 3, p. 48). Developments that can lead to paradigmatic shifts include the move towards quantification, the greater advent of information technology (IT) with enhanced modelling potential, particularly important in relation to climatology, and the development of specific theories such as chaos theory. The data explosion, particularly utilizing remote sensing but also encompassing geographical information systems, has been complemented by the possibilities offered by the Internet as explained by A. Perry (1998). In relation to data for the climatologist, he noted that the Internet has emerged as one of the most significant forces offering access to vast global collections of international computer networks, which must be the first choice from which researchers acquire data and exchange information. He quotes the fact that in a recent survey of the 13 most popular words entered into one of the most widely used search engines, 'weather' was at number 10, while the previous nine all referred to sex and pornography! A further strand in the intellectual climate of the late twentieth century has been the search for relevance and the desire for further applications of physical geography research. To the forefront of this has been scientific interest in global warming and global change. Awareness of events such as the Kyoto Climate Conference in December 1997, described as the world climate summit, is important, and Hulme (1998) contended that this was the single most important event bringing together science and the politics of global climatic change.

Two other influences particularly evident in recent years concern funding and future careers. Funding has become of increasing concern because, particularly with the expansion of higher education, it has not been possible to fund all disciplines and their research at the same unit level. In the UK, as the number of students in higher education increased from approximately 8% of the age group in the 1970s to some 33% of the age group in the late 1990s, there have been cuts, or so-called efficiency gains, imposed upon universities, which basically means that the unit of resource affecting the teaching and research of physical geography decreased substantially. One way of redistributing funding resource in the UK has been based on the Research Assessment Exercises, whereby every four years all departments in every university are invited to submit the details of their research-active academics, and the ratings of research are employed to apply funding differentially. Similar pressures have affected research funding organizations, who have not been able to match the increases in demand for research funds.

Greater awareness of the need to link to the requirements of external organizations has arisen for two reasons. Firstly, established physical geographers have looked to consultancy and to sources other than research councils and national academies to fund their research investigations. This has often been a very positive development because it has encouraged research on applied topics for which organizations could justify funding. A second reason is that students, when confronted with choice between courses, may have tended to vote with their feet and select for advanced study those courses that they perceive to be the ones that will inculcate the most useful skills and perhaps lead them towards the best employment prospects. The influence of such student choice has affected the pattern of university courses provided in physical geography, and hence the balance of academic staff expertise.

The way in which physical geography is perceived may change, and Gregory (1999) differentiated perceptions by physical geographers themselves, by human geographers and geography in general, and by other disciplines. The major trends are that physical geography was once unevenly balanced, but this has been changed to some extent, particularly by greater concentration on human impact and on studies that are not restricted to one of the three major branches but are more concerned with physical environment as a whole; physical geographers have become more in tune with the climate of the times in their research agendas; and the standing of physical geographers as viewed by other disciplines has been enhanced, so this gives considerable potential for development in the future (Gregory, 1999).

Just as Lowenthal (1961) argued in the field of behavioural geography that the world of experience of an individual is very parochial and covers a small fraction of the total available, so in physical geography it is the *personal viewpoint and experience* of particular environments, training in specific disciplines, or exposure to particular developments that can all condition the way a single individual approaches physical geography or a particular research problem within it. There are many examples of such emphases in the present and past writings of physical geographers. The landscape features and deposits of Poland were major considerations in the development of periglacial geomorphology in that country in the 1950s, and in the creation of the new journal *Biuletyn Peryglacjalny*, first published in Lodz in 1954, as a result of the leadership of the distinguished geomorphologist Professor Jan Dylik. A further example was provided in a review of American field geomorphology where Graf (1984, p. 78) suggested that spatial bias is a major element in geomorphological theory development because, with a relatively small number of researchers, publications by a few individual scientists can affect the development of the subject. By analysing the geomorphology papers in nine journals published in the US, Graf (1984) showed how 472 field localities for geomorphological research papers from 1817–1945 have a substantial bias, whereas the 469 field localities for 1946–80 showed a different spatial bias which was less pronounced. Graf concluded that modern American geomorphological theory was likely to be spatially biased and that the classic humid cycle of geomorphological theory was based on the assumption that many landscapes were similar to New England conditions or to those encountered in the western federal surveys (Fig. 1.1). It therefore seemed to Graf (1984, p. 82) that:

The Southeast, South, and the Basin and Range Province outside California and Oregon remain the invisible regions of American geomorphology . . . Field localities for future research opportunities might serve science best if they are located in some geomorphologic blind spots.

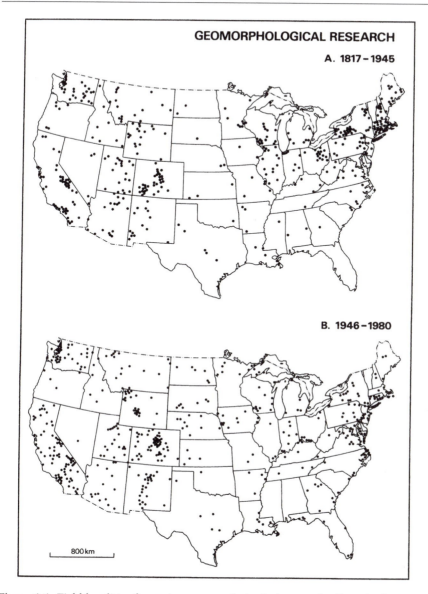

Figure 1.1 Field localities for major geomorphological research efforts in the US, as summarized by W.L. Graf (1984).

In any branch of physical geography it is therefore important to think what areas have been investigated and who has done the investigations; Gregory (1996) showed the world locations of published studies of palaeohydrology, underlining how palaeohydrological interpretation has been very much driven from the US and the UK.

1.3 An approach to physical geography

Physical geography has experienced a number of recent changes and continues to do so, and in common with all disciplines has had to confront a great explosion in the availability of information and research activity. One way of responding to this development is for each physical geographer to focus upon less and less (the reductionist or fissiparist approach), which can mean that more is known about less and less, with major world problems and issues absent from the physical geography research agenda. Alternatively it is necessary for every physical geographer to have some appreciation of the present range of physical geography, perhaps to specialize in a part of the discipline, but also to be in a position to appreciate how the subject should be developed to do justice to the major problems of the Earth's surface and the need for sustainable development.

To achieve such an understanding, a simple way of describing the purpose and objectives of physical geography was attempted by Gregory (1978a) using what was described as a physical geography equation, the idea of which has been accepted by a number of writers including Yatsu (1992). Physical geography is concerned with morphological elements or results of the physical environment (F), processes operating in the physical environment (P) and the materials (M) upon which processes operate over periods of time *t*. The physical geography equation was devised to indicate how physical geography focuses upon the way in which *processes* operating on *materials* over time *t* produce *results* which may be expressed as a landform, for example. Thus in the soil system, pedogenic processes operate on parent materials influenced by soil-forming factors to produce a soil profile or a soil landscape. In the same way a plant community can be visualized as the result of ecological processes in a particular environment. The approach is particularly appropriate for geomorphology as Yatsu (1992) agreed, and a landform such as a drumlin reflects not only the subglacial processes operating but also the materials available and the time involved. The application is somewhat different in the case of meteorology and climate because in those cases the result is the weather type or the climatic regime. However, in just the same way they are the result of atmospheric processes acting upon the available materials to produce a result.

For those who prefer the idea in equation form it can be expressed as:

$$F = f(P, M)\, dt \qquad (1.1)$$

This approach categorizes the types of investigation undertaken by physical geographers (Gregory, 1978a) into four levels:

- *Level 1: study of the elements or components of the equation* – study of the components in their own right. Some studies by physical geographers can be focused on the description (which may of course be quantitative) of landforms, climate types, soil character, or plant communities. The way of describing environmental components can require considerable innovation but is often preparatory to other levels.

- *Level 2: balancing the equation* – study of the way in which the equation balances at different scales and in different branches of physical geography. At the continental level this can involve demonstrating how the energy balance operates by relating available energy for environmental processes to radiation, and moisture received in relation to materials available locally. Studies of this kind focus upon contemporary environments and upon interaction between processes, materials and the resulting forms or environmental conditions.

- *Level 3: differentiating the equation* – includes all those studies that analyse the way in which relationships in the equation change over time and therefore how one equilibrium situation is disrupted and eventually replaced by another. This can be thought of as differentiating the equation, and these studies depend upon reconciliation of data obtained from different time-scales and upon some conceptual approach to adjustments of environment over time. Included is the impact of climatic change and also of human activity, because this may be the regulator that has created a control system.

- *Level 4: applying the equation* – when the results of physical geography research are directed to applied questions and problems. This very often builds upon the extrapolation of past trends into the future, and the difficulty is to extrapolate from particular spatial or temporal scales to other scales for which information is required to address management problems.

In 1978 it was thought (Gregory, 1978a) that all studies undertaken by physical geographers could be accommodated within this framework, but a fifth level can now be identified (see Chapter 10, p. 269):

- *Level 5: appreciating the equation.*

1.4 Method of approach

This volume is written to succeed *The Nature of Physical Geography* (Gregory, 1985), to provide an introduction to the development, present status, trends and potential of physical geography. I hope that it will stimulate young physical geographers to think about where the subject is going, or where it should be going, because they are the ones who will determine its eventual destiny. The book is structured in three major parts. The first (Chapters 2 and 3) outlines the development of physical geography by covering the century from 1850 to 1950 in Chapter 2, and the themes that emerged in the period 1950 to 1980 in Chapter 3. This provides the basis for Part 2, which includes five chapters each dealing with a major late-twentieth-century approach to physical geography, namely the environmental system (Chapter 4), environmental processes (5), temporal approaches and landscape chronology including the Quaternary (6), the impact of human activity (7), and applications and environmental management (8). Part 3 embraces trends for the millennium to bring together the earlier material in two chapters: one (9) on global physical geography, outlining the way in which there has been a welcome focus at the global level and also effective use of the materials provided especially by remote sensing; and the other (10) on cultural physical geography. The future of physical geography is briefly touched upon in a conclusion (Chapter 11). The complete bibliography is provided at the end of the volume, but several selected references are given in the further reading at the end of each chapter, selected to provide stimulating reading, and discussion issues are also offered.

It has been noted above, and referred to by many writers, that the literature available for any discipline is now vast, that the growth in the last two decades has been colossal, with a doubling rate of material available of around 5–7 years. Add to this the fact that, as A. Perry (1998) noted for climatology, the Internet can provide a major data source, then a major problem is that of selection of material. A few years ago my elder daughter gave me for Christmas a catechism of geography published in 1823 (Fig. 1.2), described as 'An easy introduction to the knowledge of the world and its inhabitants', and a volume of 70 pages was sufficient, less than two centuries ago, to reflect all the information known about the world – covering human as well as physical geography. A catechism is strictly 'instruction achieved by question and answer', and this surely epitomizes an approach that we should foster to increase learning about physical geography. It is the art of knowing what questions to ask, while appreciating what we do not know as well as what we do know, that provides a major challenge for the physical geographer.

Further reading

A useful volume for reference is:
> GOUDIE, A. *et al.* 1994: *Encyclopaedic dictionary of physical geography*. Oxford: Blackwell.

A systematic approach to many aspects of physical geography is provided in:
> CLARK, M.J., GREGORY, K.J. and GURNELL, A.M. (eds) 1987: *Horizons in physical geography*. Basingstoke: Macmillan.

A stimulating paper is:
> SHERMAN, D.J. 1996: Fashion in geomorphology. In B.L. Rhoads and C.E. Thorn (eds) *The scientific nature of geomorphology*. Chichester: John Wiley, 87–114.

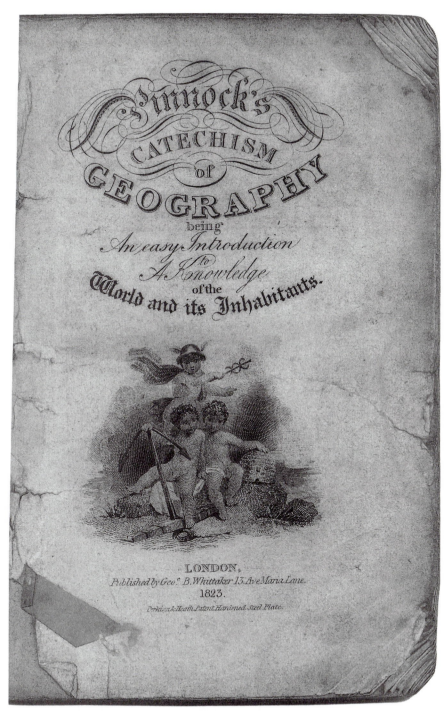

Figure 1.2 The title page of Pinnock's *Catechism of Geography*, published in 1823.

A particular perspective is provided in:

GREGORY, K.J. 1992: Changing physical environment and changing physical geography. *Geography* 77, 323–35.

An excellent perspective, largely from human geography, is provided in:

HAGGETT, P. 1990: *The geographer's art.* Oxford: Blackwell.

Topics for consideration

(1) Thinking of the position of physical geography as measured by the proportion of the academic staff, the emphasis of courses, and the amount of the curriculum devoted to physical geography in a department of higher education that you know, is the balance appropriate, has it changed, and should it change more?

(2) Is the position of physical geography within geography as a whole appropriate? Some research developments in the branches of physical geography have become so specialized that there is more contact and affinity with scientists in other disciplines than with human geographers. Is this appropriate or should the trend be redressed?

(3) Should the balance of activity within physical geography be uneven, with a heavy emphasis upon geomorphology – is this inevitable?

(4) Is the physical geography equation (p. 18) a sufficient framework to embrace all approaches to physical geography?

II

DEVELOPMENT OF PHYSICAL GEOGRAPHY

|2|

Establishing the foundation 1850–1950

This chapter outlines the development of physical geography for the hundred years up to 1950. In the nineteenth century, influential formative concepts (2.1) included uniformitarianism, evolution and exploration and survey; less rapidly assimilated were developments in science and conservation. These are outlined (2.1) and followed by a summary of the major formative developments in physiography and in the branches of physical geography (2.2), namely climatology, soil science, biogeography and geomorphology, including the impact of W.M. Davis (2.3). Trends of the mid-twentieth century are suggested before summarizing the growth which occurred between 1945 and 1950 (2.4).

Justification for this chapter may be found in the fact that understanding physical geography is just like understanding the environment: one needs to appreciate how and why the present state developed. By 1850 the discipline of geography was becoming established (see Fig. 1.2), and in the UK, physical geography, and physiography after 1877, was for a time one of the most popular high school subjects (Chapter 11, p. 285) and accounted for some 10% of the examination papers sat in English and Welsh schools (Stoddart, 1975; Holt-Jensen, 1999). The foundation of geographical societies and the creation of university professorships were clear signs of these early stages. In the early nineteenth century, many new scientific societies were founded, and the first geographical society was established in Paris in 1821, quickly succeeded by the German geographical society founded in Berlin in 1828, and the Royal Geographical Society in London in 1830. In addition to consolidating effort in an emerging discipline, the new societies were created to form an avenue for publication, each becoming associated with a major journal. Sir Clements Markham, a President of the Royal Geographical Society (1893–1905), pointed out in 1880 that, although the Royal Society had in theory published geographical work since 1662, only 77 of the 5336 papers published by that society between 1662 and 1880 could be called geographical. Each society that formed tended to develop its own distinctive focus; the Royal Geographical Society was significantly involved in exploration, but its journal was also the forum for many illuminating debates (Freeman, 1980). By 1866 there were 18 genuine geographical societies, and by 1930 the number had risen to 137. In North America the American Geographical Society was established in 1852.

University chairs in geography were created as the discipline became established, eventually forming separate university departments. In France a chair was created in Paris at the Sorbonne in 1809, and by 1899 there were chairs at five other French universities (Harrison Church, 1951), and most of the more important German universities had a professorship in geography. In the UK the first professor was Captain Alexander Maconochie at University College London from 1833–36 (Ward, 1960), and subsequently Halford Mackinder was appointed to the University of Oxford in 1887 and Yule Oldham to Cambridge in 1893. In the US the first professor of geography was Arnold Guyot appointed at Princeton in 1854, and by 1900 12 universities offered courses in geography, although not all had established geography as a part of their curriculum because physical geography was frequently taught in departments of geology (Tatham, 1951). As these new chairs were established the presence of a physical geographer could be very influential; in Germany between 1905 and 1914, physiography (see Section 2.2) was very prominent because of the influence of A. Penck who held the prestigious chair at the University of Berlin (van Valkenburg, 1951).

Although the growth of the discipline was stimulated by the foundation of societies and by the establishment of chairs and departments in universities, it was often the contribution of particular individuals that significantly affected the direction and rate of growth of the subject. W.M. Davis's contribution is outlined in Section 2.3, but many others had a significant influence upon American geography; of the 79 individuals included in a volume on leaders in American geography and geographical education (Barton and Karan, 1992), at least 11 had a significant influence upon the development of physical geography, including C.O. Sauer (1889–1975) who first enrolled for graduate study at Northwestern University in geology but quickly found that the emphasis on petrology did not meet his interest in the Earth's surface. He therefore transferred in 1909 to the University of Chicago where he was influenced by R.D. Salisbury (1858–1922) who since 1903 had organized and headed the first sizeable geography department in any American university, although the courses that he personally gave were listed in the university catalogue under geology (Barton and Karan, 1992). Sauer studied under H.C. Cowles (1869–1939), a pioneer of plant ecology in the US, who led students to interpret vegetation in the field just as Salisbury taught students to interpret landforms. Although Sauer became regarded by some as the greatest geographer of his time, geomorphology was not his forte

(Stoddart, 1996). However, his training illustrates the way in which many physical geographers were influenced: not just by the establishment of societies and the creation of university departments, but also by personal interaction with other disciplines. The effect of the prevailing intellectual climate, the 'climate of the times', could be as influential as the way in which some physical geographers received their training in Europe and then, on moving to the US, or later to Canada and Australia, were responsible for the dissemination of prevailing ideas which were often modified and adapted to the new environment. Sometimes the influence of a group of individuals could be particularly important, and the foundation of the Institute of British Geographers in 1933 was a response by a group of individuals to a perceived need (Wise, 1983). Similarly the brief appearance of a *Journal of Geomorphology* (1938–1942) was the result of the efforts of a number of enthusiasts including D.W. Johnson, and its eventual successor *Geomorphology* (1990) was initiated by Professor Marie E. Morisawa.

2.1 Influences upon physical geography

To understand the early development of physical geography it is necessary to appreciate the context after 1850 when several very significant influences affected general scientific thinking about the Earth's surface. It has to be remembered that earth science was then emerging from the prevailing beliefs that the Earth was created in 4004 BC and that its major surface features had been the consequence of catastrophic events. Therefore the concept of uniformitarianism and the theory of evolution were particularly influential, subsequently complemented by results of exploration.

2.1.1 Uniformitarianism

Perhaps the most dominant influence upon physical geography, and especially upon geomorphology, was gradual acceptance of the content of the *Theory of the Earth* published as two volumes by James Hutton in 1795 and subsequently clarified by Playfair (1802) in his *Illustrations of the Huttonian Theory of the Earth*. This theory rejected catastrophic force as the explanation for environment and gave rise to a school of uniformitarianism whereby a continuing uniformity of existing processes was regarded as providing the key to an understanding of the history of the Earth. The arguments required, and the perseverance necessary to overcome the catastrophists and those such as Dean Buckland who believed that the world began in 4004 BC, provides a fascinating story comprehensively described by Chorley *et al.* (1964), who also described how Charles Lyell, who published his *Principles of Geology* in 1830, came to be regarded as 'the great high priest of uniformitarianism'. Uniformitarianism not only largely replaced catastrophic ideas of landscape formation, but also promulgated the idea that 'the present is the key to the past', a very important step towards appreciating that the Earth's present surface can provide information about landscape processes and mechanisms, which in turn can aid our understanding of the past. It must not be assumed, however, that the present rate of operation of landscape processes can automatically be extrapolated to past environments. As the notion of uniformitarianism came to exercise an important influence in geology and then in physical geography, there were occasions when, as a doctrine, it may have been taken too far. Sherlock (1922) reached this conclusion after considering the geological and biological effects of human activity (see Chapter 7, p. 171).

2.1.2 Evolution

In the early 1860s, the concept of evolution, the relatively gradual change in the characteristics of successive generations of a species or race of an organism, came to pervade physical geography as a whole after publication in 1859 of Charles Darwin's *Origin of Species*. The impact on geography was resolved by Stoddart (1966) into four components. First was the idea of *change through time*, reflected in the way that evolutionary sequences of landforms were developed following Darwin's own 1842 study of the evolution of coral islands. This influenced W.M. Davis who proposed the cycle of erosion in 1885 and described it as a 'cycle of life'. In plant geography and in ecology a similar influence was detected; Clements, a man who held a position in ecology similar to that achieved by Davis in geomorphology, proposed plant succession as the 'universal process of formation development ... the life history of the climax formation' (Clements, 1916). The conceptual similarity of plant succession and the cycle of erosion was stressed by Cowles (1911) who advocated 'physiographical ecology' by amalgamating Davisian geomorphology and Clementsian ecology (Stoddart, 1966). Secondly, the idea of *organization* led to emphasis upon the inter-relationships and connections between all living things and their environment. This found particular resonance with European researchers who were concerned with community structures and functions and eventually with the idea of the ecosystem as expressed by Tansley (1935). The third idea of *struggle and selection* and the fourth of *randomness and chance* did not find such a clear and immediate reflection in physical geography, except that Darwinism was interpreted in a deterministic rather than a probabilistic sense. The unique contribution of Darwin's theory, that of random variation, was neglected in geographical circles (Stoddart, 1966) and was not really reflected in work by physical geographers until much later, in the 1960s. Evolution imposed a historical perspective upon physical geography that became a predominant influence upon geomorphology, upon the study of soils and of biogeography but also found parallels in the study of clima-

tology for at least 100 years. Perhaps the combined force of uniformitarianism and of evolution encouraged many manifestations of the historical
approach.

2.1.3 Exploration and survey

A third influence arose because basic exploration occurred in the nineteenth
and well into the twentieth centuries. Results obtained by expeditions continued to provide new data which could be of great value for physical geography. In his review of the first 150 years of the Royal Geographical Society,
Freeman (1980) suggested that the initial purposes prominent in the society
were exploration and map-making. Exploration not only provided data
about areas hitherto relatively unknown, but also enabled progress to be
made in particular branches of physical geography, such as the study of
coral islands made possible by the travels of Charles Darwin (1842) and
J.D. Dana (1853 and 1872).

It was as a result of travel to particular areas that the glacial theory
became established, with the idea that glacier ice rather than diluvialism or
icebergs could explain features and deposits of glaciated landscape, with the
works of L. Agassiz and De Charpentier paving the way for the glacial
theory. Exploration provided familiarity with new environments and so
stimulated fresh ideas and opinions. The explorers of the American West
including John Wesley Powell, G.K. Gilbert (see Chapter 5, p. 118) and C.E.
Dutton were notable (Chorley *et al.*, 1964), and provided material and ideas
which later became incorporated into geomorphology, often via the cycle of
erosion. It was concluded (Chorley *et al.*, 1964, p. 591) that:

> not since the time of Lyell had a new body of thought had such an
> immediate effect on geomorphic thinking in general ... particularly
> true in the United States ... Most landscape studies began to admit
> readily the power of sub-aerial erosion.

The foundation of the Ordnance Survey in Britain in 1795 and the
Geological Survey in 1801 were the beginnings of surveys which provided
basic data and information for physical geography. Important concepts
were established, and the ideas in Geological Survey memoirs still provide
testimony to the perceptive abilities of the early geological surveyors.
Subsequent mapping of other aspects of the physical environment included
national soil surveys, which in the UK dated from 1949. At that time the
only vegetation surveys available in the UK were embraced within the first
(1930s) and second (1960s) land-use surveys. An automatic weather station
was shown to the Royal Society by Robert Hooke in 1679 (Rodda *et al.*,
1976) and the first records made of river stage included those for the Elbe at
Magdeburg from 1727–1869 (Biswas, 1970). Environmental monitoring in
Britain included the permanent records made by tide gauges at coastal sites

since 1860, rainfall records kept at Burnley since 1677, and continuous river discharge measurements on the Thames since 1883; by 1935–36 there were 28 river gauging stations throughout Britain, by 1945–53 this had increased to 81, and by 1975 there were c. 1200 gauging stations. Approximately half of the area of the continental US was covered by soil maps from 1899–1935 (Barnes, 1954), and by 1950 the US Weather Bureau had 10,000 regular and cooperative stations measuring precipitation. Systematic and continuous measurements of streamflow in the US began in 1900 and the basic network of gauging stations was established during the period 1910–1940. By 1950, observation occurred regularly at about 6000 points. Such growth in environmental monitoring emphasizes how recent has been the acquisition of data on the physical environment and how data availability affects the areas that could be studied, the methods used and the concepts developed. More recently the advent of new techniques of data capture (Chapters 3 and 9, p. 68 and p. 235) has made possible levels of research investigation that could not have been anticipated, even in the 1980s.

2.1.4 Positivist background and developments in scientific method

Influences considered so far had a very clear effect upon the development of physical geography, whereas some influential ideas in what Osterkamp and Hupp (1996) called the basic sciences, namely physics, chemistry and biology, did not have the same impact, and certainly not at the same time, within what they referred to as the composite sciences, such as geomorphology and ecology (see Chapter 3, p. 49). Perhaps physical geography ignored for too long the developments taking place in the scientific world, including the influence of positivism. Established as a concept by Auguste Comte during the 1830s in France, this was conceived to supersede free speculation or systematic doubt as defined by René Descartes (1596–1650) and was characterized by Comte as the metaphysical principle (Holt-Jensen, 1981). Positivism is essentially the philosophical system that recognizes only positive facts and observable phenomena; positivist approaches came to be the foundation of what was widely known as the scientific method and depended upon making empirical generalizations, statements of law-like character, which relate to phenomena that can be empirically recognized (Johnston, 1983b, p. 11). Because positivism depends upon the use of empirical generalizations, it embraces the verification principle because it demands testing of the empirical hypotheses proposed, leading to verification or falsification. The aim is to achieve general laws that are not specific to a given set of circumstances. Positivism was given enormous impetus by the advent of Darwin's *Origin of Species* and the concept of evolution.

In Vienna in the 1920s, a group of scientists known as logical positivists extended the fundamental principles of positivism by arguing that formal logic and pure mathematics, as well as the evidence of the senses, provide knowledge, and they opposed all unverifiable phenomena. The verification principle was central to the work of logical positivists and subsequently this promoted further debate which could have been relevant to physical geography. However, Comte and positivism do not feature in the history of the study of landforms (Chorley *et al.*, 1964), reflecting the fact that the approach that became so fundamental in science was not used by physical geographers, and also that any influence in geography as a whole was associated with the determinist–possibilist debate (Harvey, 1969). This was because Comte not only believed that the natural sciences seek the laws of nature, but also that scientific investigations of societies would discover the laws governing society. From the mid-nineteenth century the sciences were influenced not only by the Darwinian theory of evolution but also by the development of equilibrium theory (Osterkamp and Hupp, 1996). Particularly important was Le Chatelier's principle, which states that 'if to a system in equilibrium a constraint be applied, then the system will readjust itself so as to minimize the effect of the constraint'. Much science was based upon a methodology that required the development of hypotheses by models, theories or laws, and then the testing of these hypotheses using empirical data, so that in 1953 the function of science was expressed by Braithwaite (1953, p. 1) as:

> To establish general laws covering the behaviour of the empirical events or objects with which the science in question is concerned, and thereby to enable us to connect together our knowledge of separately known events, and to make reliable predictions of events as yet unknown.

One of the main approaches in science involved the notion of equilibrium, and the implications of Le Chatelier's principle had to wait until greater attention was accorded to environmental processes (Chapter 5, p. 104). The positivist approach, although not really explicitly assimilated in physical geography until the late 1950s and 1960s, was already under fire in the scientific world, and the subsequent advent of quantum mechanics and quantum theory, although initiated in science in the 1920s, did not begin to have an influence in physical geography until 50 years later.

2.1.5 Conservation

Concern for conservation of environment also began in the mid-nineteenth century but had comparatively little impact on physical geography until the twentieth century. The conservation movement (Mumford, 1931) is generally thought to have started with the work of George Perkins Marsh in his book entitled *Man and Nature* (Marsh, 1864), written to suggest the possibility and the importance of the restoration of disturbed harmonies and the material improvement of waste and exhausted regions; and, incidentally, to illustrate the doctrine that man is, in both kind and degree, a power of a higher order than any of the other forms of animated life, which like him are nourished at the table of bounteous nature.

This book, which included physical geography in its subtitle, proved to have a great impact on the way in which people visualize and use the land (Lowenthal, 1965), but it was many years before physical geography fully appreciated and utilized the lead given by Marsh (Chapter 7, p. 172). In the twentieth century, experience of certain sensitive environments prompted field investigations, and major publications that provided information on the effects of use of the environment. Some of the earliest details of the physical geography of China published in the West focused upon the erosion taking place in the loess lands of the middle Huang He Basin, and this was referred to in a book on the *Rape of the Earth* (Jacks and Whyte, 1939). Other books produced in the 1930s and subsequently in the US – often as a result of experience of soil erosion produced by importing Old-World farming methods into a New-World environment – described the measures designed to deal with erosion problems (Bennett, 1938). Also influential in the decades before 1950 were a series of US Department of Agriculture Yearbooks, and these mines of factual information included *Soils and Men* (Bennett, 1938) and *Climate and Man* (Kincer *et al.*, 1941). It is a curious paradox that the major problem of soil erosion did not attract the attention of physical geographers to any great extent at the time, because of the lack of real focus on landscape processes, until the 1960s when the influence of R.E. Horton became evident in relation to hydrology (see Chapter 5, p. 122).

2.2 Emergence of branches of physical geography

In the late nineteenth century, T.H. Huxley's *Physiography* (Huxley, 1877) was important not least because it endeavoured to present an integrated view of the physical environment through physiography, defined as 'the study of the causal relationships of natural phenomena or a consideration of the "place in nature" of a particular district'. The term 'physiography' had been common currency in eighteenth-century Scandinavia, and was in regular usage in the English-speaking world in the nineteenth century. In an assessment of Huxley's book, Stoddart (1975) analysed the enormous success of this work, described as 'one of the best books read for many a long day' and 'a real service to the human race' (quoted in Stoddart, 1975, p. 21). Its success was due partly to the way in which the book began by considering the London Basin, and argued from the local and familiar to the unfamiliar. Physiography in Huxley's sense was particularly appropriate for the expansion of popular education in the decades of rapid industrialization, population growth and social awareness in the wake of *The Origin of Species*, holding a dominant position in British education for nearly a quarter of a century before being partly displaced by science subjects in their own right in the schools, but it never gained a central position in university geography because of the influence exercised by geology. The new science of geomorphology, although still called physiography in the US, was supplied with a new unifying principle, the cycle of erosion; a new technique, the historical analysis of landforms; and a new field of regional analysis, while at the same time abandoning climate, oceanography, biogeography and the study of human geography to other disciplines (Stoddart, 1975, p. 32). Physiography was evident in Germany from 1905 to 1914, and it survived in Britain in certain university departments such as Cambridge (see Haggett, 1990, p. 2), but usage of the term declined despite the fact that in the Huxley sense it was very appropriate to signify the holism of physical geography. Occasional more recent uses of the term include Professsor Ann Henderson-Sellers's (1989b) suggestion that quantitative synthesis based upon appropriate description of process and spatial analysis is urgently needed in contemporary climatology, and that 'geographers have much to offer to both "the atmospheric physiography" and "meteorological modelling" aspects of modern climate study'. Establishing physical geography required books such as that by Huxley (1877), and in France, Emm de Martonne had produced a *Traite de Geographie Physique* in 1909, a book used world-wide, later translated into other languages and running to six editions.

Despite such books, which presented a comprehensive physical geography, there were two trends of particular note. Firstly, it was not always clear whether a particular scientist was a physical geographer or not. The boundary with other disciplines remained blurred because not only in the US were many physical geographers to be found in geology departments, but also in Germany. Von Richthofen, who followed Humboldt and Ritter, had trained as a geologist, and Ratzel came to geography from early training in geology, zoology and comparative anatomy. Not only did some parts of physical geography appear under the umbrella of other disciplines, but some physical geographers were trained in other disciplines. In North America C.F. Marbut (1863–1935) was primarily a soil scientist who had trained in geology and made a very significant contribution in developing the study of soil science, but he also wrote a number of papers in geographical periodicals and was President of the Association of American Geographers (1925). Marion Newbiggin (1865–1934), who trained as a biologist, was a significant figure in the development of biogeography in the UK. Secondly, it is clear that separate branches of physical geography were beginning to emerge, partly because of the progress being made by other disciplines, especially where there was a need to undertake national surveys.

This is very well exemplified by the growth of *soil science*, where important progress was made by the Russian school headed by V.V. Dokuchaev. The prevailing perception of the soil had been influenced by agricultural chemists such as J. von Liebig in Germany, who had evolved the so-called 'dustbin theory', visualizing soil in a static way as a closed system into which could be replaced what was extracted by crop production. This notion of a layer of soil as a skin, almost independent of above and below, was superseded by a view of the soil as the product of soil-forming factors. Dokuchaev together with his students, particularly Sibirtsev (1860–99), contributed the zonal theory of soils published in 1900 (Bridges, 1970), which was particularly apposite in the USSR where soils existed in broad zonal groups.

Whereas zonal soils were determined predominantly by climate, intra-zonal soils were produced where other factors such as water, rock type or topography were dominant, and azonal soils were relatively immature and had insufficient time for development to profiles of one of the other two types. This conception of broad belts of soils which were zonal and related to broad patterns of climate and vegetation was a natural development against the background of the latitudinally arranged physical environments of the USSR. Other Russian soil scientists included K.D. Glinka (1867–1927) who distinguished soils amenable to climatic influences as being ectodynamomorphic, as distinct from endodynamomorphic ones which had inherent characteristics enabling them to resist, or at least to modify, the influence of outside factors. There was a delay before the Russian ideas of soil formation diffused to Western Europe and North America, where soil was still regarded in a geological way. This diffusion was retarded because the translation of K.D. Glinka's work by C.F. Marbut, who was then head of the US Soil Survey, was based upon a 1914 edition of Glinka's book and not published until 1927. In assessing some people involved in the development of pedological thought, Tandarich *et al.* (1988) quoted Simonson (1986), who showed that Russian ideas had been introduced as early as 1901 through a translation of a Sibirtsev paper. In addition, the Imperial University of St Petersburg had been represented at the World's Columbian Exposition in Chicago (1 May–31 October 1893), and in the catalogue for the Russian section there was listed a soil exhibit prepared by Dokuchaev and Sibirtsev, together with a short statement about Russian soil work. Despite their availability before 1900, there was no evidence of incorporation of such Russian ideas into pedological thinking until the Marbut translation appeared (1927). This illustrates how important the diffusion of ideas could be and how important the presence of a receptive scientist was. Marbut was well placed to amalgamate Russian and American ideas and published a work on the soils of the US in 1935, which reconciled the needs of a national classification with the field survey scale. The emphasis that the Russian school placed on factors of soil formation was not fully utilized until 1941 when Jenny produced a book entitled

Factors of Soil Formation (Jenny, 1941) and used a central notion that any soil character (S) was a function of climate (Cl), organisms (O), relief (R), parent material (P) and time (T) as well as other unspecified factors in the form:

$$S = (Cl, O, R, P, T, \dots).\qquad(2.1)$$

For constant climate, organisms, parent material and topography, Jenny (1941) concluded that the soil profile would be solely a function of time. The implication, suggested by Osterkamp and Hupp (1996), is that there could be orderly changes of soil variables through time in a way analogous to the Davisian cycle of erosion, to the Clementsian succession, and to Darwinian evolution; and this early work on soil development later led to concepts of chronosequences and chronofunctions of soils. A further view, not readily accepted, was put forward by C.C. Nikiforoff, a pedologist who emigrated from Russia to the US and joined the Soil Survey. His paper on soil formation (Nikiforoff, 1942) considered soil processes as fluxes of matter and energy, but his work attracted little attention at the time, just as G.K. Gilbert's notion of equilibrium was overlooked (Osterkamp and Hupp, 1996), although both were subsequently recognized with the inception of systems thinking (Chapter 4, p. 85).

Studies of the ***atmosphere*** received a great impetus from the output of the Meteorological Institute at Bergen, and it has been suggested that few groups have ever dominated a scientific field so completely as the Bergen group (Hare, 1951a). Stimulated by the lack of information for the 1914–18 war zone and facilitated by the dense network of stations that had been established especially in Norway, the Bergen group essentially placed fronts on the weather map. The central themes of the Bergen school centred upon the realization that the atmosphere is composed of large bodies of fairly homogeneous air separated by gently sloping boundaries or frontal surfaces. The life history of a frontal wave cyclone was developed by J. Bjerknes and Bergeron in 1928, and in 1938 they proposed a classification of air masses which provided the foundations for dynamic climatology, although the possible explanatory description of world climates in terms of air mass theory was not pursued, to the detriment of the physical geographer (Hare, 1951a). A number of climatic classifications were produced at this time, including the Koppen classifications dating from 1900–36, and the approach towards a rational classification was produced by Thornthwaite (1948). Also of significance in climatology was the progress towards air-mass climatology (e.g. Lamb, 1950; Belasco, 1952; Barry and Chorley, 1976) but the leads embodied in these approaches were not immediately followed up by physical geographers. Studies of the atmosphere were lifted to a different plane with the advent of the radiosonde, which meant that small radio transmitters sent up by balloon could illuminate the data record up to 50,000 feet, so that upper air synoptic charts and the significance of upper air investigations began to assist the interpretation of synoptic situations.

Thornthwaite's (1948) work on the water balance was extremely important, influencing the development of hydrology as well as climatology, and Hare (1968) contended that Thornthwaite, Budyko and Penman transformed the way in which we look at climate as an element of the natural environment. Such transformation was certainly needed – Henderson Sellers (1989b) quoted F.K. Hare as saying that in the 1940s climatology was a word hardly ever heard professionally, whereas the situation was very different by the 1970s, possibly largely as a result of global climate studies and awareness of global change. In the early days emphasis was placed upon a compilation of statistics with little attempt at elaboration of the underlying dynamic or thermodynamic phenomena that control climate, the dynamic climatology approach being introduced by Tor Bergeron in 1929. He proposed (Rayner and Hobgood, 1991) that the concepts of air masses and fronts could be developed to form a comprehensive dynamic climatology. However, although the basic physics of the atmosphere was known by the mid-1850s, it was not until the 1950s with the advent of the digital computer that solutions to major world climatic problems could be attempted. Oliver (1991) showed how classification of climate proceeded from the earliest approaches, based upon a single variable; to the late-nineteenth-century approaches which were more quantitative, using more than one variable; to those where vegetation distribution underpinned the classification derived, as exemplified by the Koppen approach; and were then succeeded by the use of climatic indices and the development of the moisture balance concept by Thornthwaite (1948). Landsberg (1941) emphasized that whereas descriptive/regional climatology answers the question of 'what is where', physical climatology should tell us 'why it is so' and it should 'analyze the observational data and abstract therefrom what is to be regarded as typical, what the interplay of cause and effect is . . .'.

The fields of physical and boundary layer climatology provide a rich heritage for geographer–climatologists according to Brazel *et al.* (1991), who summarized the early history of physical climatology as follows:

- It should be heavily weighted with mathematics, physics and meteorology.
- The ultimate objective should be to make maps of the heat and moisture budget on a topoclimate scale.
- Thornthwaite, Penman and Budyko transformed the way we look at climate as an element of the natural environment.
- Energy budget studies must form the central part of investigations into all scale relationships – there is therefore a need for synoptic climatic investigations of energy transfers and the degree of horizontal variation of specific microclimates.

Biogeography was dependent upon the emergence of ecology, plant communities being initially regarded in a static way, with the vegetation of large areas usually described by compiling species lists or by developing simple relationships between vegetation and climate (Merriam, 1894). This early approach was analogous to the dustbin theory approach to soils, succeeded in biogeography by the major influence of Clements (1916) who developed plant succession. Osterkamp and Hupp (1996) showed how Henry Cowles was one of the first to emphasize change through time, and how he may have interacted with W.M. Davis and was strongly influenced by his teaching and writing, so that the ecology developed by Cowles was parallel to the cycle of erosion model. Clements (1916) fostered a dynamic approach to ecology which embodied the concept of climax vegetation, in which the plant community passed through all phases of its succession and reached a state of equilibrium with the climate of the region, and during the vegetational sequence there was a series of transitional stages called seres. The early ideas of Clements may have fostered a somewhat unsatisfactory analogy between the development of a plant community over time and the life cycle of an organism (Harrison, 1980). Much attention was devoted to the concept of succession by biologists and then by biogeographers, with a number of features emphasized such as the distinction between primary succession on new bare areas, and secondary successions on areas previously supporting vegetation, where human activity or natural hazards modified the former vegetation and initiated the secondary succession. Harrison (1980) suggested that the initial view of succession was followed by one in which succession is viewed as a process resulting from plant-by-plant replacement, where the patterns generated by this replacement process have routine statistical properties. Whittaker (1953) stressed the evolutionary significance of succession and visualized the climax as a pattern of species abundances which is locally constant but varies from place to place, and Harrison (1980) concluded that this view was accepted by many biogeographers. Initially it was thought that the climax community was climatically controlled, but it was subsequently accepted that climax communities are relatively permanent, stable, and adjusted to a particular blend of environmental and biotic conditions.

In the first half of the twentieth century, not everyone was persuaded that succession was the only approach for ecology and biogeography. Questions arose not only because the processes were insufficiently known, but also because a more quantitative approach was required, and field experience in particular areas stimulated other distinct approaches. Osterkamp and Hupp (1996) cited the work of Forrest Shreve who worked at the Carnegie Institution's Desert Laboratory, Tucson, from 1907 to 1940 and who suggested (Shreve, 1936, p. 213) that it was necessary to

> weave together the separate threads of knowledge about the plants and their natural setting into a close fabric of understanding in which it will be possible to see the whole pattern and design of desert life . . .

distribution of a plant species reflects its tolerance for a range of environmental conditions, thus few if any species have identical distributions.

For similar reasons the separation of ecology into autoecology, dealing with the environmental relations of individual plants, and synecology, concerned with the environmental relations of plant communities, became influential. The concept of the ecosystem developed in 1935 by A.G. Tansley (Tansley, 1935) was subsequently influential in relation to the systems approach (p. 83), provided a conceptual framework for biogeography and incorporated human activity as an integral component (Tivy, 1971).

2.3 Geomorphology and the impact of W.M. Davis

The ideas of W.M. Davis (1850–1934) were clearly detectable in the development of ecology and particularly biogeography, and he was undoubtedly the dominant influence upon physical geography in this period. Although his ideas were eventually challenged and placed in a much broader context, physical geography would have advanced more slowly had he not written more than 500 papers and books. Of the three published volumes on the history of the study of landforms, a whole volume of 874 pages is devoted almost exclusively to Davis (Chorley *et al.*, 1973).

The essential focus of Davis's geomorphology was that the normal cycle of erosion could be used to classify any landscape according to the stage that it had reached in the erosion cycle, whether youthful, mature or old age, and in addition he offered a trilogy for the understanding of landscape in terms of structure, process and stage or time reached in an erosion cycle. Whereas the normal cycle was the work of rain and rivers, arid landscapes were fashioned under the arid cycle of erosion, and in the marine cycle particular attention was given to shorelines of emergence or of submergence. In addition there were two principal accidents to the normal cycle: volcanic activity and the glacial accident. A striking feature of Davis's work was the clarity of the supporting sketches and illustrations which assisted in the acceptance of his approach. In addition to his very numerous papers, a volume of his essays was subsequently published, and his physical geography was compiled from his later lecture notes (King and Schumm, 1980). Of the many geomorphologists who adopted the Davisian message, Cotton (1942, 1948) was particularly notable. The great success of the Davisian approach and its subsequent popularity was attributed by Higgins (1975) to 12 reasons, namely:

(1) simplicity – in particular the initial uplift;
(2) applicability by students to a wide range of erosional landscapes;
(3) presentation in a lucid, compelling and disarming style – the style of writing and the numerous line drawings and sketches;

(4) apparent basis of careful field observations – although no measurements were made, they did relate to actual examples;
       particularly in relation to the geological community:
(5) it filled a void – and complemented uniformitarianism;
(6) it synthesized contemporary geological thought – it incorporated concepts that had been introduced by others including base level (Powell, 1834–1902), graded stream (Gilbert, 1843–1918), and the profile of equilibrium advocated by French engineers;
(7) it provided a basis for prediction and historical interpretation – this enabled geomorphology to use landform study as a tool for deciphering the later stages of earth history and to function as a part of historical geology;
       in terms of popularity:
(8) it was rational – and appealed to positivists;
(9) it was consistent with evolution;
(10) it appeared to confirm stratigraphic thought at the time – namely a tectonic model of rapid diastrophism followed by a long period of coastal stability and rest;
(11) it set humid temperate as 'normal', and this was attractive to many earth scientists;
(12) the cyclic approach was also attractive to many earth scientists.

Thornbury (1954) noted that 'Geomorphology will probably retain his stamp longer than that of any other single person', and this stamp survived in many textbooks, sometimes without a clear indication that the approach was Davisian. Davis 'both organized and systematized geography in the United States and won recognition for the subject as a mature science and as an academic discipline' (Chorley *et al.*, 1973, p. 734). Not only had a means of explanatory description been provided, but also a new terminology had been furnished with over 150 terms and phrases credited to Davis and probably at least a further 100 generated by his students. Reasons for the continued acceptance of the ideas of the Davisian cycle were reviewed by Bishop (1980), who argued that the cycle is not a scientific theory because it is irrefutable in relation to the concept of stage, and also because it was modified in an *ad hoc* manner as objections were made to it. The concept that underlies Popper's definition of science, as including falsifiable hypotheses of high information content, caused Bishop (1980) to conclude that the Davisian hypotheses could have been of more value had they been expressed in such a way as to enable testing by falsification. It is not easy to appreciate the extent to which Davisian ideas took hold and how difficult it was in the early twentieth century for such ideas to be challenged or replaced, but an extremely good summary of the situation and the way in which it was questioned was provided by Stoddart (1997b), complementing that by Beckinsale (1997) in the same volume. The impact of Davis was reflected throughout geomorphology, and Church (1996, p. 149) concluded that:

For more than half a century it remained the dominant template for landscape interpretation. It portrayed landscape as a staged sequence of erosional transformations of an initially elevated landmass.

Sherman (1996, p. 107) saw Davis as the prototype geomorphological fashion dude:

> He was a propagandist and heckler. He was a prolific writer and public speaker, and he was a powerful political advocate for his own ideas and for the stature of our discipline ... it is commonly accepted that the paradigm governing the discipline through the early decades of the twentieth century was the Davisian model of the geographical cycle ... Certainly this was the first general landscape model to receive widespread, international acceptance ... and it provided the touchstone for geomorphological development and debate for decades. It is also commonly accepted that this paradigm was rejected sometime around the middle of the century and replaced by process (or quantitative/dynamic/systematic) geomorphology.

Osterkamp and Hupp (1996, p. 422–424) concluded that

> [an] identity for geomorphology occurred with William Morris Davis ... Davis patterned his model after Darwin's concept of evolution and therefore subscribed to uniformitarianism ... Davis as eloquent lecturer and writer dominated geomorphology for half a century using Darwin's exemplar.

When Davis retired to the west coast of North America, he envisaged a basis different from the one generated in most of his publications derived from the more stable eastern area of the country, and wrote (quoted in Chorley *et al.*, 1973, p. 647):

> ... the scale on which deposition, deformation and denudation have gone on by thousands and thousands of feet in this new made country is 10 or 20 fold greater than of corresponding processes in my old tramping ground.

Had Davis experienced the west coast of North America at an earlier stage in his career, then the development of geomorphological thinking could have been rather different.

The impact of the Davisian approach should be thought of not just in terms of Davis's direct contribution, but also in terms of developments, alternatives and objections. *Developments* occurred in research, and many textbooks (e.g. Wooldridge and Morgan, 1937; von Engeln, 1942) continued the Davisian ideal. Whereas these views reflect the North American perspective, Beckinsale and Chorley (1991) concluded that other areas temporarily embraced historical geomorphology and regional geomorphology – the latter including morphoclimatic approaches (see Chapter 5, p. 127).

However, some studies of regional geomorphology employed Davisian ideas. D.W. Johnson (1931), a disciple of Davis, produced *Stream Sculpture on the Atlantic Slope;* Wooldridge and Linton (1939) produced a Davisian interpretation of southeast England in *Structure, Surface and Drainage in South East England*; and Cotton's (1922) *Geomorphology of New Zealand* is another exemplar of the way in which Davisian ideas were applied to specific areas. The *Physiography of the Eastern United States* (Fenneman, 1938) and its sister volume for the western US (Fenneman, 1931) also utilized the regional approach. *Alternatives* arose because some approaches, such as that of Walther Penck (1924), depended upon an essentially different basis for geomorphology but one that was not so readily understood and applied, one that took much longer to become widely available in the English-speaking world (Simons, 1962; Tinkler, 1985), and one that was not accompanied by such a great volume of publications as Davis's ideas. *Objections* developed as some geomorphologists began to see possibilities other than those offered by a purely Davisian approach. Thus Kirk Bryan followed Davis to a large degree, but also took account of Penckian views and, together with his students, made contributions to the newly developing fields of periglacial geomorphology and arid geomorphology. Peltier (1954) argued that in the US a new era in geomorphology began about 1940, heralded by descriptive investigations particularly focused upon mapping; by dynamic studies of fluvial, solution, marine, periglacial, aeolian and volcanic processes; and by applied studies.

Glacial theory had already developed to some extent independently of Davisian ideas and *The Alps in the Ice Age* by Penck and Bruckner (1901–9) was distinct. In this field the erosional effectiveness of glaciers was first established by L. Agassiz, but it was not easy to gain acceptance of the notion that glaciers could erode, one view being that glaciers could no more erode the landscape than custard could erode the custard dish! Flint (1957, 1971) explained how the development of the idea that glaciers were formerly more extensive dated from the second quarter of the nineteenth century. An address by Louis Agassiz to the Helvetic Society in 1837 pictured a 'great ice period' caused by climatic changes and involving a sheet of ice from the North Pole to the Alps and central Asia. Ten years later he recognized that the former glaciers of northern Europe were separate from the former Alpine glaciers, and that the Alpine ice postdated the uplift of the Alps themselves. Agassiz became a professor at Harvard in 1846 and so started the disciplines of glacial and Quaternary geology and of glacial geomorphology in North America. Elsewhere ideas took time to become accepted, and the important steps to be taken included acceptance that extensive glacial drift was the product of glacial deposition rather than short-lived submergence beneath the sea; that several former glaciations had occurred; that far-travelled erratic rocks could indicate the direction of movement of glacier ice; that different pluvial climates had existed in certain areas; that sea-levels had changed; and that land uplift could occur after glaciation. The majority of papers on glacial topics between 1900 and the 1960s concerned the classification and interpretation of specific landforms (Tinkler, 1985).

2.4 Approaches established by 1945–1950

The view of Church (1996, p. 150) that 'In geomorphology the nineteenth century ended about 1950' can be applied to physical geography as a whole. By 1945 notable features of the position established were as follows:

- Physical geography had in some countries evolved an integrated coherent approach, sometimes styled 'physiography'.
- Distinct branches of the subject had started to emerge, with geomorphology, climatology, soil science and biogeography constituting the quartet often seen as the core of physical geography.
- Despite the emergence of such branches, the unifying approach of W.M. Davis, although most influential in geomorphology, also permeated other branches because they were affected by general scientific developments, of which Darwinian evolution was most notable.
- Other approaches had been developed and not everyone perceived the Davisian model as robust enough to fulfil all the needs of physical geography.
- Philosophical developments in science (see p. 48) had not yet been acknowledged.

Considerable growth of geography, higher education and research occurred after 1945 because after the end of the Second World War there was a sudden increase in the number of geographers, expansion of universities, growth in student numbers and in the courses provided, and new journals such as *Erdkunde* (1945). Several influential papers published around 1950 promised much for physical geography for the next decade. As well as more students being available, people from the armed forces completed their degrees and provided an influx of practical expertise for university staff. Many had gained experience in a variety of areas including weather forecasting, aerial photography interpretation and terrain analysis (e.g. Tinkler, 1985) and so were able to bring a pragmatic approach, and Hare (1951a) noted that:

> Climatology in a university Geography department is now very likely to be taught by a competently trained meteorologist with a much deeper understanding of his field than was generally the case before the war.

New techniques such as the advent of radar afforded new lines of investigation and had a salient effect on the development of climatology. The advent of an approach towards a rational classification of climate (Thornthwaite,

1948) complemented earlier classifications proposed by Koppen and A.A. Miller. Whereas nineteenth-century physical geography often encompassed oceans and hydrology, oceanography became more separate, due in part to the growth in the science of oceanography (e.g. Sverdrup *et al.*, 1942), although Burke and Eliot (1954) advocated a new approach to the regions of the oceans despite describing most of the professional geographers in the US as land-lubbers. Developments in hydrology had a greater impact, however, and work on evaporation by Penman (1948) produced a clearer understanding of spatial variation (Penman, 1950), while new books on hydrology such as that by Linsley *et al.* (1949) supplemented the edited volume that had been available for nearly a decade (Meinzer, 1942). Such progress, supplemented by publications by R.E. Horton (1945), had a significant effect when processes began to attract greater attention (see p. 000), but Meigs (1954) noted that, although an interest in water supply problems was growing, hydrology was being underestimated as oceanography had been. Although not firmly established until the 1960s, the links with hydrology proved to be particularly fruitful (see p. 122) and complemented other emerging branches of physical geography.

Not all writers agreed about the content of physical geography: Davis preferred the term 'physical geography' to 'geomorphology', and Wooldridge and East (1951) in their book on *The Spirit and Purpose of Geography* headed a chapter 'Physical geography and biogeography' so that it was not clear whether physical geography should include biogeography or not. In addition to plant geography, biogeography could include animal geography and zoogeography, and when recognizing the three approaches of regional, historical and ecological, Stuart (1954) noted that it is unfortunate that the aims and methods of zoogeography have never been clearly formulated, and that:

> if scholars trained in geography and thoroughly grounded in the methods of regional study were as adequately trained in systematic zoology and palaeontology, they could be in a position to render important service to zoogeography. (Stuart, 1954, p. 449)

In geomorphology the Davisian school, as denudation chronology, was upheld vehemently by some including Wooldridge, and involved more precise specification of the techniques, for example mapping and analysis of river long profiles (Brown, 1952), leading to debates about whether erosion surfaces were marine or subaerial in origin (e.g. Balchin, 1952).

Alternatives to the Davisian approach appeared because of the supremacy of Davisian geomorphology within geomorphology and within physical geography as a whole, and criticism was particularly evident in the US, where Strahler (1950a, p. 209) commented:

> Davis's treatment appealed then, as it does now, to persons who have had little training in basic physical sciences, but who like scenery and

outdoor life. As a cultural pursuit, Davis's method of analysis of landscapes is excellent; as part of the basis for the understanding of human geography it is entirely adequate. As a branch of natural science it seems superficial and inadequate.

In a retrospective assessment it was suggested that 'to some the Davisian method had become a stranglehold or at least a sedative' (Chorley *et al.*, 1973, p. 753). Davis had acknowledged some new developments, and King and Schumm (1980), editing lecture notes for courses presented by Davis at the University of Texas in 1927 and at the University of California Berkeley in 1929, showed how Davis had accepted parallel slope retreat and pediments. Criticism was growing, much of it summarized in a special issue of the *Annals of the Association of American Geographers* in 1950, although Wooldridge (1949) in his discussion of taking the 'ge' out of geography expressed his fear that 'the baby might be thrown away with the bath water'. Other models of landscape evolution were being considered, and results of early work on slopes, stimulated by the ideas of Penck (1924), were followed by Wood (1942) and were then incorporated into the work of L.C. King, who was already working on arid and semi-arid landscapes of Africa (see p. 142). In the *Annals of the Association of American Geographers* for 1940 and 1950, papers pointed towards alternatives, including approaches to slopes (Strahler, 1950a), and a year earlier it had been argued (Russell, 1949, p. 3–4) that:

> the geomorphologist may concern himself deeply with questions of structures, process and time, but the geographer wants specific information along the lines of what and where and how much.

Alternatives to the Davisian approach arose with the study of other environments, and the significance of periglacial processes in past landscapes came to be appreciated following the work of Bryan (1946) and developed into the periglacial cycle by Peltier (1950), although the diffusion of important work on the European continent (Dylik, 1952; Poser, 1947) into the English-speaking literature took until the early 1960s.

In glacial geomorphology (in addition to detailed work on specific areas by Pleistocene geologists), research on glacier fluctuations (Ahlmann, 1948), material embraced in *Glacial and Pleistocene Geology* (Flint, 1947) and in *The Pleistocene Period* (Zeuner, 1945) presented a perspective of potential interest to physical geographers, although some, such as W.V. Lewis, had already established themselves in this field (Lewis, 1949). In studies of the Pleistocene new techniques were becoming available (Zeuner, 1945) and these also began to influence the study of biogeography, where pollen analysis was to become a major technique for environmental reconstruction. The area of coastal geomorphology was also developing somewhat independently of Davisian ideas, as exemplified by J.A. Steers's *Coastline of England and Wales* (Steers, 1948).

In a re-evaluation of the geomorphological system of Davis, and a prelude to new approaches, Chorley (1965) drew attention to three major criticisms:

(1) It led to a dogma of progressive, irreversible and sequential change, signifying that the amount of energy for the transformation of landforms was simple and a direct function of relief or angle of slope. This method of analysis was not supported when it was shown that, for example, drainage densities, erosional slopes and river meanders do not necessarily evolve cyclically.
(2) The emphasis was upon historical sequence rather than upon functional associations, which were more dependent upon process investigations.
(3) The approach was highly dialectical and semantic.

A dominant aproach had not emerged even as late as 1967, when Chorley (1967, p. 59) drew attention to:

> ... many national preoccupations, some of long standing with particular geomorphic objectives. Thus the development of the American style of 'dynamic-process' geomorphology, the Franco-German climatic geomorphology, the British denudation chronology/geological approach, the Polish Pleistocene-dominated geomorphology, the Russian applied geomorphology, the Swedish studies of process almost per se, the Eastern European morphological mapping, and the central European tectonic bases have created a Godot-like atmosphere of articulate introspection.

It was in this conceptual vacuum or lack of an integrated approach to physical geography that quantification began to grow, and what Chorley (1965) described as American-style dynamic process geomorphology was stimulated by Strahler and the Columbia School, building upon the contributions of Horton, and by Leopold (see p. 119). In a retrospective view of quantitative/dynamic geomorphology at Columbia from 1945–1960, Strahler (1992) admitted that the first paragraphs in his 1950 paper (Strahler, 1950a) were Davis-bashing, but argued that the paper as a whole had subsequently been grievously misconstrued as a denouncement of the fundamental validity of the Davisian denudation 'cycle'.

The impression may be created that physical geography by 1950 was becoming a disparate field of enquiry, with new branches like hydrology complementing existing ones, and few signs of integrative study of the physical environment. Signs did exist, however, and although mapping and classification of areas had become a feature of climatology, geomorphology, pedology and plant geography, all epitomized as taxonomic approaches by Chorley (1978), there were some attempts to address inter-relationships of an environmental character, perhaps most noticeably in the USSR. The influence of Dokuchaev and the school of soil science had prevailed, so that in 1950 L.S. Berg produced a volume dealing with the natural landscape zones and the inter-relationships within these zones (Berg, 1950), and this was the foundation for later attempts to integrate physical geography (e.g. Suslov, 1961) and landscape science (see p. 206). Such approaches had been attempted elsewhere by *Forest Physiography* (Bowman, 1922) and also at different scales. The small individual slope envisaged by Wooldridge (1932) as the atoms from which landscape was built was reconciled with information from other spatial scales in an influential paper on the delimitation of morphological regions (Linton, 1951), which acknowledged the influence of concepts of the site (Bourne, 1931) and the catena (Milne, 1935). These two concepts were to influence the progress of soil survey investigations and the way in which soils surveyed in the field related to larger soil bodies and to other environmental characteristics. Two other trends towards more integrated approaches were land systems (Christian and Stewart, 1953), and relationships between rainfall and runoff (Langbein *et al.*, 1949) over the US which became a stepping stone for palaeohydrology and environmental change.

The prevailing emphasis throughout the 100 years to 1950 had been upon evolution of environment and its classification. However, the focus in evolutionary studies had usually been upon millions of years and the Tertiary, rather than upon thousands of years and the Pleistocene, with very few studies devoted to the last few hundred years. Similarly in classification, the focus had been upon static patterns rather than upon the dynamics of environment. An exceptionalist perception of physical geography as fulfilling an integrating role linking more specialist sciences together was difficult to sustain in 1950, and even more implausible as techniques multiplied and diversified in subsequent decades (Chapter 3).

Further reading

An interesting view in relation to geomorphology is provided by:
TINKLER, K.J. 1985: *A short history of geomorphology.* London: Croom Helm.

The definitive statements which merit careful attention are:
CHORLEY, R.J., DUNN, A.J. and BECKINSALE, R.P. 1964: *The history of the study of landforms, vol. I: geomorphology before Davis.* London: Methuen.

CHORLEY, R.J., BECKINSALE, R.P. and DUNN, A.J. 1973: *The history of the study of landforms, vol. II: the life and work of William Morris Davis.* London: Methuen.

BECKINSALE, R.P. and CHORLEY, R.J. 1991: *The history of the study of landforms or the development of geomorphology, vol. 3: Historical and regional geomorphology 1890–1950.* London: Routledge.

In relation to the contribution of Davis, and the way in which the geomorphology of long-term landscapes has been complemented by other approaches, a thought-provoking paper written from a particular perspective is:

BAKER, V.R. and TWIDALE, C.R. 1991: The reenchantment of geomorphology. *Geomorphology* **4**, 73–100.

An interesting viewpoint for the future and for Chapters 3 and 11:

HENDERSON-SELLERS, A. 1989: Atmospheric physiography and meteorological modelling: the future role of geographers in understanding climate. *Australian Geographer* **20**, 1–25.

Topics for consideration

(1) The way in which the concept of evolution pervaded all developments in physical geography up to the 1950s.
(2) The way in which the physical, economic and intellectual environment in particular countries conditioned approaches taken to physical geography.
(3) Davis changed his views once he retired to the west coast of North America. Had Davis's contribution not existed, would physical geography have developed in a very different way? (See Baker and Twidale, 1991, noted above.)

|3|

Developments 1950–1980

Developments in the philosophy of science were given insufficient attention until 1950, but subsequently came to have a significant influence upon thinking in physical geography. The physical geography literature prior to 1950 contains little allusion to scientific thought, to epistemology (the theory of the development of knowledge), or to positivism. Greater awareness after 1950 is approached by outlining relevant developments in science (3.1), by summarizing the impact of quantification (3.2), by highlighting some major developments in branches of physical geography (3.3), and by pinpointing the overall position achieved by physical geography by the 1980s (3.4). The nature of physical geography changed quite substantially in the period 1950–1980 as techniques were revolutionized. Publications in that 30-year period and subsequently appreciated that 'practitioners of physical geography stand to gain clarity, unity and strength from philosophical and methodological awareness' (Rhoads and Thorn, 1994).

3.1 Positivism and post-positivism

The positivist approach in science (Chapter 1, p. 14) embraced measurement as a necessary feature of scientific investigations. The vision proclaimed by Kuhn (1962) saw science not as a well-regulated activity, whereby each generation automatically builds upon the results of earlier workers, but instead as a process of varying tension in which tranquil periods, characterized by a steady accretion of knowledge, are separated by crises which can lead to upheaval in the subject and to breaks in continuity. Paradigms (Chapter 1, p. 13) were 'universally recognized scientific achievements that for a time provide model problems and solutions to a community of practitioners'. Kuhn (1962) saw the development of science taking place in a series of phases in which a *pre-paradigm phase*, characterized by conflicts focused around individuals, is succeeded by *professionalization* when definition of the subject is acute, and then by a series of *paradigm phases*, each characterized by a dominating school of thought and each separated by a crisis phase when revolution occurs because problems accumulate that cannot be solved by the prevailing paradigm. Kuhn thus visualized scientific activity as seeking solutions within generally accepted, but often unspecified, rules and conventions, and such puzzle-solving is characteristic of what he called 'normal science'. During the paradigm phase the scientist accepts established theories, uses them as a framework for puzzle-solving, and is not normally engaged in trying to overthrow theories or develop new ones. The way in which a theory is verified, according to Kuhn (1962), is more a matter of faith than of logic, and the procedures of verification and confirmation are an integral part of the rules that the scientific community associates with the prevailing paradigm. Any anomalies that arise may then accumulate to form the basis for the next crisis phase whereby revolution will engender the next paradigm phase. Use of the term *paradigm* received some criticism, so later Kuhn (1977, p. 297) substituted *exemplars* as 'concrete problem solutions, accepted by the group as, in a quite usual sense, paradigmatic'. Osterkamp and Hupp (1996, p. 436) contended that composite sciences, such as geology, ecology and geomorphology, matured in a manner different from that of the basic sciences such as physics, chemistry and biology. They believed that a principal reason for the difference is that the basic sciences are defined by exemplars, but the composite sciences are too complex to be thus defined (see also Stoddart, 1986, chapters 8 and 11). Physical geography as a whole could be thought of as a composite science in the sense intended by Osterkamp and Hupp (1996), who observed that Kuhn scarcely referred to composite sciences, so that his model is defined uniquely for the basic sciences and cannot easily be extrapolated to the composite ones.

These approaches began to appear in the bibliographies of books and articles by physical geographers. The ideas of Kuhn were an influence upon *Models in Geography* (Chorley and Haggett, 1967) which advocated a model-based paradigm for geography as a discipline; paradigms were generally embraced within physical geography, applied to continental drift and plate tectonics and to Davisian geomorphology. However, the existence of a number of prevailing paradigms led to the realization that a Kuhnian view of science may not be so readily appropriate for physical geography. Stoddart (1986, pp. 6–18) analysed the way in which such paradigms were received in geography and concluded:

Why ... is the paradigm idea so popular, not only in human geography (and other social sciences) but in physical geography too? I suggest that a major reason lies in the way in which the concept of revolution bolsters the heroic self-image of those who see themselves as innovators and who use the term paradigm in a polemical manner, coupled with the fact that Kuhn's terminology supplies an apparently 'scientific' justification for the advocacy of change on social rather than strictly scientific grounds.

Students of physical geography who have studied some science but little philosophy were addressed by Haines-Young and Petch (1986), who provided a survey of approaches to science followed by aspects of practice, including theorizing, modelling, classifying, measuring and experimental design in the context of the critical rationalist tradition, and this encouraged physical geographers to address issues previously insufficiently confronted. They showed how the classical approach to science, contending that scientific knowledge is secure because it rests on experience, involving the idea of inductive argument, and using repeated observations to produce general statements, was flawed because of the absence of sound principles of verification or induction and because observations are actually theory-dependent. They proceeded to the critical rationalist view, whereby a rational basis for scientific knowledge is provided by deducing the consequences of theories and then attempting to expose their falsity by critical testing. Critical rationalism developed in opposition to logical positivism after the 1930s as a consequence of the work of K.R. Popper. Popper's view was that falsification replaces verification, so that a theory is assumed to be true until it is shown to be false. The scientist proposes trial solutions which are then evaluated critically, the trial solutions being speculative theories set up in an attempt to solve the particular problem to hand. Falsification is justified as a procedure because, whereas no finite number of facts can verify a universal proposition, a single fact can demonstrate the proposition to be false. Scientific statements are therefore conceived as being falsifiable whereas non-scientific ones are not. This critical rationalist stance of falsification was criticized by Kuhn because, if a single failure is the basis for theory rejection, then all theories ought to be rejected at all times. From a critical rationalist position, three objections to the logical positivist thesis (p. 30) were summarized by Haines-Young and Petch (1980):

(1) Facts are not objective because they are observations which are perceived in a particular way, according to the technology available for measurement and observation, and cannot therefore be shown to be true.
(2) The principle of induction involves a logical error because no number of apparently confirming observations can show that a general proposition is true.
(3) Observations cannot be made independently of theory because the variables perceived to be important and thus selected for measurement are chosen in the light of some preconceived theory.

Lakatos (1978, p. 2) suggested that it is easily demonstrable that there can be no valid derivation of a law of nature from any finite number of facts, so that if all scientific theories are equally unprovable, 'what distinguishes scientific knowledge from ignorance, science from pseudoscience?' The emphasis upon logical positivism or logical empiricism or upon critical rationalism was important into and throughout the 1960s. Whereas the logical positivists or logical empiricists believed that all knowledge must be grounded in direct phenomenal experience and so tried to verify hypotheses, the critical rationalists tried to falsify them (Rhoads and Thorn, 1994).

In *On Geography and its History*, Stoddart (1986) included other influences such as: the analysis of the social and political relationships between individuals within institutions, and the importance of particular groups in society, illustrated by the Royal Geographical Society; the external characteristics of the subject, departments, degrees and publications, and citation structures; the biographical approach by studying the contribution of important individuals, or histories of ideas and themes. All these have to be accommodated, and:

> Each of these approaches has its own validity and its own strengths. I claim here only that the history of geography is an infinitely richer and more varied landscape than acquaintance with the standard works on the subject would suggest. (Stoddart, 1986, p. 27)

The philosophy of science in this century was divided (H.I. Brown, 1996) into two periods, namely positivist and post-positivist. Whereas positivists dominated the field until the late 1950s, and maintained that scientific theories should be evaluated solely on the basis of observational data in accordance with a set of formal rules, H.I. Brown (1996, p. 18) contended that 'Positivists took physics as the paradigmatic science and – since the thesis that methodology is the same for all sciences was one of their guiding assumptions – they did not see any need to examine other sciences'. A discipline can be in a multi-paradigm state for considerable periods of time simply because it takes years to develop a new paradigm, and heretical thinking must go on for a long time before paradigm change can occur (Watkins, 1970). This multi-paradigm development was embraced within the methodology of scientific research programmes advocated by Lakatos (1970) to solve some of the problems which both Popper and Kuhn failed to solve. Lakatos claimed that the typical descriptive unit of great scientific achievement is not an isolated hypothesis, but rather that each is a research programme, and that each research programme is supported by a heuristic, problem-solving machinery which, with mathematical techniques, can digest anomalies and convert them into positive evidence. The concept of a research programme was developed by Haines-Young and Petch (1986) who concluded that the ideas do not avoid the criticism of relativism in the operation and the problem of choice of research programmes. Feyerabend (1970, 1975) claimed that the only rule in science is that anything goes, and argued that scientists should adopt a pluralistic attitude embracing all ideas because they can all potentially add to knowledge. This was certainly the case found by Slaymaker (1997) when he considered that a real, unified geomorphology requires pluralism and that 'there is no recognizable central concept in geomorphology, and we have no problem focus'.

Significant contributions to understanding the position of geomorphology and of physical geography in relation to the philosophy of science were made by Rhoads and Thorn (1994, 1996). The logical positivist picture of science, which had attempted to avoid instrumentalism and to provide theories with a realist interpretation, was by the late 1960s known as the 'received view' and considered by many philosophers of science to be fatally flawed. Three alternative perspectives were expressed by Rhoads and Thorn (1994):

(1) Post-positivist empiricism – rejects the strict instrumentalist view of scientific theories and holds that the aim of science is *to generate empirically adequate theories*, and that there is a high degree of conformity between directly observable phenomena.

(2) Social constructivism – emphasizing that observation is theory-laden, it involves the notion that the aim of science is *to solve problems* and not to seek the truth.

(3) Scientific realism – which derives from the contributions of C.S. Peirce at the end of the nineteenth century, and of Karl Popper since the 1930s. Critical realism adopts the perspective that the aim of science is *to seek the truth*, not merely to solve problems. It readily acknowledges that all aspects of scientific enquiry are theory-laden, that current theories are approximately true and are the foundation for scientific progress.

These perspectives share four commonalities (Rhoads and Thorn, 1994) which are:

- acknowledgement that the methods of science are theory-dependent;
- increased emphasis on naturalistic philosophical perspectives that fully recognize the constraints imposed upon theories of science, both by the judgements on which scientists base their claims to knowledge, and by the natural history of science;
- elimination of the notion of theory as axiomatized, hierarchical, mathematical constructs, and proliferation of less formalized views of scientific theories;
- recognition that no clear distinction exists between the context of discovery and the context of justification.

In a stimulating editorial on 'real geomorphology', Richards (1990) advocated a realist approach as one alternative to geomorphological sociology (Church *et al.*, 1985), and although greater involvement in current philosophical debate was welcomed by Bassett (1994), Rhoads (1994) showed that several brands of realism exist (see also Chapter 5, p. 124).

In the context of geomorphology, but actually applicable to physical geography as a whole, Richards (1996) suggested that general explanations commonly demand a methodology in which a complex migration occurs between the poles that are represented in typical dichotomies. Thus different individuals researching a particular problem may be simultaneously at opposite poles of the dichotomy, but as research evolves communities may shift position. Thus in Fig. 3.1A (Richards, 1996), the left-hand column summarizes a broadly empirical and positivist approach to research concerned with observational and experimental evidence, whereas the right-hand column summarizes a realist approach which employs methods appropriate to a world-view, or an ontology, in which a distinction is drawn between three levels of a phenomenon (Richards, 1994): the underlying mechanisms and the intellectual structures representing them; events caused by those mechanisms in particular circumstances; and observations of those events. Such methodological distinctions are important (Richards, 1996, p. 174) because they influence the way in which investigations are undertaken (see also Chapter 5, p. 110). A conventional positivist approach leads to an emphasis upon sampling theory, large *n* studies, statistical methods and empirical generalization. Alternatively, intensive research may involve the study of a single case or small number of cases with the objective of developing an explanation of the mechanisms generating observed patterns in an extensive investigation (Fig. 3.1A). Richards (1996) used inverted commas around 'positivism' and 'realism' in order to signify the fact that, in

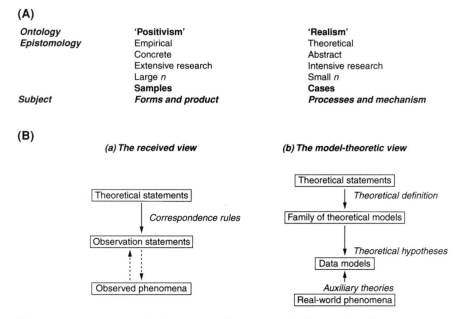

(A)

Ontology	**'Positivism'**	**'Realism'**
Epistomology	Empirical	Theoretical
	Concrete	Abstract
	Extensive research	Intensive research
	Large *n*	Small *n*
	Samples	**Cases**
Subject	***Forms and product***	***Processes and mechanism***

(B)

(a) The received view   *(b) The model-theoretic view*

Theoretical statements
                    │ *Theoretical definition*
                    ▼
Family of theoretical models
                    │
                    │ *Theoretical hypotheses*
                    ▼
Data models
                    ▲
            Auxiliary theories
Real-world phenomena

Theoretical statements
                    │ *Correspondence rules*
                    ▼
Observation statements
                    ▲
                    ▼
Observed phenomena

Figure 3.1 Summary of the 'received' view and the 'model-theoretic' view. (A) Richards's (1996) view of some linked apparent dichotomies in the scientific methodologies employed by geomorphologists. (B) Rhoads and Thorn (1996) demonstrated contrasting philosophical perspectives on the structure of scientific theories.

practice, research moves backwards and forwards between the two poles, because the labels attached to the methodologies are semantic devices rather than rigid definitions.

Considerable effort was devoted to developing an alternative to the received view, and this was known as the 'semantic' or 'model-theoretic view' (MTV) of theories. According to the MTV a theory is specified by defining a family of abstract structures as its models; Fig. 3.1B shows how Rhoads and Thorn (1996, p. 127) contrasted philosophical perspectives on the structure of scientific theories. In the 'received' view (Fig. 3.1B(a)), theoretical statements are connected to observational statements via correspondence rules, which are explicit or partial definitions. Observational statements are directly testable (verified or refuted) by comparing these statements with observations. However, in the 'model-theoretic view' (Fig. 3.1B(b)), basic theoretical statements define families of theoretical models that represent abstract, idealized representations of some domain of real-world phenomena. Raw data on real-world phenomena, along with auxiliary theories governing data reduction and analysis, are then used to develop data models which provide the basis for evaluating theoretical hypotheses that individually make claims about the real-world system being a system of the type defined by the theory (Rhoads and Thorn, 1996).

Relevant to this debate was the contribution made by Charles S. Peirce, who is now regarded as 'the one truly universal mind that nineteenth century America produced' (Dusek, 1979; quoted in Baker, 1996a). Baker (1996a, 1998) suggested that American geomorphologists including G.K. Gilbert, T.C. Chamberlin, and even W.M. Davis, showed the influence of Peirce's philosophy in their papers on the nature of geomorphological reasoning. Peirce's vision of scientific hypothesizing was associated with a kind of scholastic realism which holds that universals, including theories, general concepts and hypotheses, exist independently of our perceptions of them. Alternatively there exists the nominalism whereby *generals* do not refer to something real but rather to names that we attach to things (Baker, 1996a). Nominalism, and the logical empiricist scientific philosophy that derived from it, hold hypotheses to be useful fictions. Baker (1996a, p. 81) quoted Kitts (1980) who suggested that it is possible to argue that the proponents of naturalism and realism in the hypothetical reasoning underpinning geomorphological inference were philosophically naïve. However, earlier physical geographers were operating in the philosophical tradition of the time, and Baker (1996a, p. 82) concluded that:

> All science relies upon regulative principles and other metaphysical notions. It is precisely because we know so little of them that they should be thoroughly criticized and alternatives proposed that may seem more in accord with our actual practices of reasoning.

Rhoads and Thorn (1996, p. 117) drew attention to the main tasks of contemporary philosophy of science as outlined by Shapere (1987) to include a critical function, an overview function, and a detailing function (Box 3.1), and there will continue to be a range of stances reflecting the approaches that can be taken. Approaches in physical geography have to reflect the range of approaches available in science, although Rhoads and Thorn (1994, p. 98) commented:

> Because geomorphology is concerned with distinctive types of natural systems that include synergistic physical and biological elements and employs characteristic investigative methods, it cannot be reduced to the underpinning disciplines.

Box 3.1 Main tasks of contemporary philosophy of science *(after Shapere, 1987; quoted by Rhoads and Thorn, 1996)*

Critical function
- To continue its traditional task of exposing confused or mistaken interpretations of science.

Overview function
- To provide an overview of the rationale (or lack thereof) of scientific change.
- To determine how specific beliefs develop and change.
- To ascertain how certain beliefs are considered knowledge (i.e. are judged free from specific and compelling doubt).
- To formulate concepts of scientific reasoning and knowledge.

Detailing function
- To conduct detailed studies of science, including case studies within specific disciplines, to determine how important presuppositions, beliefs, methods, criteria, goals, and so forth have developed and changed, and to demonstrate important commonalities and differences among these factors across the various domains of science.

The outcome of these debates is that there are great pressures in favour of a realist approach as an alternative to an empiricist stance. However, Harrison and Dunham (1998) concluded, after drawing lessons from quantum mechanics and related concepts of decoherence and entanglement, that the empiricist and realist approaches are both misguided in their attempts to view the researcher as in some way 'detached' from the system under investigation. They therefore advocate an idealist approach to geomorphology, one that recognizes the primacy of consciousness, as being better equipped to appreciate the unpredictable and probabilistic nature of the world. Emphasis on 'local' circumstances and interdisciplinarity are seen (Richards *et al.*, 1997) as commending a realist perspective. There is no doubt that a pluralist approach should be maintained, as suggested in geomorphology (Slaymaker, 1997), and this was reinforced by the advent of quantification.

3.2 Quantification in physical geography

When opening his chapter on bases for theory in geomorphology, Chorley (1978, p. 1) wrote, 'Whenever anyone mentions theory to a geomorphologist, he instinctively reaches for his soil auger'. This statement had such resonance that it was used as the introductory quotation for two separate chapters in the volume compiled as a tribute to Dick Chorley on his retirement (Stoddart, 1997a, b). In addition it was suggested (Rhoads and Thorn, 1996, pp. 115–16) that 'the mention of philosophy is perhaps the surest way to get a geomorphologist into the field posthaste!' Although it was not until the 1980s that theory and philosophy (Section 3.1) were accommodated in physical geography, quantification was embraced earlier. This was facilitated, firstly, by the production of several appropriate books including statistical techniques (Barry, 1963; S. Gregory, 1963) and mathematical methods (Wilson and Kirkby, 1974). Initially many physical geographers did not fully utilize the opportunities afforded by mathematics or acknowledge the differences between mathematical and statistical methods, and the essentially descriptive quality of both groups of methods, but researchers quarried the material presented for earth scientists as a whole (Miller and Kahn, 1962; Krumbein and Graybill, 1965; Davis, 1973). A second type of contribution to the general quantitative environment was provided by books in branches of physical geography (e.g. Barry and Perry, 1973) and those that included aspects of physical geography in a quantitative manner. *Frontiers in Geographical Teaching* (Chorley and Haggett, 1965) was followed by *Models in Geography* (Chorley and Haggett, 1967), and by *Network Analysis in Geography* (Haggett and Chorley, 1969). In *Network Analysis in Geography*, inclusion of branching networks extended the focus of attention in terms of spatial structures, and exemplified the way in which an intellectual environment was created by proposing a reorganization of geographical knowledge, by collating and sometimes importing material which would otherwise have been difficult to trace, and hence providing a stimulating and research-provoking work. Networks, although very refreshing and original, were heavily associated with certain parts of physical geography, particularly with developments from R.E. Horton (see p. 122), and the book had a structure at variance with the established subdivisions of physical geography into geomorphology, climatology and biogeography. A decade or so later, similar collections of material, spanning physical as well as human geography, further reflected progress of quantification (Wrigley, 1979; Wrigley and Bennett, 1981).

Against this background and the developments in scientific thinking (Section 3.1), the discipline was ready for the thought-provoking approach provided by David Harvey's *Explanation in Geography* (Harvey, 1969). In that book he cited the work of Caws (1965) and related the subjective pole of experience (S in Fig. 3.2A) to sense perceptions or *percepts*, to mental constructs or images which provide *concepts*, and to linguistic representation which gives *terms*. Harvey (1969) also reviewed the routes to scientific explanation including the Baconian route (Fig. 3.2B) and the alternative, which acknowledges dependence upon an *a priori* model (Fig. 3.2C). Adoption of the first route (Fig. 3.2B) can be dangerous because acceptance of the interpretations may depend upon the standing and charisma of the scholar involved (Moss, 1970). The second method begins with the researcher perceiving a pattern of some kind in the real world; an experiment is then formulated which can test the validity of the *a priori* model. The first route (Fig. 3.2B) is effectively inductive by proceeding from unordered facts towards generalization, whereas the second is deductive because it relies upon an *a priori* model which is perceived at an early stage to allow for manipulation of data and for conclusions to be drawn about some set of phenomena even if a complete theory is not available. An alternative *a posteriori* model expresses the notions contained in the theory in a different form such as mathematical notation, but these have not been used as extensively in physical geography as have *a priori* models. Once a model has been tested, by the route shown in Fig. 3.2C, this may lead to a theory, a set of sentences, expressed in terms of a specific vocabulary, to facilitate discussion of the facts that the theory is to explain. Whereas a theory cannot necessarily be shown to be true or false, a hypothesis is regarded as a proposition which can be shown to be true or false (Harvey, 1969) and is more restricted in science. Braithwaite (1960, p. 2) contended:

> A scientific hypothesis is a general proposition about all the things of a certain sort. It is an empirical proposition in the sense that it is testable by experience: experience is relevant to the question as to whether or not the hypothesis is true, i.e. as to whether or not it is a scientific law.

From the late 1960s a range of models were employed in the route (Fig. 3.2C) towards scientific explanation and in addition to the continuation of existing approaches, Werritty (1997) advocated further enquiry into chaos theory.

In addition to the intellectual environment created across geography and over related parts of the earth sciences, general statements were made in physical geography. A very influential stimulus was the systems approach (Chorley, 1971; Chorley and Kennedy, 1971) as outlined in Chapter 4. Chorley (1971) drew attention to the ever-deepening dilemma confronting

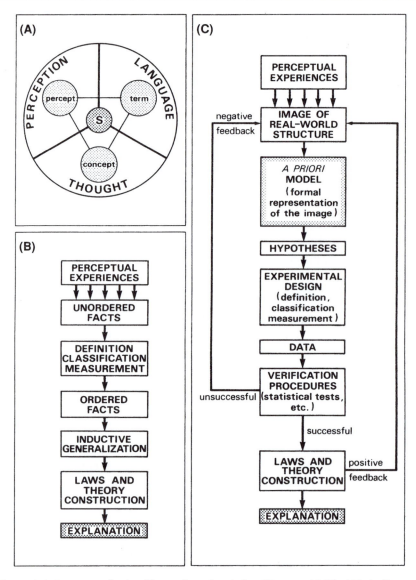

Figure 3.2 Patterns of scientific explanation (after Harvey, 1969). (A) A diagrammatic representation of the relationship between percepts, concepts and terms; (B) the 'Baconian' route to scientific explanation; (C) an alternative route to scientific explanation employing an *a priori* model.

physical geography. On the one hand physical geography had become responsible both for advanced research and for basic teaching in many earth science subjects, and on the other hand it was required to play a relevant role in an increasingly economically- and socially-oriented human geography. Although both roles had been fulfilled efficiently in the past, it was suggested that the increasing technical demands made upon researchers in the earth sciences, and the increasing preoccupation of human geographers with spatial socioeconomic matters, was making the position of the physical geographer increasingly difficult, and this was pointedly characterized (Chorley, 1971, p. 89) as:

> ... rather like that of a tightrope walker attempting to walk simultaneously on two ropes which are becoming more and more separated. Will the acrobat eventually commit himself wholly to one or the other of the ropes, thus either removing into the earth sciences the last vestiges of what used to be called the 'physical basis of geography', or continuing to feed forcibly increasingly unreceptive human geographers with a diet of what seems to them to be physical irrelevancies? On the other hand, will a painful schism occur rupturing traditional physical geography into two parts in which, for example, the 'geographical geomorphology' and descriptive climatology of 'true' physical geography are divorced from the geological geomorphology and dynamic climatology of the earth sciences?

In this penetrating analysis it was suggested (Chorley, 1971) that physical geography must be sensitive to the changing objectives of geography as a whole and also structured within a sufficiently viable intellectual framework. Three frameworks were offered based upon: model building, which had been profitable but would not provide the kind of fusion needed between physical and human geography; resource use, which did not offer a sufficiently comprehensive intellectual base for academic physical geography; and systems analysis, which was the most attractive (Chapter 4, p. 83). Systems analysis provided a framework for a summary of statistical applications in physical geography (Unwin, 1977) because it was suggested that traditional subdivisions of physical geography were increasingly irrelevant to what physical geographers actually did, and prompted the conclusion that greater emphasis had been placed upon deterministic than upon stochastic approaches; that quantitative advances and applications in the 1970s had not continued the momentum generated in the 1960s; and that this loss in momentum contrasted starkly with the continued growth of statistical modelling in human geography.

Complementing the emphasis upon statistical methods, the increasing utility of computing (Mather, 1976) and the general availability of mathematical methods (Sumner, 1978) provided two additional props for quantification. Considerable scope remained for more extensive use of mathematical methods when Wilson and Kirkby (1974) concluded that:

> ... there are many analytical problems associated with geographical systems of interest and many 'real world' planning problems, which could benefit from a more mathematical treatment.

Some 20 years after the conference that led to *Models in Geography* (Chorley and Haggett, 1967) a further volume on *Remodelling Geography* was produced (Macmillan, 1989a). Many of the chapters relate developments in the branches of physical geography up to the late 1980s, and Unwin (1989) concluded that when models were first introduced in 1967 they were broadly acceptable and intelligible to all. However, he contended that 'Looking at the variety of modelling in physical geography undertaken since 1967, it is evident that this work is now much less accessible', and furthermore, 'this lack of accessibility to the average student I regard as a major challenge to education in physical geography'. He concluded that the implications for the future of physical geography are either that we become the last generation of geographers able to contribute to modelling research, or that we pay much more attention to the problem of training the next generation of modellers than we have of late. This theme was amplified by Henderson-Sellers (1989a), and Unwin himself showed how energy balance simulation can be used for teaching a model-based climatology (Unwin, 1981). Four trends that changed the style and subject matter of modelling in the 1980s were characterized by Thornes (1989) as:

(1) the relaxation of subject boundaries within the earth sciences – this has seen the consequent fruitful invasion of physical geography by other scientists anxious and well equipped to do the job;
(2) the consequent development of modelling at regional and global scales;
(3) the development of evolutionary models which revitalized flagging interest in long-term behaviour;
(4) some areas previously neglected in mega-models, such as vegetation cover, have now been taken up.

To take advantage of the opportunities available it is necessary to develop rigorous approaches, and Kirkby (1989, p. 271) analysed the situation as follows:

> ... physical geography is in great danger for lack of a consistently rigorous approach to theory, to models, or to experimental procedures. In many areas of research we are increasingly under threat from other scientists who are directing their interests to the physical environment, and we frequently suffer in comparison with them.

He proposed the following solution (Kirkby, 1989, p. 256):

> ... it becomes increasingly important for physical geographers, and others involved, to have an effective training in the physical, chemical and biological principles which underlie their system of interest. We are already in danger of losing our grasp on these different topics, and only improved scientific training and the recruitment of interested scientists can maintain our hitherto substantial contribution.

Despite the strength and popularity of geography as a school subject in Britain

(p. 285), it is not easy to see how this training can be sufficient when the sciences are less popular. The search for improved models has continued, the lack of modelling has been redressed, but it still remains important to scrutinize the data quality and its appropriateness if GIGO (garbage in, garbage out) is to be avoided. In any modelling situation, decisions are required about the parameters that are used and how they are obtained, what qualitative models and knowledge underpin the models that are built, and how much validation or testing is undertaken to verify whether the model is effective or not. We are still dominantly empiricists or theoreticians and need to avoid becoming intransigent, looking at a problem from one perspective only.

3.3 Developments in branches of physical geography

Results of the advent of quantification included the following:

- new statistical techniques were sought and incorporated into physical geography, sometimes with insufficient concern for data quality and programme specification where a technique was in search of a problem;
- batteries of techniques were presented to the geographical or physical geography audience, sometimes with no real conceptual frameworks (e.g. Cole and King, 1968);
- conceptual progress by models, systems or networks was sometimes the vehicle for the selective presentation of techniques;
- the impact of quantification was not equal throughout all branches of physical geography, as indicated below.

Study of the *atmosphere* was not exclusively, or even predominantly, the domain of physical geographers, and E.H. Brown (1975) drew attention to the imbalance of physical geography, noting that, of the physical geography papers submitted to the 1972 International Geographical Union Meeting in Montreal, 44% were concerned with the land, 27% with air, 9% with water, 10% with plants and animals, and 10% with soils (cf. Table 11.1). In Britain it was argued (Unwin, 1981, p. 261) that:

> Climatology has always sat uneasily within British geography departments yet to date it has not been convincingly taken up by any other discipline. In consequence it has been depressed, a sort of Cinderella to the ugly sisters of geomorphology and biogeography.

An inspection of nine English-language journals over the decade 1970–80 revealed on average fewer than one climatological article per journal per year (Atkinson, 1980). Internal reasons, embracing the expense and skills required, and external reasons, particularly the existence and role of government or national organizations, were suggested to account for this situation.

Quantitative methods became especially pronounced in studies of the atmosphere, and the attitude to climatology as a form of book-keeping, involving the numerical record of mean atmospheric conditions at particular places, provided the basis for descriptive and physical climatologies that were effectively statistical in nature (Atkinson, 1980). By 1957 the distinction was clearly drawn (Court, 1957) between complex (analysis and presentation of climatic information related to practical applications); dynamic (explanatory description of the atmosphere (Hare, 1957)); and synoptic climatology (obtaining insight into local or regional climates by examining the relationship of weather elements to atmospheric circulation processes). Climatology, visualized as a subject for the geographer, was dependent upon models in climatology primarily statistical in character, and also on meteorological models which are physical and mathematical, relying on the basic laws of physics and hydrodynamics. Theoretical analyses of climate were almost exclusively undertaken in university departments of physics, geophysics, mathematics, meteorology or oceanography, or in government institutions such as the British Meteorological Office (Atkinson, 1980).

Contributions by geographers in climatology were resolved (Atkinson, 1980) into four fields:

- regional–physical concerned with classification and particularly following the rationally based approach to classification inspired by the work of Penman (1948) and Thornthwaite (1948), and also with heat and water budgets of global or continental areas. Hare (1968) had suggested that Thornthwaite, Budyko and Penman had transformed the way that we look at climate as an element of the natural environment;
- synoptic climatology, concentrating upon data methods and applications, to link global understanding of the atmosphere and knowledge of local and regional-scale phenomena, and well exemplified in the book by Barry and Perry (1973);
- boundary layer climates which include the physical, topo-, local, meso- and regional climates of the lowest kilometre of the atmosphere, and embrace both natural and man-modified climates (Oke, 1978);
- climatic change has also been an arena for work by physical geographers.

In each of these four contributions quantitative methods were a necessary tool, readily utilized by physical geographers, but although the focus of research on the atmosphere by geographers centred upon climatology, it was argued (Hare, 1966) that, as the shift occurred away from parameters such as temperature and humidity towards the measurement of fluxes, the geographer should be exclusively concerned with climate as environment and therefore also with the direct impact of climate on human health, efficiency and psychology, and indirect impact on human activity (see p. 217). W.H. Terjung (1976, p. 222) noted a similar opportunity:

> Because of a predilection for self-contemplation and sometimes almost suicidal academic isolation, geography has a history of missing the boat on vital social and environmental issues.

A particular reason for the failure of physical geography to grasp the opportunity was identified (Henderson-Sellers, 1989b, p. 1):

> The relationship between geography and climate has been prey to bad timing and lack of mutual understanding and the villain of this piece is the digital computer. Just when budding partnerships with other disciplines were developing in the 1950s and 1960s, the revolutionized and quantified geography was ignored by atmospheric scientists because they had need to focus their full attention on the difficult task of solving the equations of motion fast enough for weather forecasters to beat the real world events.

This wealth of data, the acceleration of data collection, and the increased speed of forecasting followed attempts in the late 1950s to increase the quality and quantity of observations by improvements at existing weather stations and from upper-air radiosonde stations. The World Weather Watch (WWW) led to activities under a Global Atmospheric Research Programme (GARP), aiming to provide a more fundamental understanding of atmospheric circulation and of climate systems as a whole (Atkinson, 1980). Development of the use of remotely sensed information from satellites also increased data availability, and the initial GARP Global experiment (FGCE) was the first occasion on which observation for one year (1 December 1978 to 30 November 1979) included observations of the Earth's atmosphere from a truly integrated system of satellites.

Quantitative methods were necessary in climatology in view of the way in which the fundamentals of meteorology employed basic laws of physics and hydrodynamics. Great advances in the mathematical modelling of atmospheric circulation occurred at several scales, and meteorological models were dominant within climatology (Barry, 1967). General circulation models (GCMs) were of interest to physical geographers, but models were developed later in climatology than they were in geomorphology (Henderson-Sellers, 1989a) because observational data had to await global coverage by satellites, and the concept of climate as an ensemble of mean weather could not become established until meteorological models were shown to work. In addition, climatology seemed to be exclusively directed towards the atmosphere, rather than the troposphere, with insufficient recognition of water vapour greenhouse feedback and cloud feedback (Henderson-Sellers, 1989a).

In *geomorphology* an influential paper, triggering direct and indirect quantitative developments (Horton, 1945), was not the first in which Horton developed concepts that would later assume great significance, but the 1945 paper was important, firstly because it provided the foundation for the Horton runoff model (see p. 123) and hence stimulated emphasis upon processes, and secondly because it provided the basis for a quantitative approach to morphometry. Horton's approach was quantitative, it offered a way of relating form to process, and it appeared to be more related to contemporary landscape and its problems than did denudation chronology (see also p. 141). The Strahler school at Columbia (p. 119) stimulated work on drainage basin morphometry and, after nearly a decade, it was appreciated that stream ordering and drainage composition reflected statistical relationships rather than deterministic ones and were unrewarding (Werritty, 1972), but the bonanza of developments was summarized by Strahler (1964) and retrospectively by Strahler (1992). In an extensive review, Smart (1978) also concluded that network topology could not yet be beneficially utilized in relation to hydrological prediction models.

Whereas the most cited development of geomorphology benefiting from quantification was drainage basin morphometry, it is arguable with hindsight that the approach initiated by G.K. Gilbert in the nineteenth century should have been more widely adopted (see Chapter 5, p. 104). Other developments were either conscious of the research inspired by Horton or developed in parallel. A singular approach (Scheidegger, 1961, 1970, 1990) to *Theoretical Geomorphology* was developed from the standpoint of geodynamics, but did not receive the acknowledgement it deserved because of reliance upon a mathematical theoretical foundation and did not cover the range of geomorphology completely. It was not until 1988 that *An Introduction to Theoretical Geomorphology* was provided (Thorn, 1988) as a book to commend geomorphology to students as a 'thinking before digging (or looking) discipline'. Thorn saw theory as an integral part of being human, whether applied to daily life, geomorphology, or any other human endeavour, and argued (1988, p. 115) that it is necessary to have a clearly defined objective and a well-developed grasp of relevant theory as the way to appreciate not only strengths but also limitations and weaknesses, so that although applied geomorphology was doing much to transform the discipline by restricting it to a human, management scale (Thorn, 1988, p. 118):

> geomorphology ... needs to exhibit better theoretical articulation so that it may initially present a more persuasive case to the scientific community at large, and, thereby, subsequently become more visible to society in general.

Slopes also offered scope for a new approach to geomorphology, originating in some of the work of the Columbia school (Strahler, 1950b); in the Netherlands (e.g. Bakker and Le Heux, 1946) developing mathematical models also undertaken by Scheidegger (1961); and exploration of the potential of morphological mapping (Waters, 1958; Savigear, 1965). The last approach concentrated upon the entire landscape, but methods for the analysis of slope data were sought including the use of data processing (Gregory and Brown, 1966), and deductive models of slope evolution developed (Young, 1963). Whereas these approaches were focused primarily upon morphological data, alternatives were geomorphological mapping as in applied physical geography (p. 204), and general geomorphometry defined by Evans (1981, p. 31) as 'analysis of the land surface as a continuous, rough surface, described by attributes at a sample of points or arbitrary areas'.

Although employing quantitative methods for large amounts of data collection, studies of slopes did not advance significantly until a more realist approach focused upon slope mechanics was developed (see p. 108). Sedimentology, especially of coastal environments, also provided additional opportunities for quantitative approaches, and these were directed particularly towards the description of sediment characteristics so that they could be the basis for deduction about environment and landscape change. Appropriate statistical techniques for description and comparison of sample frequency distributions were necessary for grain-size characteristics of sediments, for the possibilities introduced by till fabric analysis of glacial deposits (Holmes, 1941), and for the analysis of the orientation of coarse particles in tills (Andrews, 1970b). Similar methods were subsequently employed for fluvial deposits, and all these advances were reflected in the content of quantitative texts produced (e.g. Miller and Kahn, 1962; Krumbein and Graybill, 1965). Particle shape could be assessed by simple visual estimation (e.g. Krumbein, 1941), but quantitative measures of flattening (Cailleux, 1947) and of roundness (Tricart and Schaeffer, 1950) were the basis for environmental interpretations from specific deposits and later developments (e.g. Fleming, 1964; Gregory and Cullingford, 1974).

In geomorphology, therefore, the decades of the 1950s and 1960s were characterized by an expanding range of subjects for attention with the import of quantitative techniques (Doornkamp and King, 1971), but it took time to appreciate the precise applications of these techniques. Even in 1981, Thornes and Ferguson (1981, p. 284) concluded that the main application of quantitative methods was still the 'description and analysis of field data on particular sites, events or areas in a mainly inductive framework'. Some progress had been made towards spatial analysis in geomorphology (Chorley, 1972), influenced by two distinct quantitative dynasties: one concerned with the application of quantitative techniques to geology, pioneered particularly by Krumbein (e.g. in Krumbein and Graybill, 1965); and the other being spatial analysis in human geography beginning in the later 1950s (Chorley, 1972). Delay in adopting spatial analysis techniques was adduced to be because: many central geomorphological problems had not traditionally been expressed in quantitative terms; many innovators of quantitative techniques had insufficient mathematical background to develop their use in geomorphology; in the US and Sweden, where spatial analysis was most developed in geography, the links between human geography and geomorphology were traditionally rather weak compared with those in Britain; work in automated cartography largely centred around technical problems of data storage and retrieval rather than use of data in model-building; and spatial model building proceeded most rapidly where ideas of systems analysis had already been assimilated.

It is very difficult to separate quantitative developments from conceptual progress in other fields, and from advances especially relating to landscape processes (Chapter 5, p. 109). However, several papers exerted important influences upon, and to some extent were generated by, the inception of quantitative approaches. These papers included the concept of entropy in landscape evolution (Leopold and Langbein, 1962); probability concepts in geomorphology (Scheidegger and Langbein, 1966) advocating a statistical or probabilistic viewpoint; and the concept of minimization of effects (Williams, 1978). A further important paper referred to association and indeterminacy (Leopold and Langbein, 1963). Association was identified as the basis for geological reasoning and was useful to indicate sequences of events in time; to extend from local to general; and to indicate processes, although it was concluded that 'in geomorphologic systems the ability to measure may always exceed ability to forecast or explain' (Leopold and Langbein, 1963, p. 191). The principle of indeterminacy, although long recognized in physics, was new to geomorphological thinking, and referred to those situations in which the applicable physical laws could be satisfied by a large number of combinations of values of interdependent variables, so that the result of an individual case is indeterminate, and although the range of uncertainty should decrease as more is learned about the factors involved, the uncertainty will never be removed. Leopold and Langbein (1963, p. 192) concluded:

> The measure of a research man is the kind of question he poses ... Geomorphology is a field of enquiry rejuvenated not so much by new methods as by recognition of the great and interesting questions that confront the geologist.

It is debatable whether the full import of these challenging developments was realized in geomorphological research, and Kirkby (1989) suggested that we should beware of developing new principles exclusive to geomorphology or to other branches of physical geography, because landscape entropy, enunciated as a principle by Leopold and Langbein (1962), is at least in part a restatement of the principle of least work.

Comparatively few quantitative contributions to **biogeography** occurred before the 1970s because biogeographers were relatively few in number and were directing energy towards historical aspects (p. 149), so that they tended to be more concerned with plants and animals in evolutionary time than with ecological time aspects (Simberloff, 1972, p. 94). This paucity is reflected in the fact that there was no frontier described for biogeography in *Frontiers in Geographical Teaching* (Chorley and Haggett, 1965), and the chapter in *Models in Geography* was directed towards organisms and the ecosystem (Stoddart, 1967b), and eventually separated from the physical chapters when the book was reprinted in 1969. Pleas that biogeography deserved a more significant place in geography (e.g. Edwards, 1964) curiously ignored the quantitative advances being made, as well as the greater attention accorded by ecologists to biological production by ecosystems, emphasized by the inception of the International Biological Programme in 1965. Biogeography tended to emphasize spatial aspects and environmental relationships, with insufficient early recognition of the potential application of quantitative methods, although by the 1980s ecosystems provided a more significant approach (Simmons, 1987). Quantitative advances in ecology embraced two groups of developments. First was theoretical deduction from mathematical models, such as the theory of island biogeography, which involved a relationship between the area of an island and the number of species of a given taxon. Second were developments in the statistical analysis of vegetation composition, with a wide variety of numerical approaches, generated after the mid-1950s, and developed by 1970 (Moore *et al.*, 1970). Some of these approaches were later acknowledged in the researches of biogeographers, and five trends isolated by Watts (1978) were:

(1) investigations of soil–vegetation–environment complexes;
(2) relationships between major vegetation types and particular animal species;
(3) analyses of distributions of individual species and of the influencing processes;
(4) Quaternary community or ecosystem change;
(5) mankind–ecosystem–community relationships.

Although the seeds of quantitative techniques were beginning to germinate in the work of biogeographers, there was some delay before productivity was very apparent. Stoddart (1986, p. 271, Chapter 12), following earlier concerns about taking the 'ge' out of geography (Wooldridge, 1949) by researchers concerned with morphology rather than origin of landscape, argued for putting the geography back in the bio- and for a spatial approach to be clearly central in biogeography. He felt (Stoddart, 1986) that there had been too many distractions from classical concerns of distribution, ranges and limits of particular plants and animals in favour of ecosystem principles and concepts, organisms as components of patterns, and nutrient cycling in specific ecosystems, and that only Stott (1981) had maintained the theme that the core of biogeography lies in the analysis and explanation of pattern and distribution.

Quantitative approaches required an even longer gestation period in the case of *soil geography*, although this may have been because biogeography tended to subsume the study of soils until the 1970s when advantages were perceived for studying soil geography as a new and vigorous branch of the subject (Bridges, 1981). The essentially quantitative perspective provided through the factors of soil formation (Jenny, 1941), although often cited, was not developed by physical geographers until the 1970s with the systems approach (Chapter 4, p. 85). The quantitative approach to the soil profile was also somewhat delayed, although there were attempts to rationalize data collection and mapping, and to codify soil profile properties so that each soil profile could provide up to 200 items of information for computer analysis (Webster *et al.*, 1976). In soil geography the process of data acquisition, especially by mapping, and the dominance of researchers other than physical geographers, together with the data-rich character of the soil itself, necessarily retarded the development of quantitative methods. When techniques for analysis of soil distribution were reviewed by Courtney and Nortcliff (1977), they noted that, despite the great attention devoted to soil survey procedures, most investigators paid scant attention to spatial relationships. Although there is no single soil property that typifies the soil as a whole, classification, ordination, regression and analysis of variance methods were now available, meriting further consideration in relation to studies of soil boundaries and soil variability (Courtney and Nortcliff, 1977). By the 1980s Trudgill (1987) argued that attention should be paid to 'basic processes, developments in soil survey, soil properties, soil management for agriculture, and linkages with other environmental components', thus illustrating the way in which quantification had been absorbed as a technique useful for thrusts in this and other branches of physical geography.

3.4 Physical geography in 1980

The advent of quantification, greater awareness of the need for theory, and the impact of modelling approaches affected each branch of physical geography after 1950; similar issues had to be addressed, albeit with slightly different timings, and the choice between a holistic view and increasingly reductionist approaches had to be made. This required appreciation of the range of scale of systems (Fig. 3.3A) and the consequences for understanding the range of complexity (Fig. 3.3B). As was the case in other fields (Shapere, 1974), the core branches of physical geography, such as geomorphology (Rhoads and Thorn, 1996), were identified fairly readily whereas the peripheries were fuzzy and rather less clear. This fuzziness around the edges and the need for pluralist approaches characterized individual branches (Chapters 4 to 8), which were aided by developments in remote sensing, computing and geographical information systems; they were also affected by relationships with other disciplines.

One of the greatest impacts in relation to data quantification and the transformation of physical geography was from enhancement of sources of *remote sensing*. In 1960 remote sensing simply referred to 'observation and measurement of an object without touching it', but now usually refers to 'the

Figure 3.3 Size of systems and gradients of complexity. (A) The size of nested systems and 'life-span' of systems, after Huggett (1991), with permission from the author and from Springer-Verlag. (B) Gradients of time, space and biological complexity from Simmons (1993a) to show the difficulty of measuring impacts upon ecosystems and the predictability of impacts upon systems.

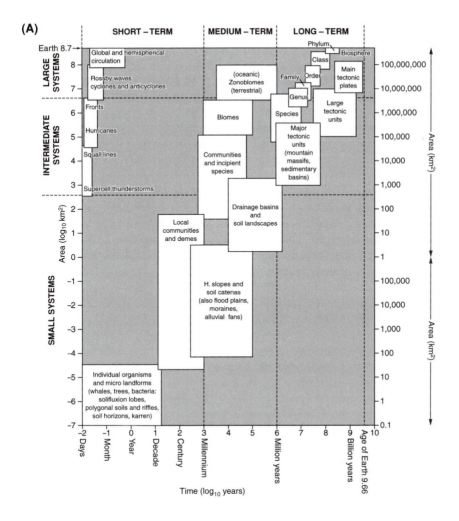

(A)

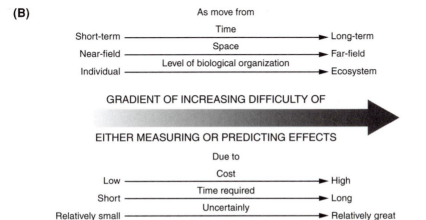

(B)

As move from

	Time	
Short-term	——————————————→	Long-term
Near-field	Space ——————————————→	Far-field
Individual	Level of biological organization ——————————————→	Ecosystem

GRADIENT OF INCREASING DIFFICULTY OF

EITHER MEASURING OR PREDICTING EFFECTS

Due to

	Cost	
Low	——————————————→	High
Short	Time required ——————————————→	Long
Relatively small	Uncertainly ——————————————→	Relatively great

use of electromagnetic radiation sensors to record images of the environment which can be interpreted to yield information' (Curran, 1985). Terrestrial remote sensing encompasses all those techniques used to obtain information about the Earth's surface and its atmosphere by sensors located on a specific platform recording radiation from the electromagnetic spectrum. Many such platforms are available and include aircraft, balloons, or even a large tower. Remote sensing from aircraft obtaining black and white and subsequently colour aerial photographs had long been employed in physical geography as a particularly important ingredient for terrain resources evaluation and had been used in relation to landscape ecology, land systems and more specific land evaluation techniques (p. 205). However, more radical developments occurred as more wavebands were used from the electromagnetic spectrum, and the number of platforms increased. The electromagnetic spectrum includes the photographic bands of visible and near infra-red, used for more than 30 years, but it also includes linescan (visible and infra-red), active microwave (side-looking radar) and passive microwave (Hardy, 1981); investigations explored how sensing of all parts of the electromagnetic spectrum could be used to enhance physical geography research. Enhancement of the number of platforms available was achieved because not only can satellite platforms survey large areas of the world, but they also provide frequent and repeated surveys of the same area. Thus, the problems of environmental data collection and of the need for repeated observations were alleviated by remote sensing from satellites. The multispectral scanner on the Landsat satellite system available since 1972, following a sun-synchronous orbit, repeated a pass along a given track every eighteenth day, so providing data every 18 days at a particular location and giving information on deliberate or incidental environmental change. In the 1970s it was possible to deduce, from Landsat imagery, the progress of dam construction along the Huang He in China (Smil, 1979). Aerial photograph interpretation contributed to land and resource inventory, but satellite imagery provided even more (Allan, 1978; Townshend, 1981a). In the field of satellite climatology significant developments were made (Barrett, 1974), including prediction of rainfall intensities from satellite data (Barrett, 1970); estimation of rainfall (Barrett and Martin, 1981); and the climatology of clouds (Henderson-Sellers, 1980). The fields to which satellite data analysis could be applied were legion, and Lulla (1983) collected together the applications of Landsat data, resolving them into those concerned with:

- aquatic environment and ecosystems, including chlorophyll and particulate suspended solids monitoring; wetland and coastal ecosystems including wetland biomass estimation and measurement of coastal primary productivity;
- terrestrial ecosystems including leaf area index estimations, applications to agricultural disease prediction and to forest to estimate timber volume, and to terrestrial biomass and primary productivity estimates;

- scarred landscapes including desertification and studies of drought impact.

Of the many implications of remote sensing data sources for physical geography, the most dramatic are (Lulla, 1983):

- the possibility of taking a global view;
- obtaining frequently repeated data from imagery;
- obtaining images that allow interpretation of patterns such as plant disease which are not feasible by other methods;
- analysis of results in what is essentially a non-destructive method for obtaining estimates of bioenvironmental parameters such as biomass, leaf area index, and canopy coverage.

Box 3.2 shows just some of the exciting ways in which information from remote sensing sources revolutionized the data available to physical geography, the types of problems that could be addressed, and the issues where a spatial approach is indeed a vital one. Two problems outstanding were the resolution available and the interpretation of the remotely sensed information. Whereas Landsat 4 in the early 1980s afforded ground resolution of 30 m, Landsat 5 launched in 1984 carried the thematic mapper which recorded information at six discrete wavelengths for successive areas 30 m^2 on the ground, together with the emitted radiation in the thermal infra-red waveband in every 120 m^2, and the information was recorded every 16 days for nearly all the Earth's surface. In 1985 the first of a series of French Earth observation satellites (SPOT-1) provided 10 m resolution panchromatic and 20 m resolution multispectral data (Briggs and Jackson, 1984). In summarizing the taxonomy of remote sensing satellites Curran (1985, Fig. 5.1, p. 130) identified 14 major sources including Landsat and SPOT, and by the 1980s the rate at which data about the biophysical environment were collected from satellites exceeded that from ground level (Townshend, 1987).

The quantity and quality of information extracted and interpreted (Townshend, 1981b) was subsequently aided by developments in information technology, although visual interpretation of aerial photographs undertaken for many years meant that one expected to see a visual photographic image, but the digital processing of remotely sensed data does not necessarily involve visual images. Remote sensing information collected for picture elements (pixels) is stored as digital data, and since a single scene from the thematic mapper covering an area of 185 × 185 km comprises about 300 million separate items of information, and because many such scenes may be required in analysis for one purpose, the enormity of the information processing task is evident. Thus Townshend (1981c) concluded that the most important future development was likely to be the integration of remotely sensed information with other data sources to provide geographical information systems. Image processing systems for the analysis of

Box 3.2 Recent remote sensing developments. *This summary lists all the recent developments cited by White (1998) in relation to specific areas of physical geography.*

Developments in technology such as compact solid-state imaging instruments have further enhanced applicability of Earth observation data to environmental monitoring, and general examples include:

- buried waste from thermal images as waste trenches tend to be cooler and damper than surroundings;
- atmospheric haze assessed by comparison with haze-free images of the same area;
- effects of atmospheric pollution from metal smelters on vegetation mapped;
- pollution from mine tailings discharged into rivers detected by mineral mapping from hyperspectral data;
- data from airborne surveys of effluent outfalls compared with models to identify inadequate outfall operations;
- locating opium poppy crops;
- mapping urban areas from night-time images;
- slope and aspect data, potential increased through increasing availability of multiple-view angle data;
- magma levels monitored in deep conduits in volcanic areas.

Aspects of vegetation:
- different vegetation communities;
- potential drought conditions;
- forest resources, inventory;
- biomass estimation;
- forest fire hazards;
- hurricane damage;
- growth of fires in real time from thermal systems;
- advance warning of fire risk.

Soil characteristics:
- estimating soil salinity;
- soil moisture estimation;
- soil colour, from optical RS;
- iron oxide concentrations in surface layers, mapped from space;
- geochemical mapping of sand seas based on iron oxide content;
- cyanobacterial crusts on surface (role in preventing soil erosion) mapped from RS data due to higher blue reflectance compared with surrounding surfaces;
- crop residue amounts (inhibit erosion) can be remotely sensed.

Geomorphological aspects:
- mapping of alluvial fans on the basis of mineral differences between fan surfaces of different ages;
- rock roughness related to age of surface – can be detected by radar backscatter;
- buried fluvial features detected by surface penetration of microwave radiation through dry materials;
- interaction of fluvial and marine processes at river mouths.

Ice and snow:
- position of snow line;
- grounding line of shelf ice;
- growth rates of aufeis (overflow icings) determined from SAR interferometry;
- changes in tundra vegetation, from multiple-view angle data;
- permafrost mapping.

satellite digital data were developed as microprocessors and microcomputers advanced in capability and decreased in size. This was the culmination of developments that proceeded from the use of computers for analysis and simulation of environmental problems to computer cartography and thence to databases. The *database*, as computer-stored information which replaces the published map or data-set, became increasingly valuable because it could be continuously updated and provide the basis for the most up-to-date map or list, issued as and when required.

In cases where there is a requirement for large volumes of spatial data of different kinds, it is possible to process information which may or may not be derived from remote sensing, and to develop an *automated geographical information system* (GIS), stored by referencing each item to a set of geographical coordinates (Heywood *et al.*, 1998). Many geographical information systems were originally devised in response to the needs of government programmes, in the way that the Coastal Zone Management Act of 1972 led to the development of an automated geographical information system for the State of Carolina, subsequently the basis for a review of the data needs of the State (Cowen *et al.*, 1983). The impact of geographical information systems has been massive, and the 1980s saw a revolution in the use of computers in research and application in geography which Rhind (1989) suggested to have the following primary characteristics:

- enormous growth in design, creation and use of databases;
- move from parametric descriptors and inferential statistics;
- number of academic geographers is miniscule compared with this area in other disciplines;

- the commercial sector, not governments or academia, has produced new technical developments and fostered new applications;
- existing turn-key systems are effective;
- much of the research work is produced in conference proceedings;
- the main thrust is the design, creation and exploitation of databases;
- the rapid decrease in the cost of computing power with micro-computers.

The impact of technology was illustrated by Rhind (1989, p. 180) who noted that in 1973 the most sophisticated system in British universities (a NUMAC IBM 360/67) had a memory and raw processing power which was equivalent to that bought at the end of the 1980s for less than £5000. The way in which technology facilitated access to computing power to enhance the speed and extent of analysis inevitably meant that the boundaries between groups of techniques became less clear, and Macmillan (1989b, p. 310) concluded:

> The separate categories of statistical, mathematical and data-based modelling are already becoming redundant. It is easy to foresee them being superseded by the development of enhanced GIS . . . with sophisticated spatial analysis and mathematical modelling tool chests.

For many physical geographers, textbooks may be the only contact with the field of GIS; Maguire *et al.* (1991) identified 13 textbooks up to 1991, of which the first was published in 1986 (Burrough, 1986), albeit with a land resources and computer cartography orientation. It was in 1989 (Aronoff, 1989) that the first textbook appeared that expressed the spirit and purpose of GIS. From the multiplicity of definitions of GIS available, Maguire (1991) concluded that 'A geographical information system is best described as an integrated collection of hardware, software, data and liveware which operates in an institutional context'. GIS can be categorized as map, database and spatial analysis views, because GIS are a special case of information systems but are differentiated by the general focus upon spatial entities and relationships, together with specific attention devoted to spatial analytical modelling operations. Maguire (1991) described the evolution of GIS as a three-stage process whereby:

(1) the early stage is oriented towards data collection and inventory operations;
(2) after about 3–5 years the emphasis shifts to a more analytical operation;
(3) maturity is achieved in a further 3–5 years when there are true decision support systems. Spatial analytical and modelling operations are routinely employed.

In their editorial comment for the first issue of the *International Journal of Geographical Information Systems*, Coppock and Anderson (1987, p. 3) represented the rapidly developing field of GIS, lying at the intersection of many disciplines including cartography, computing, geography, photogrammetry, remote sensing and statistics surveying, as being concerned with handling and analysing spatially referenced data. That journal, initiated in 1987 with four issues per year, had increased to eight per year in 1997 and 1998.

Potential applications of GIS continue to be explored, but the timeliness was stressed by Maguire *et al.* (1991, p. 6–7):

> Perhaps the ubiquitous concern of scientists, politicians and other human beings at large in the late 1980s and early 1990s is that for the fate of the environment ... GIS have already made useful contributions to the collection, analysis and understanding of information on the environment and promise much more success.

Environmental databases have grown rapidly, but not many benefited initially from GIS technology because of the investment required to convert existing analytical data to digital format, and also because of the need to accommodate 3- and 4-dimensional data, which is not easy for GIS. Townshend (1991) collated examples of geographically referenced data holdings and suggested that the diverse features of data-sets include:

- locational referencing – two-dimensional is satisfactory for some but others, including marine and atmospheric data, need three dimensions and time;
- longevity of databases – some have old data which are still important;
- types and formats of data – there are many different forms, raster and vector forms and digital and analogue forms;
- availability of digital data – information is not all readily available in digital form;
- degree of integration of data-sets – some, such as meteorological, are good internationally, others are good at the national scale, and some not complete at that level.

However, Townshend (1991) also recognized certain similarities which were:

- growth in size of data-sets;
- increasing digitization of data holdings;
- mismatches between the requirement for the use of data and GIS technology;
- increased interlinking and spatial integration of data-sets.

GIS found very ready application in relation to soil information systems and, although introduced since the mid-1970s, the impact was gentle but profound as soil scientists moved from a descriptive to a quantitative science. Burrough (1991, p. 165) commented:

> As a result of being able to handle much larger volumes of data, soil scientists have come to grips with the difficult problems of describing the spatial variation of soil and they are now providing useful infor-

mation services to a wide range of different kinds of land user ranging from urban planners in western countries to land resource experts in developing lands.

The increasing potential of GIS could be illustrated in many ways, such as employing a regression-based methodology for mapping traffic-related air pollution within a GIS environment using data obtained by mapping NO_2 in Amsterdam, Huddersfield and Prague (Briggs *et al.*, 1997).

One further development is the growing potential of *artificial intelligence*, computer programs that in some way mimic intelligent human activity. The sub-fields of most interest are expert systems and image understanding (vision) (Fisher *et al.*, 1988). An expert system contains 'the knowledge of a number of known human experts in a particular field or domain in computer usable form' and is already the basis for consultation systems, for example to assist in the management of fire in a national park; for applications in remote sensing; and for applications in map interpretation and geographical information systems.

Great technical strides made in remote sensing, computing and information processing were of significance for all the environmental and earth sciences and were sustained by *external developments* (Chapter 1, p. 14) including:

- greater concern for global environmental change, which had been suspected in the mid-twentieth century, was subsequently debated sometimes quite fiercely, and was then accepted (Chapter 9, p. 234);
- increasing awareness of physical environment including the Gaia hypothesis (Lovelock, 1979) as an integral part of the cultural trend (Chapter 10, p. 254);
- greater awareness of funding mechanisms and limitations, in common with all disciplines. This arose as a consequence of growth in numbers of students and of researchers, was characterized by greater scrutiny of the purposes of research funding and by greater difficulty in achieving it, and stimulated the move towards more directed research programmes which, if not actually applied in character, became more grey skies than blue (Gregory, 1997).

Despite great strides in technology and changes in the overall scientific environment contributing to the revolutionary changes after 1950, the influence of particular individuals was very significant. For example, there is no doubt that Dick Chorley contributed to the creation of a new atmosphere in British geography and was an immensely successful catalyst of change and of research (Goudie, 1994). Stoddart (1997b) commented that it was perhaps difficult to recognize the scale and scope of the changes brought about by Chorley's work and how rapidly they were achieved (e.g. Fig. 3.4). However, all individuals operate within particular physical environments, the character of which pervades the way in which investigations are under-

Figure 3.4 A representation of Professor Richard Chorley in the early 1960s (from Stoddart, 1997b, with permission from Routledge publishers). This was described as the way 'a local wag' viewed Dick Chorley constructing one of his many works from that time.

taken. Thus in introducing papers on Australian physical geography, Thom (1988, p. 157) noted that 'longevity of landforms involving the impact of discrete events is a theme which runs through Australian geomorphology'. This theme was developed by Tooth and Nanson (1995) who showed that Australian fluvial systems often contrast with Anglo-American observations in view of the key features of the Australian setting, including low long-term denudation rates, absence of extensive Quaternary glaciation, and the predominance of low-gradient fluvial systems over much of the continent. The spate of books and articles produced in the period 1982–94 collectively could suggest that the discipline, if not in crisis, was experiencing acute growing pains (Rhoads and Thorn, 1996), and approaches to the discipline in the next five chapters reveal how some of these growing pains built on the quantitative, theoretical and technological foundation.

Further reading

CURRAN, P.J. 1985: *Principles of remote sensing*. London: Longman.

MACMILLAN, B. (ed.) 1989: *Remodelling geography*. Oxford: Blackwell.

THORN, C.E. 1988: *An introduction to theoretical geomorphology*. Boston: Unwin Hyman.

Topics for consideration

(1) While thinking of particular physical geography problems studied before the late 1960s it is informative to consider how the approach taken would be altered by statistical and quantitative methods, and by the advent of technological developments.
(2) Hindsight is important and it is instructive to think how specific examples of investigations undertaken in the 1950s and 1960s might have been dramatically changed (or not) had they had the benefit of a better conceptual basis or a sounder theoretical foundation.

PART

III

CURRENT APPROACHES IN PHYSICAL GEOGRAPHY

|4|

The environmental system

The unifying methodology provided by the systems approach became accepted with varying degrees of success throughout all areas of physical geography after 1970; the approach may not have offered all that was initially hoped for, and adjustments of attitude and method were demanded. This chapter outlines how systems were developed in science (4.1), how the approach developed in physical geography (4.2), the manner in which the impact of the approach was reflected in different parts of the subject (4.3), suggests how energetics can provide a focus for integration (4.4), and finally demonstrates how systems have become related to models as an integral part of research and of conceptual approaches (4.5).

4.1 Systems in science

The concept of a system was not new since Newton had written on the solar system, biologists had been concerned with living systems, and geographers had implicitly used notions of the systems concept since the early days of the subject. However, as Harvey (1969, p. 449) noted, systems concepts, although old, had tended to remain on the fringe of scientific interest, acting more as constraints than as subjects for intensive investigation. *General systems theory* was proposed by the biologist Ludwig von Bertalanffy as an analytical framework and procedure for all sciences. He perceived the theory as a way of uniting all sciences, but the academic world did not readily accept the theory as presented at a philosophical seminar in Chicago in 1937 (Holt-Jensen, 1981). Although physics was inclined towards general theory at that time, the trend elsewhere in the first half of the twentieth century was to emphasize detailed investigations and to avoid general theories. A change of attitude by the 1950s was emphasized by the development of new fields including cybernetics, information theory and operations research. In the late 1940s cybernetics was a new branch of science formed to study regulating and self-regulating mechanisms in nature and technology. It was primarily concerned with control mechanisms in systems and with communication processes that determine their successful operation, and part of its mathematical basis is found in information theory (Holt-Jensen, 1981). The philosophical aspects of general systems theory that von Bertalanffy published in the 1950s and 1960s could be approached using cybernetics. Von Bertalanffy (1972) distinguished three aspects of the study of systems: firstly, systems science which deals with the scientific investigation of systems and with theory in various sciences; secondly, systems technology concerned with applications in computer operations and theoretical developments such as game theory; and thirdly, systems philosophy involving reorientation of thought and world view consequent upon the advent of system as a new scientific paradigm. A system has been defined as:

- a set of elements with variable characteristics;
- the relationships between the characteristics of the elements;
- the relationships between the environment and the characteristics of the elements.

In analysis, attention has been accorded to the structure of the system; its behaviour which involves energy transfer, its boundaries, its environment, its state whether transient or equilibrium; and its characteristic parameters which are unaffected by the operation of the system.

The systems approach was necessarily identified with a positivist approach, and as such has been less resilient in human (Johnston, 1983a) than in physical geography. Thus Willmott and Gaile (1992, p. 182) concluded:

> Contemporary social theories disdain strict empiricism or positivism and formal modelling . . . Critics focus on what models do not do; they 'cannot replace evocative description' or 'answer the important questions' . . . These critiques of positivism cannot be dismissed, but neither should they be accepted in total. Critical social theorists often fail to come to grips with physical geography, wherein positivism has been especially fruitful. They also find positivistic methods useful while rejecting the philosophy that underlies them.

Although most of the ideas of general systems theory are certainly valuable, it has been argued that they were applied without formal knowledge of theory (e.g. Jennings, 1973, p. 124). Nevertheless, systems thinking was probably responsible for a comprehensive review of many environmental situations. The theory is primarily inductive in nature and so lacks explanatory value, but it helped to counter the trend towards specialization in science; one expression of this by Medawar (quoted in Coffey, 1981, p. 30) is that 'in all sciences we are progressively relieved of the burden of singular instances, the tyranny of the particular. We need no longer record the fall of every apple.' A basic two-fold hierarchical evolution was perceived by Lazlo (1972b) in the many systems of the universe. Firstly, the entities of astronomy present a macrohierarchy embracing galaxy clusters, galaxies, star clusters, stars, and planets and subsidiary bodies. Secondly, in a microhierarchy are the terrestrial entities of physics, chemistry, biology, sociology; and international organizations which are atoms, molecules, molecular compounds, crystals, cells, multicellular organisms and communities of organisms. The basic components of each hierarchy are atoms or their constituent elementary particles which are composed of quarks. Huggett (1976b) extended Lazlo's (1972b) bipartite scheme to include another microhierarchy in order to provide an evolutionary link between atoms and planets in the macrohierarchy, namely the hierarchy of planetary and geological systems. He also noted that the systems of the atoms-to-planet and the atoms-to-societies hierarchies commingle to produce a third hierarchy which is the hierarchy of environmental systems. He construed geography as the science that deals with systems at the uppermost levels of this environmental, or atoms-to-ecologies, hierarchy, although noting that some geographers may delve into systems at a more detailed and fundamental level. Anuchin (1973) in his view of theory in geography noted the uncertainty which characterized much of geography's growth and its attempts to become an established science, when many scientists who were substantially geographers (he cites V.V. Dokuchaev as an example) have often preferred to give their activities another name, and:

> When one recalls that for decades geography was not recognized as a scholarly discipline, one cannot perhaps be surprised that some scientists have unsuspectingly spent their whole careers studying geography – rather like the well-known character of Molière who did not know that he had been speaking prose all his life. (Anuchin, 1973, p. 44)

4.2 Developing a systems approach in physical geography

A systems way of thinking encouraged a comprehensive focus on all aspects of a particular problem, and was adopted successively by biogeography, soil geography, climatology and geomorphology from 1935 to 1971 when *Physical Geography: A Systems Approach* (Chorley and Kennedy, 1971) was published. In geography as a whole, Stoddart (1967b, p. 538) concluded: 'Systems analysis at last provides geography with a unifying methodology and by using it geography no longer stands apart from the mainstream of scientific progress'.

Ecosystem was a term proposed by the plant ecologist A.G. Tansley in 1935 as a general term for both the biome, which was 'the whole complex of organisms – both animals and plants – naturally living together as a sociological unit', and for its habitat. Tansley (1946, p. 207) further expressed the notion that:

> All the parts of such an ecosystem – organic and inorganic, biome and habitat – may be regarded as interacting factors which, in a mature ecosystem, are in approximate equilibrium: it is through their interactions that the whole system is maintained.

In his review of organism and ecosystem as geographical models, Stoddart (1967b, p. 523) showed how Tansley's concept broadened the scope of ecology beyond the purely biological content and gave formal expression to a variety of concepts covering habitat and biome which date back to the late nineteenth century. Extending Tansley's definitions and relating to the island ecosystem, Fosberg (1963, p. 1) avowed that 'such partial concepts of nature as climate, vegetation, biota, soil environment, and even community, though very useful for analytical purposes, are not especially conducive to synthetic thinking or integration'.

Although major ecosystem developments were largely external to biogeography until the 1960s, Stoddart (1967b, p. 524) argued that the ecosystem concept had four main properties which commended it for geographical investigation. Firstly, it is monistic and brings together environment, mankind and the plant and animal world within a single framework in which interaction between the components could be analysed. Because emphasis is on the functioning and nature of the system as a whole, this should dispense with geographic dualism. Secondly, ecosystems are structured in a more or less orderly, rational and comprehensible way, and therefore provide an approach which requires identification of the structures present and the links between the structural components. Thirdly, ecosystems function as a result of throughput of matter and energy, and in ecology the identification of trophic stages and quantification of food webs and of productivity are exemplars of the way in which the function can be utilized. Fourthly, the ecosystem is a type of general system and therefore can be visualized as an open system tending towards a steady state under the laws of open system thermodynamics. Ecosystems in steady state possess the property of self-regulation and this is analogous to mechanisms such as homeostasis in living organisms, feedback principles in cybernetics, and servomechanisms in systems engineering.

The systems approach was very appropriate in biogeography, and Simmons (1978) for example, in discussion of the ecosystem scale, distinguished two approaches often made: one which is synoptic and develops from intuitive perception of an ecosystem to studies of ecological cohesion including, where relevant, the significance of human activity; and another approach which is more analytical whereby measurements are made of the flow and partitioning of energy through the ecosystem and of the cycles of mineral nutrients within the system. The latter approach embraces production ecology as the study of the rate of production of organic matter in an ecosystem, and population ecology which studies changes in population numbers of the species of the ecosystem.

In soil geography it is generally suggested that the systems approach was formally applied by Nikiforoff (1959), although earlier he had distinguished accumulative and non-accumulative soils, therefore implicitly involving an open system attitude (Nikiforoff, 1949). Earlier still Jenny (1941) had provided a basis for *Factors of Soil Formation* (Chapter 2, p. 35), and subsequently (Jenny, 1961) advanced a systems-oriented approach whereby the initial state L_o represents the assemblage of properties at time zero, P_x is the combined result of inputs and outputs which provides the flux potential, and *t* is the age of the system. This provided a revised general state factor equation where ecosystem properties (l), soil properties (s), vegetation properties (v) and animal properties (a) are combined in an equation of the form:

$$l, s, v, a = f(L_o, P_x, t) \tag{4.1}$$

This enabled him to suggest five broad groups of factors, namely climofunction, biofunction, topofunction, lithofunction and chronofunction, and although it is very difficult to analyse soil systems quantitatively using this approach, a number of attempts have been made to solve the state factor equations (see Yaalon, 1975). In the generalized theory of soil genesis proposed by Simonson (1959) there was a formal introduction of systems thinking into soil science, with soil visualized as an open system with inputs and outputs. The catena approach extended to the drainage basin was used as the basis for a model of the soil system by Huggett (1975) when he attempted to simulate the flux of plasmic material in an idealized basin. This model is really an extension of the nine-unit hypothetical landsurface model and adduces definable flow lines of material within soil–landscape units. Different constituents will move through the system at different rates, and one time step in the simulation could represent a day for mobile salts but as much as a millennium for a relatively immobile element such as aluminium.

Systems were readily embraced within climatology and meteorology, led to the emphasis upon global models and upon general circulation models (GCMs) and were fundamental to systems generated for other parts of the physical environment. Thus Huggett (1991, p. 9) concluded: 'It has become clear recently that central to the elucidation of relations within the climate system is an understanding of the connections between systems of different scales within the biosphere'. Application of systems theory in geomorphology was clearly and explicitly achieved by Chorley (1962), acknowledging earlier contributions by Strahler (1952) and by Hack (1960), including the statement (Strahler, 1952, p. 935):

> Geomorphology will achieve its fullest development only when the forms and processes are related in terms of dynamic systems and the

transformations of mass and energy are considered as functions of time.

Chorley (1962) contrasted the open system view and the closed system view, which was at least partly embodied in Davis's view of landscape development. In a closed system the given amount of initial free energy becomes less readily available as the system develops towards a state with maximum entropy, where entropy signifies the degree to which energy has become unable to perform work. Open systems, however, were portrayed as those that need an energy supply for maintenance and preservation and as being maintained in an equilibrium condition by the constant supply and removal of material and energy. Open systems can import free energy (or negative entropy) into the system and they can behave equifinally whereby different initial conditions can lead to similar end results. The value of the open system approach to geomorphology was summarized thus (Chorley, 1962):

- to depend upon the universal tendency towards adjustment of form and process;
- to direct investigation towards the essentially multivariate character of geomorphological phenomena;
- to admit a more liberal view of morphological changes with time to include the possibility of non-significant or non-progressive changes of certain aspects of landscape form through time;
- to foster a dynamic approach to geomorphology to complement the historical one;
- to focus upon the whole landscape assemblage rather than those parts assumed to have evolutionary significance;
- to encourage geomorphological investigations in those areas where evidence for erosional history may be deficient;
- to direct attention to the heterogeneity of spatial organization.

This very influential paper (Chorley, 1962) was succeeded by other ideas (Chapter 5, p. 105), but not everyone welcomed the systems approach. Smalley and Vita-Finzi (1969) considered it unnecessary in the earth sciences and responsible for confusion rather than clarification in empirical investigations, and Chisholm (1967) dismissed the approach for geography as a whole as 'a jargon-ridden statement of the obvious'.

Developments had appeared within particular branches of physical geography, but *Physical Geography: A Systems Approach* (Chorley and Kennedy, 1971) applied to physical geography as a whole. It showed how physical geography could be rationalized and perhaps given new coherence in terms of systems theory, and 'by avoiding the usual *pot pourri* of information about the Earth and its atmosphere which had traditionally been termed physical geography', it was devoted to the identification and analysis of some of the more important systematic relationships with which modern physical geographers were concerned. It did not purport to deal with all the subject matter of the field and it intended 'to view traditional material in physical geography in a new light'. The dual purpose of the book was, firstly, to present a view of landscape and processes in terms relevant to the student of human geography, by indicating ways in which socioeconomic and physical systems interlock and interact, and secondly, by showing how far knowledge of the physical world and its processes is compatible with the ideas of systems theory, to demonstrate areas in which research might profitably be concentrated. In achieving these aims it was undoubtedly enormously successful and must rank as one of the most influential physical geography textbooks of the twentieth century, although not as well cited by 1984 as might have been expected (Wrigley and Mathews, 1986). The book distinguished four types of medium-scale system, each showing distinct but complementary properties, and giving a progressive sequence to higher levels of integration and sophistication. The four types of system (see Fig. 4.1) were:

(a) *Morphological systems* which comprise morphological or formal instantaneous properties integrated to form recognizable operational parts of physical reality, with the strength and direction of connectivity revealed by correlation analysis.
(b) *Cascading systems* composed of chains of subsystems which are dynamically linked by a cascade of mass or energy, so that output from one subsystem becomes the input for the adjacent subsystem.

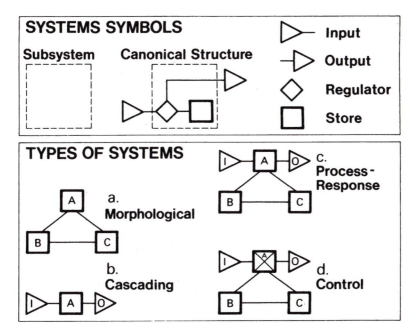

Figure 4.1 Systems terminology and classification (after Chorley and Kennedy, 1971).

(c) *Process-response systems* formed by the intersection of morphological and cascading systems and involving emphasis upon processes and the resulting forms.

(d) *Control systems* where intelligence can intervene to produce operational changes in the distribution of energy and mass.

Subsequently, a text applied to geography as a whole (Bennett and Chorley, 1978) attempted, firstly, to explore the extent to which systems theory provides an interdisciplinary focus for environmental matters, and to what extent systems technology provides an adequate vehicle; and, secondly, to ascertain the manner in which systems approaches aid in the development of an integrated theory relating social and economic theory to physical and biological theory. The approach employed used *hard systems*, which are capable of specification, analysis and manipulation in a more or less rigorous and quantitative manner, and *soft systems* which are not tractable by mathematical methods. In their conclusion, Bennett and Chorley (1978, p. 541) suggested that systems methods have illuminated thought, clarified objectives, and cut through the theoretical and technical undergrowth during the third quarter of the twentieth century in a most striking manner.

Whereas the typology advocated by Chorley and Kennedy (1971) was based on form and process facets of systems, a typology can be based on the level of system complexity (Huggett, 1985). For geography as a whole Wilson (1981) characterized systems analysis as concerned with handling complexity, with identifying and understanding systemic effects, with seeking methods that are applicable to a wide range of systems classified into certain types, and with providing tools that aid planning and problem-solving. This approach was introduced into geomorphology by Thornes and Ferguson (1981), and Huggett (1985) suggested that it recognized three main kinds of system. *Simple systems* include the solar system or a few boulders resting on a hillslope. The conditions required to dislodge the boulders and the subsequent processes can be predicted from mechanical laws involving forces, resistances, and equations of motion, in much the same way that the motion of planets around the sun can be predicted from Newtonian laws. *Complex but disorganized systems* are those in which a vast number of objects are seen to interact in a weak and haphazard manner, exemplified by the vast number of particles in a hillslope mantle, which could be regarded as a complex but rather disorganized system. Whereas these two concepts of systems have long been recognized, the third, *complex and organized systems*, are those in which objects are seen to interact strongly with one another to form systems of a complex and organized nature. Most biological systems and ecosystems are of this kind, and Huggett (1985) suggested that this could include a hillslope represented as a process-response system or similar approaches to soils, rivers or beaches. Nested systems called holons and a series making a hierarchy (Huggett, 1991) can be employed in landscape research (Haigh, 1987).

Systems have now been applied throughout the discipline and Huggett (1980) distinguished a strategy of systems analysis applicable to both a theoretical (at either a micro- or macro-scale level) and an experimental mode of analysis (which observes the nature of relationships between system parts). Huggett (1980) distinguished four phases:

- the *lexical phase* which involves identification of system components;
- the *parsing phase* which involves establishing relationships between system components;
- the *modelling phase* which requires expression of relationships in the context of a model and then calibration of the model;
- the *analysis phase* in which there is an attempt to solve the system model.

If not successful the procedure is repeated with a modified model.

4.3 Adoption of the systems approach

The systems approach was rapidly reflected in the organizational structure of textbooks (e.g. Strahler and Strahler, 1976; King, 1980; Dury, 1981). One concise book, published before the Chorley and Kennedy (1971) volume, presented the geosystem as a single planetary system (Rumney, 1970) in which land, sea and air are dynamically integrated, and specifically included the sea, an area frequently ignored by physical geographers in the latter part of the twentieth century. Much more explicit was *Environmental Systems: An Introductory Text* (White et al., 1984) which was organized to explain why a systems approach was employed and how matter, force and energy were central to it, followed by treatments of systems and their components as a basis for systems and change and for mankind's modification of environmental systems. A particularly useful approach was their Fig. 1.5 which offered seven different pathways through the book, structured according to the type of course being followed: this could be broad or general (environmental science–broad global approach, environmental science–comprehensive approach) or more specific towards a branch of physical geography (geological/geomorphological; pedological; climatological; biogeographical; or ecological bias). Written as a successor to *The Ecology of Natural Resources* (Simmons, 1974), *Earth, Air and Water: Resources and Environment in the Late 20th Century* (Simmons, 1991) took a broader perspective, utilized the notion of systems and extended them to resource systems and management and action frameworks. Not all texts adopted an overt systems approach, however, and at least one American text (Wallen, 1992) makes brief mention of systems but does not refer to the conceptual basis or give any references to the development of the approach. Perhaps by the early 1990s some thought that it was no longer necessary! However, *Environmental Geology* (Keller, 1996, 7th edn) introduced systems as one of four basic concepts of environmental science, the other three being population growth, sustainability and limitation of resources.

Perhaps one of the greatest advantages of a systems viewpoint was to cement the branches of physical geography more closely; pedagogically it offered an approach that directed the student's attention to the holism of the Earth's environment and to its dynamics. In a volume produced in 1992 on pervasive themes in American geography, movements, cycles and systems were summarized as one of the themes, although it was suggested that 'Geographers have yet to exploit the full potential of systems thinking' (Graf and Gober, 1992). The approach was utilized in branches of physical geography, as in the approach taken in the textbook *Geomorphology* (Chorley *et al.*, 1984), and in presenting a geomorphological approach to glaciers and landscape, Sugden and John (1976) utilized a systems approach as a vehicle for the explanation of complicated ideas because they believed in 'the value of a systems framework as a powerful explanatory tool'. Subsequently Sugden (1982) utilized a systems framework for his synthesis of the character of the Arctic and Antarctic. In these cases the system was used as a collective vehicle for an open-systems approach involving input, storage and output relationships; for system hierarchies; and for concepts such as thresholds and relaxation times (see Chapter 6, p. 156).

Systems approaches influence research investigations, and in geomorphology in general Thornes and Ferguson (1981) followed Weaver (1958) and Wilson (1981) in recognizing three kinds of system. First were *simple systems* which involve no more than three or four variables and can be handled by relatively simple techniques including regression models and partial differential equations, possibly extending to finite difference methods. Second were *systems of complex disorder* with large numbers of components and therefore of variables, but only weak linkages between them, which are handled by probabilistic methods of statistical mechanics. This includes probabilistic approaches to soil creep and to stream networks, coastal spit simulation and Box Jenkins models. Third were *systems of complex order* with a large number of components so that simple analysis techniques cannot be employed; catastrophe theory and perturbation analysis are examples of appropriate analysis procedures. In geomorphology, however, Thornes (1989, p. 6) argued that 'systems analysis is a popular concept enjoyed by many but whose deeper ramifications are understood by relatively few', and he suggested that this was in strong contrast with hydrology in which both the theory and practice of systems analysis are well embedded. The systems approach was easily assimilated in hydrology because the hydrological cycle and water resource systems were so appropriate for systems representation. Approaches to hydrological model-building tended to be either via physical hydrology, which investigates the components of the hydrological cycle to achieve a full understanding of the mechanisms and interactions involved, or by systems synthesis investigations, which attempt complete simulation of drainage basin operation by adjusting the components and the parameters of the model until outputs of the model agree with empirical results from known inputs. Physical geographers emphasized the physical hydrology approach using models for subsystems such as evaporation, infiltration, surface runoff or groundwater, or overall catchment models. In a review of hydrological models in geography, More (1967) concluded that despite large areas of overlap between hydrology and geography, the two disciplines had developed quite separately, that many of the hydrological models developed had not been geographic in their inception, but that geographers should not ignore the implications of the fast-moving science of hydrology. For the hydrological system a convenient distinction was made (Amorocho and Hart, 1964) between parametric and stochastic hydrology, and in subsequent research geomorphological and hydrological models interacted together (e.g. Anderson, 1988).

In climatology the system was adopted as a suitable framework and appeared as the introductory foundation for *Causes of Climate* (Lockwood, 1979a), where it was argued that the application of systems theory and mathematics had completely changed the subject of climatology. The three types of systems distinguished were *isolated systems* whose boundaries are closed to the import and export of both mass and energy; *open systems* in which there is exchange of both matter and energy between the system and its environment such as clouds; and *closed systems* in which there is no exchange of matter between the system and its environment, although there is in general an exchange of energy. The atmosphere, oceans, and landsurface were considered as a series of cascading systems connected by flows of mass or energy. In a review of climatology as it relates to geography, Terjung (1976) reviewed the position of climatology in geography teaching and research and suggested:

> Instead of drumming the trivial facts of physical geography of yesteryear into the heads of reluctant freshmen, introducing classes should teach the concepts of systems analysis and the flows of energy, mass, momentum, and information through various environments of our planet.

A systems approach was basic to much subsequent thinking in relation to processes and change (p. 105). Soil and vegetation systems were considered together by Trudgill (1977) who dealt with each component of soil and vegetation in turn to build up a sequential picture of the whole system energy flows prominent in his book. In his alternative biogeography, Gersmehl (1976) also focused upon the circulation of mineral elements modelled as a system of compartments and transfer pathways.

Figure 4.2 A view of the impact of models and systems and the new approach to geography in the 1960s (Henderson-Sellers, 1989a). The caption suggested that Mr Middle Aged Geographer contemplates the possibilities of climatic determinism, the 'minor' truth of climax vegetation and the expense of a climate model. On the whole he's safer with good old climatology.

4.4 Energetics as a focus for the environmental system

Ready adoption of the systems approach in physical geography texts certainly retarded, and perhaps even reversed, the trend towards greater specialization, and the separation of the branches of physical geography from each other and from human geography. However, adoption as a textbook framework, although producing some stimulating texts, was not immediately paralleled in research investigations and research programmes. Slaymaker and Spencer (1998, p. 19) concluded that 'Unfortunately, many have dismissed [these] ideas as theoretical systems without real world application', so that they advocated global environmental change as a more appealing and effective alternative (see Chapter 9, p. 232). It is important to distinguish between systems analysis, which can mathematically optimize some attribute of the links within and between systems, and general systems theory which claims that all systems can be understood by the application of systems principles. Adopting a systems approach uncritically has the danger that it is assumed to be sufficient to identify system structures and to portray the multitudinous variables involved in a particular system, reinforcing the first law of ecology, graphically described by Commoner (1972) as the principle that everything is connected to everything else. However, Commoner (1972) also stated three other laws of ecology:

(1) Everything must go somewhere.
(2) Nature knows best.
(3) There's no such thing as a free lunch, because somebody somewhere must foot the bill.

In physical geography, especially in textbooks, the first of these four 'laws' was perhaps emphasized initially, rather than the other three. In research investigations it is essential that the ideas underlying all of these 'laws' are considered and the environment modelled as effectively as possible. The hydrological cycle was very amenable to systems representation, the diagrams produced identified the stores and pathways between the stores, but they could not so easily represent the dynamics of the runoff-producing system. Runoff dynamics have to be visualized in three dimensions because, as the areas close to water courses expand during a storm, so the runoff-producing areas expand and other types of runoff begin to contribute to the operation of, and outflow from, the system. Systems diagrams representing this situation in two dimensions do not effectively portray the way in which changes occur during a storm event.

Research applications needed to focus upon the dynamics of systems and to explore energy flows. Before explicit statements of the systems approach, Linton (1965) had identified four sources in his geography of energy, namely:

- radiant energy from the sun;
- internal energy from the Earth's interior;
- rotational energy of the whole and parts of the solar system;
- vital energy which is energy in the service of man.

Linton (1965, p. 227) expressed the hope that salient parameters, in fields as far apart as climatology and social geography, might eventually be expressed in terms of a common set of units – the watt and the calorie – and subsequent progress has been made in the physical environment. Energy exchanges in the atmosphere, where energy signifies the capacity to do work and where work is what happens when force accelerates mass over distance, were emphasized by Hare (1965), who subsequently suggested (Hare, 1966) that the outstanding change in climatology in the post-1945 years was the shift away from parameters such as temperature and relative humidity towards the measurement of fluxes. This necessitated attention to the movement and transformation of energy in the atmospheric boundary layer, in the plant cover, and in the soil so that progress could be made towards the understanding of the mechanisms of energy and moisture exchange. In his review of climatology for geographers, Terjung (1976) advocated an emphasis on process-response systems as related to mankind which occur within the planetary boundary layer, interface and substrates; he advocated study of flows of energy, mass momentum and information through the various environments of the planet Earth. Similarly, in their review of 'Climatology, the challenge for the eighties', Mather *et al.* (1980) concluded that climatology must systematically investigate the exchanges of heat, water and momentum that occur at or near the Earth's surface, and should focus upon topoclimatology as well as on transfer processes.

In searching for a more unifying theme for the physical–human geography interface, Simmons (1978) advocated the study of energy in contemporary society, and traced the relevance of the Lindeman model of ecosystems and its development by Odum and Odum (1976) in *Energy Basis for Man and Nature*. In addition to clear definitions of energy, power as the rate at which energy flows, and efficiency as any ratio of energy flows, Odum and Odum (1976) proposed three principles of energy flows:

(1) the law of conservation of energy;
(2) the law of degradation of energy, which introduces entropy as a measure of technical disorder to signify the extent to which energy is unable to do work;
(3) the principle that systems that use energy best will survive, which is the maximum power principle or minimum energy expenditure principle.

Adopting the taxonomy of energy systems employed by ecologists, Simmons (1978) suggested that geographers could use a set of ecosystem types that broadly provide a set of spatial regions which conforms to patterns identified from analysis of satellite data. Subsequently in his text on *Biogeography: Natural and Cultural*, Simmons (1979a) used energy as a key to the understanding of natural biogeography through food chains, productivity, nutrient cycles and population dynamics to provide the basis for the treatment of cultural biogeography. Stoddart (1967b, p. 537) noted that:

> Geography is clearly concerned with systems on a multitude of levels.
> A preliminary attempt to develop a science of 'geocybernetics' has
> been made in a little known paper by Polonskiy (1963) . . . The study
> of geosystems may now replace that of ecosystems in geography . . .

A focus upon energy flows is appropriate for physical geography, but Simmons (1978) cautioned:

> A case can thus be made for energy flows as linkages between man and environment, both in terms of resource uses and environmental impact. But caution is necessary, for the homogeneity of kilocalories and gigajoules may hide qualitative and cultural aspects of the flows which as geographers we are not at liberty to ignore. F.E. Egler sounded the same note when talking about the way ecological energetics – especially the trophic level concept – ignored taxonomic consideration. He said that ecological energetics was like grinding up cows to make hamburgers – you could not be sure a monkey had not slipped in somewhere.

The basis for use of energy in biogeography derives not only from work in ecology generally but also from the field of bioenergetics. Thus Broda (1975) suggested that until the significance of the First Law of Thermo-dynamics defining heat as a form of energy had been appreciated, and until energy distribution implications of the Second Law had been assimilated, then progress in the analysis of bioenergetical processes was uneven and patchy. This can be developed by regarding organisms as chemodynamical machines, by identifying three classes of bioenergetic processes, namely fermentation, photosynthesis and respiration, and by utilizing an approach to classification based upon microphysiology and biochemistry as well as upon macrostructure of organisms and upon macrophysiological processes. Broda then applied bioenergetics to early conditions on Earth and to the ecosphere in general.

In soils, energy considerations (p. 85) had already been identified by Jenny (1941, 1961), and Gerrard (1981), referring to Runge (1973), reviewed the energy status of soil systems by recognizing three components. The *decay* component is where the energy status gradually declines and eventually the system should continue to a state of virtual exhaustion; the *cyclic* component occurs because energy, and possibly material input, changes in a rhythmic manner associated with diurnal and seasonal climatic cycles; and a *random* component is provided by irregular supplies of energy such as rainstorms. In the Runge (1973) model, soil development was visualized in terms of organic matter production, time, and the amount of water available for leaching. The water available depends upon the amount of water that infiltrates and becomes available for pedogenesis, compared with that which is removed by surface flow and hence is not available. A three-dimensional view of the plasmic flux of material in an idealized basin was modelled by Huggett (1975) to provide a model of soil genesis in the context of soil–landscape systems. New models have been developed and Phillips (1989) postulated that soil ecosystems could be represented by a series of (probably non-linear) partial differential equations, and his modelling approach could apply to the biosphere (Huggett, 1991).

In hydrology, definition of the pathways of movement through the hydrological cycle necessarily provided an important focus in studies by physical geographers, but in geomorphology the use of concepts based upon energy-balance concepts was less striking. Hare (1973) suggested that this was because fluvial processes tend to be dominated by extreme events rather than balance relationships, and that the geomorphological time-scale is very long compared with that appropriate for climatic processes, although glaciology is in a very different situation. Indeed, in the field of glacial geomorphology one of the most imaginative approaches was devised by Andrews (1972), who provided an analysis of total glacier power (WT) as the product of basal shear stress and the average velocity. Effective power (WE) was determined by the proportion of the total average velocity resulting from basal sliding, so that the ratio WT/WE could vary greatly according to the proportion of basal slip to ice deformation. Andrews proposed that WT/WE is small, between zero and 0.2 for polar and subpolar glaciers, but between 0.5 and 0.8 tending towards 1.0 for temperate glaciers. The implication that followed, that the glacial erosional forms produced by arctic and by temperate glaciers differ in size and geometry, received support from the glacial geomorphology literature. A further application was developed in geomorphology by Caine (1976) when he estimated the physical work in joules represented by different types of sediment movement. More generally, energy together with forces, resistances and responses, was used by Embleton and Thornes (1979) to introduce process in geomorphology, where energy was attributed to solar radiation, atomic energy, chemical energy, gravity, and the energy of the Earth's rotation.

Moves were made in the branches of physical geography towards a focus upon energy and upon entropy as the distribution of available energy (Leopold and Langbein, 1962), but further scope exists for use of the principle of least work (Kirkby, 1989). Some attempts were made to provide an energetic approach to the subject as a whole: Krcho (1978) offered a general approach to the physical-geographical sphere and the anthroposphere by using entropy in relation to a cybernetic system, and a generally applicable energy-based approach was advocated by Hewitt and Hare (1973). Although models of paths and storage reservoirs had been developed by ecologists, geochemists and climatologists, there was a need to progress towards a multidimensional model of the entire system.

4.5 Energetics, modelling and processes

In evaluating the systems approach in physical geography there is no doubt that the approach had an enormous pedagogical impact on teaching and learning in higher education, that it provided a convenient and timely way of addressing the environment, especially in textbooks, that it fostered a more open way of thinking not inhibited by the subdisciplines of physical geography, and that it equipped researchers with a systems modelling strategy to complement and relate to the physical modelling approaches employed for subsystems (Fig. 4.3A). Huggett (1985) concluded that discussion of 'systems' and the 'systems approach' tends to fall into one of two categories, the panegyrical or the disparaging, and that:

Scholars who praise the systems approach do so in the belief that it is a powerful and precise method of study. Scholars who try to shoot it down fail to see any advantage in it; indeed many deem it pernicious. (Huggett, 1985, p. viii)

The approach engendered a need to express environmental situations in terms of canonical structures which are the schematic diagrams representing a combined statistical and physical geography interpretation of a systems model. However, what it did not do, as pointed out in the general and specific examples mentioned in Section 4.4 above, was provide a fundamental and robust foundation for physical geography. In a summative assessment of reactions to the systems approach, Thorn (1988) concluded that it provided a vehicle for what needed to be done, at least in part, and also that it was not explored as extensively as it might have been. Huggett (1985) quoted Van Dyne (1980, p. 889) who reported a facetious comment once heard:

In instances where there are from one to two variables in a study you have a science, where there are from four to seven variables you have an art, and where there are more than seven variables you have a system.

Realization of the remaining potential may require a more fundamental understanding of, and familiarity with, the scientific basis (Chapter 3), more coherent use of the systems approach and energetics, and its greater application in modelling.

A coherent approach to physical geography was attempted in *Energetics of Physical Environment* (Gregory, 1987a) using the theme of the power of nature. Power, or the rate of doing work, required focus upon five phases, namely energy sources, energy circulation and transfers, energy budget, energy related to morphology, and changes of energy distribution; these phases are shown in Table 4.1 with suggestions of applications in physical geography. The approach really focuses upon power and fluxes, where flux is defined as the rate of flow of a fluid across a given area, and requires standardization in terms of units and measures. For global energy flux it was possible to express (Gregory, 1987b) the relative importance of the different components in consistent units (Fig. 4.3A). Although varying in degree of use throughout the branches of physical geography, the five themes included in Table 4.1 could be amplified by demonstrating the manner in which they have been reflected through approaches to the physical environment (Gregory and Thornes, 1987). Physical geographers have employed

(A)

GLOBAL ESTIMATES OF ENERGY FLUX

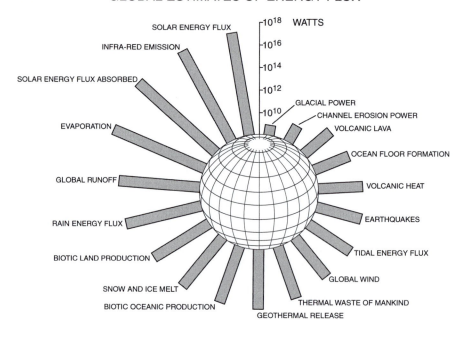

(B)

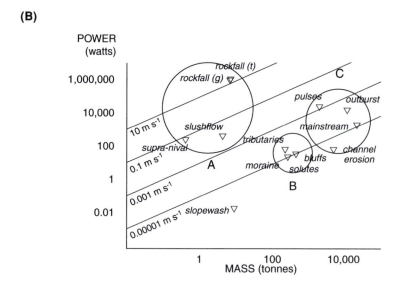

Figure 4.3 Examples of energetics. (A) Global estimates of energy flux (after Gregory, 1987b). The scale is logarithmic and the figure is intended to summarize the power of nature over the Earth's surface. Copyright John Wiley & Sons Limited. Reproduced with permission. (B) A mass–power relationship for Bas Glacier d'Arolla proglacial sediment budget processes (after Warburton, 1993).

Table 4.1 Energy phases and examples in physical geography, developed from Gregory (1987b) and Gregory and Thornes (1987)

Energy phase	Type of identification in physical geography	Atmosphere	Geosphere and hydrosphere	Pedosphere	Biosphere
Energy sources	Solar Geothermal Gravitational Rotational Vital Potential Kinetic Enthalpy	Internal energy Potential energy due to gravity Kinetic energy	Solar radiation Atomic energy Chemical energy Gravity Earth's rotation	Solar radiation Chemical energy Mechanical energy	Solar radiation Primary production
Energy circulation, transfers	Conservation of energy First Law of Thermodynamics Fluxes Energetics	Transfers, transformation, storage of energy Exchanges of heat, water, momentum Energy fluxes	Hydrological cycle Geochemical cycles	Mineral weathering Organic matter decomposition, Gibbs' free energy	Food chains Nutrient cycles Population dynamics
Energy budget, availability	Energy balance Thresholds Entropy	Energy balance	Water balance Sediment budget Erosion rates	Soil moisture budget Soil organic budget	Energy budgets at trophic levels Productivity
Energy related to morphology, rate of doing work	Equilibrium Steady state Power Efficiency	Available PE and KE Energy spectra and zonal balance	Force-resistance relations	Soil profile dynamics	Ecosystem dynamics Functional approach to ecology
Changes of energy distribution	Maximum power Minimum variance Dissipative structures Fluctuation	Climatic change	Endogenetic changes	Soil profile erosion	Evolution

energetics in three main ways which can be thought of as operational, methodological and conceptual, although they are not mutually exclusive. In the *operational* sense, energy is considered as a basic quantity which is used to define the temporal and spatial magnitudes of events resulting from its application; the efficiencies of its transformations; and its application to the management of physical environmental systems. To further this approach we need to be able to express all aspects of the power of nature in similar units. This could be important in comparing the efficacy of human activity with that of nature, and in adopting a more coordinated approach to the control and efficiency of energy expenditure in the natural environment. In the *methodological* sense energetics is used as a vehicle to organize information about the physical environment into a manageable structure, the most common format used being the budget (Table 4.1). *Conceptual* ways of deploying energetic approaches occur when transformations of energy are used as the basis for conceptualization of processes or states based upon general principles of energy expenditure. The conceptual approach has hitherto received the least emphasis, but it was concluded (Gregory and Thornes, 1987) that, if an approach to physical geography embracing space, time and mesoscale approaches, without recourse to reductionism, can be underpinned by considerations of work, power and efficiency in the language of the physical sciences, then physical geography can become more coherent and unified. The approach has been used in specific investigations; Warburton (1993) used sediment budgets as a means of characterizing the energetics of alpine proglacial geomorphic processes which enabled three basic process morphological sets to be distinguished (Fig. 4.3B) in the Bas Glacier d'Arolla proglacial fluvial sediment system. Employing a similar energetic viewpoint, Slaymaker and Spencer (1998) advocated a renewed emphasis upon biogeochemical fluxes and protection of the atmosphere, biosphere, hydrosphere and lithosphere, and argued that there is a sparsely inhabited niche between the Earth system science approach to global environmental change and the prevailing approach to physical geography (Fig. 4.4A). Their refreshing approach led to a new focus (Fig. 4.4B) on global environmental change (see p. 232, p. 285).

Other sciences successfully explored the use of an energetic approach, and Meyers (1977) showed how the approach could be used for the analysis of environmental impact in relation to water resources planning. A more comprehensive approach by Smil (1991) proceeded from planetary energetics, to human energetics and food needs, to energetic foundations of all stages of human evolution, concluding with discussion of general patterns and trends of energy use, environmental implications and socioeconomic considerations.

Modelling was accelerated by the conceptual sense (Gregory and Thornes, 1987) and use of energetics. Although in some publications the distinction between systems and models has not always been as clear as it could be, Strahler (1980) emphasized that a system is assumed to exist in the

(A)

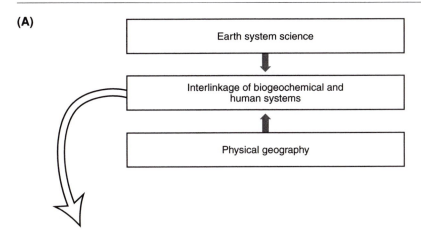

Societal context and ethical constraints

(B)

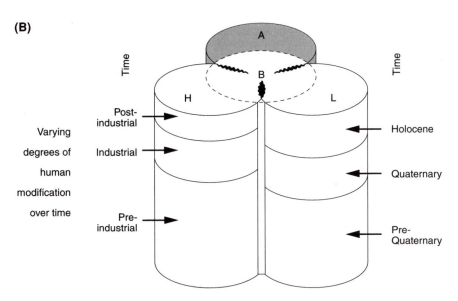

Figure 4.4 The sparsely inhabited niche. (A) This niche was suggested by Slaymaker and Spencer (1998) to occur between Earth system science and Anglo-American physical geography, and they proposed that a new and revitalized physical geography depicted in (B) can occupy the niche (see also Chapter 11, p. 286).

real world whereas a model is an attempt to describe, analyse, simplify or display a system. In his consideration of *Earth Surface Systems*, Huggett (1985) distinguished between conceptual, scale and mathematical models. A *conceptual model* is a mental image of a natural phenomenon and is always to varying degrees an abstraction or simplification of the environment. *Scale models* include physical models which are scaled-down versions of the environment, such as a laboratory flume or a wave tank, and analogue models in which some properties, as well as the size, of the system are altered, such as the use of kaolin rather than ice in modelling a glacier. *Mathematical models* are abstractions of a conceptual or scale model where the features of a system are represented by abstract symbols and subjected to the rigour of mathematical argument, and in which components and relations are replaced by an expression containing mathematical variables, parameters and constants. With particular reference to geomorphology, but of application to physical geography as a whole, Thornes (1989) drew attention to major modelling developments, largely mathematical, which have occurred, and distinguished:

- *ecological first-generation mega-models* – stemming from the development of energy flows and trophic dynamic concepts by Lindeman in 1942;
- *soil erosion second-generation mega-models* – the US Department of Agriculture (Soil Conservation Service and Agricultural Research Service) initiated a series of models in the 1980s;
- *hydrological mega-models* – include modern large-scale simulations of hydrological systems;
- *a new generation of models* – to include cross-boundary models, global and regional models, evolutionary models, modelling neglected processes (e.g. geomorphology–vegetative cover interactions), and modelling the human impact.

The interdependence of systems and modelling was stressed by Thorn (1988, Chapter 12), emphasizing that systems models, being intellectual and simplified constructions, are only models, not reality. Therefore they are tied very closely to existing theory, so that the frailties of such theory need to be kept in mind, just as in the case of positivism (Chapter 3). However, Thorn (1988, p. 192) concluded that: '. . . the fundamental case for systems modelling rests on a holistic perspective; that against the technique stems from a reductionist perspective'.

Modelling is referred to at the end of each of the next four chapters because it is now so fundamental to all modern approaches to physical geography. However, whereas mathematical models were usually broadly classified into deterministic, probabilistic and optimization models (e.g. Haines-Young and Petch, 1986), subsequent developments have extended the range of models available very considerably; Huggett (1985), for example, distinguished deductive stochastic models, inductive stochastic models, statistical models, deterministic models and dynamical systems models. Significant developments occurred within the branches of physical geography and, for example, Anderson (1988) edited *Modelling Geomorphological Systems*. In reviewing the bases of geomorphological modelling, Anderson and Sambles (1988) analysed the major themes prevalent in current geomorphology relevant to modelling research, and showed that some 45% of the papers in the period 1977–85 in the journals *Earth Surface Processes and Landforms*, *Catena*, and *Zeitschrift für Geomorphologie* related in some way to modelling. In that volume on geomorphological systems, the emphasis is upon current modelling research in hillslope and river channel processes, which is justified in view of the emphasis in research upon these areas since the 1960s. In *Process Models and Theoretical Geomorphology*, Kirkby (1994) organized two of the four sections on channel processes and valley heads, although the other two were on tectonics and general approaches and applications. In fact, some of the earliest developments in hydrological modelling were founded on nutrient cycling, implicitly embraced a systems viewpoint, approached biogeochemical cycling, and employed models for analysis, but depended upon a greater understanding of processes, which is the subject of the next chapter.

Further reading

If you can find a copy in the library it is well worth looking at:
CHORLEY, R.J. and KENNEDY, B.A. 1971: *Physical geography: a systems approach*. London: Prentice Hall.

Useful summaries of the approach are provided in:
HUGGETT, R.J. 1980: *Systems analysis in geography*. Oxford: Clarendon.

HUGGETT, R.J. 1985: *Earth surface systems*. Berlin: Springer Verlag.

This volume on models is good to look at:
ANDERSON, M.G. (ed.) 1988: *Modelling geomorphological systems*. Chichester: John Wiley.

Topics for consideration

(1) A systems approach was built upon increased interest in environmental processes, but catalysed thinking towards the regional and the global scale.
(2) Is an energetic approach (see p. 92) sufficient to apply to research investigations at all scales?
(3) In systems approaches, do we pay sufficient attention to linkages between systems and subsystems at different spatial and temporal scales?

|5|

Environmental processes

In the light of the adoption of systems and energetics (Chapter 4), it is perhaps difficult to imagine a time when processes were not a major theme in physical geography. It is important to appreciate why the need for greater emphasis upon processes was recognized (5.1), and how processes have been investigated (5.2), prior to understanding how developments and shifts in emphasis occurred in branches of physical geography (5.3) and how these had implications for patterns of process in space (5.4) and time (5.5), culminating in exciting new challenges for the investigation of process (5.6).

5.1 A focus upon environmental processes

Prior to the development of the systems approach and the surge of interest in models, it was appreciated in physical geography as a whole, and in its branches, that there was an outstanding need for the study of environmental processes. Whereas under the Davisian trilogy of structure, process and stage or time the emphasis had been very clearly upon stage, with very little upon process, alternative views had been available, such as that of G.K. Gilbert (1843–1918) in geomorphology, which was clearly directed towards the basic importance of the study of processes, but his work had been eclipsed by that of W.M. Davis. Sack (1991, p. 30) characterized Gilbert's approach:

> ... to solve geomorphic problems, Gilbert used field observation, logical classification of similar phenomena, multiple working hypotheses of explanation, and rigorous hypothesis testing, often supported by quantitative analysis ... Gilbert's hypotheses, explanations, and tests were commonly derived by analogy from physical mechanics ... He studied landforms as manifestations of geomorphic processes acting on earth materials ...

Not all physical geographers were convinced of the need for the study of process, and with reference to geomorphology Wooldridge (1958, p. 31) avowed that: 'I regard it as quite fundamental that geomorphology is primarily concerned with the interpretation of forms, not the study of processes'. This 'remarkably out-of-touch statement' (Stoddart, 1997, p. 384) contrasted with a contrary view that had been expressed in an influential paper on the dynamic basis of geomorphology by Strahler (1952, p. 924), who concluded:

> If geomorphology is to achieve full stature as a branch of geology operating upon the frontier of research into fundamental principles and laws of earth science, it must turn to the physical and engineering sciences and mathematics for the vitality it now lacks.

Strahler (1952, p. 937) suggested a programme of five steps for further research in geomorphology, adumbrated as:

(1) a study of geomorphological processes and landforms as various kinds of responses to gravitational and molecular stresses acting on materials;
(2) quantitative determination of landform characteristics and causative factors;
(3) formulation of empirical equations by mathematical statistics;
(4) building concepts of open dynamic systems and steady states for all geomorphological processes;
(5) deduction of general mathematical models to serve as natural quantitative laws.

Strahler suggested that this daunting programme was necessary because geomorphology was already half a century behind developments in chemistry, physics and the biological sciences, and his prescription for geomorphology was applicable to physical geography as a whole. Developments taking place in the 1960s and 1970s led to the production of new textbooks, often referring to form and process in their titles, such as *Hillslope Form and Process* (Carson and Kirkby, 1972) and *Drainage Basin Form and Process* (Gregory and Walling, 1973). Functional and realist approaches, identified by Chorley (1978), were both evident in the greater focus upon processes. Functional studies were essentially positivist in character and depended upon the notion that phenomena could be explained as instances of repeated and predictable regularities in which form and function were assumed to be related, hence the titles of the books referred to above. The realist approach, although developed as an extension of the functional positivist approach, attempted to probe beyond the relationships derived from observed regularities and to seek the mechanisms and underlying structures responsible for the operation of environmental processes (see Chapter 3, p. 53). Functional and realist stages tended to develop in sequence and characterized each branch of physical geography albeit to varying degrees; indeed, some branches had already proceeded as far as possible without an enhanced knowledge of processes. Thus the understanding of glacial land-forms such as glacial cirques depended upon further knowledge of the processes of ice movement and glacial erosion, and any advance of interpretation of planation surfaces produced by marine erosion required knowledge of the nature and rate of processes of coastal erosion. Whereas earlier studies in physical geography had tended to ignore the Holocene and to concentrate instead upon earlier phases of landscape development, it was now necessary to know more of processes and to understand their significance for contemporary environments.

Within geography as a whole the replacement of an idiographic by a nomothetic approach created an atmosphere in which process investigation was very germane. In their textbook, Thompson *et al.* (1986) applied the idea of the nomothetic and idiographic approaches to the river system. Whereas idios was concerned with the individual character of a particular river, how it changed, how it evolved and how it differed from other rivers, nomos made quantitative observations to test quantitative theories. The idiographic approach considered that the real world is so complicated that generalization is impossible, whereas the nomothetic approach was concerned with general theories.

Processes were also featuring much more prominently in related disciplines. This was shown in the orientation of journals and the inception of new ones. The *Journal of Hydrology* (1963) arose because in the first volume it was contended that: 'The steadily rising interest in many countries in hydrology in all its branches makes this an opportune time in which to launch a new Journal concerned with the scientific aspects of the subject'.

Other influential journals were initiated, often inspired by one visionary physical geographer, including *Earth Surface Processes* (1977) initiated by Professor Mike Kirkby, developing to *Earth Surface Processes and Landforms* in 1981, *Hydrological Processes* (1986) initiated by Professor Malcolm Anderson, *Permafrost and Periglacial Processes* (1990) inspired by Professor Hugh French, and other journals initiated after 1970 are included in Table 5.1. A further influence external to physical geography was greater public awareness of the environment, realization of the possibility of finite resources, and appreciation that spaceship Earth required concern for the wise use of its resources and reserves. Although this intellectual atmosphere or aspect of social concern (Stoddart, 1981) was particularly germane to the effects of human activity (Chapter 7, p. 171) and to applications (Chapter 8, p. 198), it also fostered a situation in which acquisition of information on environmental processes and environmental change was certainly favoured and sometimes positively encouraged by Research Councils, funding agencies and national mapping agencies. Many of the new journals initiated after 1960 included results of investigations into environmental processes, existing journals exhibited an influx of process-based papers, and *Progress in Physical Geography* (1977) included regular progress reports of which many were concerned with developments in

Table 5.1 Some journals initiated since 1970 that include papers on environmental processes: the second and third columns are not clearly distinguished, but physical geographers may have had a major role in initiating or managing those in column 2.

Year	Of major interest to physical geographers	Of interest to several disciplines
1970	*Geoforum*	*Nordic Hydrology* *Quaternary Research* *Quaternary Newsletter*
1971		*Water, Air and Soil Pollution*
1972		*Science of the Total Environment* *Ambio* *Boreas*
1973	*Catena*	*Geology* *Coastal Zone Management*
1974	*Journal of Biogeography*	*Environmental Conservation* *Landscape and Urban Planning*
1975		*Environmental Geology*
1976		*Geo Journal* *Environmental Management*
1977	*Earth Surface Processes and Landforms* *Progress in Physical Geography*	*Coastal Engineering* *Polar Geography and Geology*
1978		*Journal of Arid Environments* *Climatic Change*
1980	*Physical Geography*	*Annals of Glaciology*
1981	*Journal of Climatology* *Applied Geography*	*Soil Survey and Land Evaluation* *International Journal of Remote Sensing*
1982		*Quaternary Science Reviews*
1984	*Regulated Rivers*	*Journal of Coastal Research*
1985		*Journal of Quaternary Science*
1986		*International Journal of Geographical Information Science*
1987	*Hydrological Processes*	*Landscape Ecology*
1989	*Geomorphology*	
1990	*Permafrost and Periglacial Processes*	*Global Environmental Change* *Polar and Glaciological Abstracts* *Quaternary Perspectives* *Quaternary International*
1991	*The Holocene*	*Global Ecology and Biogeography Letters*
1993		*Biodiversity Letters* *Ecumene*
1997		*Global Environmental Outlook*

research on environmental processes. International collaboration was encouraged by the International Geographical Union by establishing commissions that included one on present-day processes, and a later one on field experiments in geomorphology (Slaymaker *et al.*, 1980). Those trends were continued by the Continental Erosion Commission of the International Association of Hydrological Sciences, and subsequently (1985) the creation of the International Association of Geomorphologists.

5.2 Approaches to investigations of process

Successful growth of research in environmental processes effectively required three developments. Firstly, techniques for the measurement of processes should be adopted and adapted and sometimes innovatively designed. Secondly, measurements should be made and data obtained of the rate of operation of actual, or simulated, processes. Thirdly, results should be analysed and interpreted using basic physical principles as advocated by Carson (1971), Statham (1977) or Williams (1982). It proved difficult to meet all three requirements. The more realist approach advocated by Strahler (1952) had the implication that not only was an approach to processes needed in terms of physical principles, but a similar approach to materials was required. After Strahler, one of the earliest movements in this direction was in *Rock Control in Geomorphology* in which Yatsu (1966) proposed a quantitative approach to the underlying mechanics of processes, and remarked:

> Geomorphologists have been trying to answer the *what, where* and *when* of things, but they have seldom tried to ask *how*. And they have never asked *why*. It is a great mystery why they have never asked *why*.

In the 1970s this *why* was addressed in an important volume on *The Mechanics of Erosion* (Carson, 1971) – a book that provided a unified introduction to the mechanics of erosional processes aimed at undergraduates in earth sciences, considered the concept of stress, the mechanics of fluid erosion, stress–strain inter-relationships, mass movement in rock and soil masses and the mechanics of glacial erosion, and concluded (Carson, 1971, p. 166) that:

> The neglect of the study of geomorphic processes by workers in geomorphology must rank as one of the most puzzling features in the development of this discipline ... There is something strange about a subject in which its research workers are willing to dabble at the application of the jargon of thermodynamics but unwilling to apply even the most rudimentary aspects of mechanics to these problems. One suspects that geomorphology will emerge as a reputable discipline only when its students have become well versed in the established principles of natural science.

Carson (1971) identified achievements already made by the mechanics-based approach, predicted that more would follow, and concluded that because erosional processes are weathering-limited to varying degrees, on a geological time-scale erosion mechanics must be closely linked with the mechanics and chemistry of rock breakdown. However, progress towards understanding the mechanics of weathering processes (e.g. Curtis, 1976) required not only an adequate knowledge of chemistry and of exchange reactions, but also of processes at a different scale, and these are required for the extensive survey of *The Nature of Weathering* (Yatsu, 1988) in which some 108 pages (17% of the book) are devoted to summarizing the thermodynamic properties of chemical species. Geomorphology was characterized as a meso-scale science in the late 1970s (Chorley, 1978), and the extent to which it was realistic to extend investigation of geomorphological processes to the micro-scale was then debatable, especially as the positivist view was no longer uncritically sustained in the natural sciences. However, geomorphology needed to overcome its reluctance to use the principles underlying landscape processes because this would not necessarily prejudice the retention of a meso-scale approach as shown in sedimentology (e.g. Allen, 1970) and in soil science. Carson (1971, p. 167) argued that an approach more firmly based in mechanics would lead to a more integrated understanding of geomorphological processes so that the student would be more impressed by unity rather than alarmed by superficial diversity.

The importance of underlying theory was appreciated fairly slowly and this is why Chorley (1978) suggested that the geomorphologist reaches for his soil augur as soon as theory is mentioned (Chapter 3, p. 56), and why greater familiarity with scientific and mathematical methods and notation was advocated (S. Gregory, 1978) especially through closer alliance with other earth and environmental sciences (Clayton, 1980a). However, the greater tendency at pre-university level to focus upon the impact of human activity and upon management of the environment, with much less, if any, emphasis on the mechanics and principles of landscape development, is rather like putting the cart before the horse. It is very difficult later to take up that study of the horse when all the previous emphasis has been placed upon the cart! In the 1980s, physical geography in general, and geomorphology in particular, was in danger of developing with insufficient understanding of basic mechanics and principles, somewhat analogous to the decline of positivism in human geography. From the standpoint of geotechnical science, Williams (1982, p. 3) asserted that:

> ... the tunnel vision of scientists is perpetuated by traditional packaging into 'subjects', whether by the effects of history or of practical ends. This is unfortunate because, whatever we study in the so-called 'derivative' or 'secondary' sciences – biology, geology, pedology, hydrology – we must always operate within established principles and laws of the primary sciences of mathematics, physics and chemistry.

Other books (e.g. Statham, 1977) followed the approach used by Carson, and excellent illustrations of the research benefits to be realized by this method (e.g. Prior, 1977) still left considerable potential for development which was gradually realized during the 1980s and further in the 1990s.

The primary objective of process measurement is to obtain data on the transfer of energy or mass or flux within the physical environment. However, there are effectively an infinite number of points in space and time, and at each point several process elements need to be measured, so that a strategy has to be adopted for sampling from what is an infinite population. In geomorphological terms Church (1980) expressed the problem in terms of mass or energy fluxes or changes in content of a storage or control volume. Thus a sequence of observations could specify magnitudes of fluxes (q_t) or successive states of a system parameter (x_t). The net amount of work is often estimated by observing the frequency of an externally applied effective stress (such as a weather- or hydrology-related stress) which is more easily measurable or is available from records. A stress could be converted into the sequence of interest using a rating calibration or transfer function which links the applied stress and material flux or strain. Thus the change in landscape over a period T = $t_2 - t_1$ could be measured either by integration of fluxes

$$\int_{t_1}^{t_2} q_t \, dt \qquad (5.1)$$

or by net displacement of system parameters ($x_{t_2} - x_{t_1}$). Church (1980) clearly identified the fact that the character of geophysical event sequences complicates simple analysis, and three significant features are:

- trend, which includes well-defined cyclic behaviour;
- persistence, when a particular value in a sequence is constrained by adjacent values;
- intermittency, whereby there is a tendency for non-periodic grouping of like values over long periods of time, an effect called the 'Hurst phenomenon'.

These complications are illustrated in Fig. 5.1 together with an example of variance spectra. Data on fluxes or changes in storage volume in relation to a specific process have been obtained in four main ways and the data obtained (cf. Chapter 3, p. 57) then have to be interpreted using a sequence such as that in Fig. 3.2. Firstly are theoretical approaches that depend upon some conceptual framework to apply established relationships such as the continuity equation. Secondly are empirical measurements, thirdly are experimental investigations, and fourthly are historical techniques. These four basic approaches correspond to the major types of models employed in physical geography, and experience has shown that in any particular environmental problem the best results are often obtained where reliance is not placed upon a single approach alone.

Theoretical approaches have the great advantage that they do not require the establishment of a monitoring framework and the long periods of time necessary for the collection of empirical data. Furthermore, theoretical models are not confined simply to the small areas that are monitored but can be applied to large areas or whole systems. However, they depend upon available basic conservation equations including energy equations and water balance equations and they necessarily require simplifying assumptions. They have been used most effectively where there are relatively simple morphological sequences to deal with and where there is a clear relationship between the existing system and the processes operating upon it. Field studies might appear more expensive than computer modelling, and less guaranteed to produce results; however, Richards (1994, p. 280) argued:

> If modelling appears cheap, is this because the start up costs associated with sophisticated models that capture in a rigorous manner the fundamentals of process have been discounted? ... If it is thought that modelling can take place without very detailed field study, both to identify the boundary and initial conditions and to accumulate spatially distributed data to validate and test model runs (particularly in · terms of internal system state predictions), is this not a strange concept of modelling?

Results of direct *empirical measurements* are provided by national monitoring agencies including measurements of meteorological elements and of river discharge. Such measurements may not be available with the spatial and temporal frequencies that meet the requirements of research programmes, so that additional measurements often have to be made related to an *a priori* hypothesis and possibly derived from small experimental areas. Fluvial processes were a major focus for field experimentation after the mid-twentieth century, and the Vigil network scheme initiated in the mid-1960s involved the careful measurement of processes at individual sites and in selected study basins. The intention to maintain measurements over a decade or more was realized at some sites (Leopold and Emmett, 1965). A range of objectives for watershed experiments was identified by Ward (1971) which extended from specific black-box approaches to comprehensive studies in which there is an attempt to monitor many of the processes operating within a small area. The Hubbard Brook experimental basin (15.6 ha) in New Hampshire was monitored in great detail, including the effects of complete clearing of the woody vegetation and treatment with herbicides for three years. Some small instrumented areas were selected as representative of particular conditions, whereas others were experimental, in which change due to logging or other forms of land-use change could be monitored either by change in one basin or by comparison of several areas in which changes were taking place to different degrees. The enthusiasm for

(A)

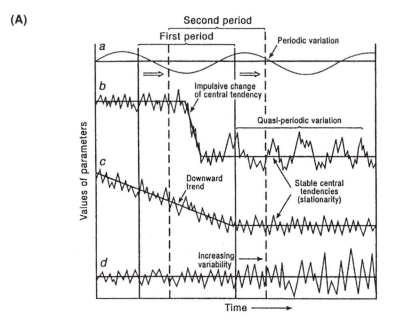

(B)

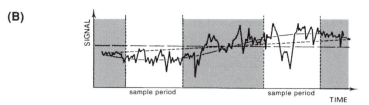

(C)

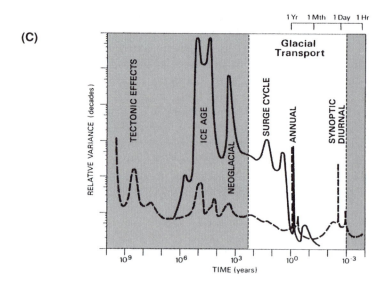

small instrumented areas produced important data and Slaymaker (1991) classified over 300 field experiments, but there were also a number of problems including lack of control and replicability of measurements, insufficient representativeness, unreliable data, and problems of finding suitable methods of analysis for the data collected. Thus Ackermann (1966) writing about the US suggested that 'small experimental watersheds have cost this country uncounted millions of dollars and unfortunately have yielded a small return on the investment of time and money'.

However, the small instrumented area approach galvanized the development of process measurements and the foundation provided has subsequently been built upon in several ways (Burt, 1994). Conceptual problems have been specified more clearly in the initial stages; experimental design has been more flexible, involving measurements at different scales; measurements have benefited from technological developments in instrumentation and especially the use of electronic devices; there has been closer liaison with long-term national monitoring networks; and analysis has been more robust. As physical geography research students were often dependent for their doctoral research upon field data collected from experimental investigations, the merits of research teams and groups became very evident because the group could offer a long-term monitoring framework into which specific doctoral research projects were accommodated. This is exemplified at the University of Exeter where since the 1960s research has been undertaken on the Exe Basin (see p. 123 and Fig. 5.2)

Experimental investigations embrace a range of approaches that experimentally reproduce a section of the physical environment, and range from field plot experiments through laboratory hardware models which attempt to use scaled-down versions of the real world, to analogue models which employ a different medium for investigation (e.g. Kirkby *et al.*, 1987). A kaolin glacier was an example of the last-named (Lewis and Miller, 1955), but much more extensively used have been measurements of rainfall simulators, especially in relation to erosion experiments, flumes, wave tanks and wind tunnels. The potential available was considerable in geomorphology (Mosley and Zimpfer, 1978), once the problems of scale and of relating observations to geophysical event sequences (e.g. Fig. 5.1) were overcome.

Figure 5.1 Variations over time. (A) An idealized time-series (curves a to d) of a representative parameter of a climatic element that is continuous in time, such as temperature or pressure (after Hare, 1996, with permission from Routledge publishers). (B) shows an event sequence and (C) indicates a conjectural variance spectrum (after Church, 1980). In the event sequence is shown the long-term mean, the linear trend, the well-defined cycle and the intermittent signal, so that a typical sample period as shown will produce statistics which are biased with respect to the whole sequence. The conjectural variance spectra for glacial transport is standardized on the annual cycle.

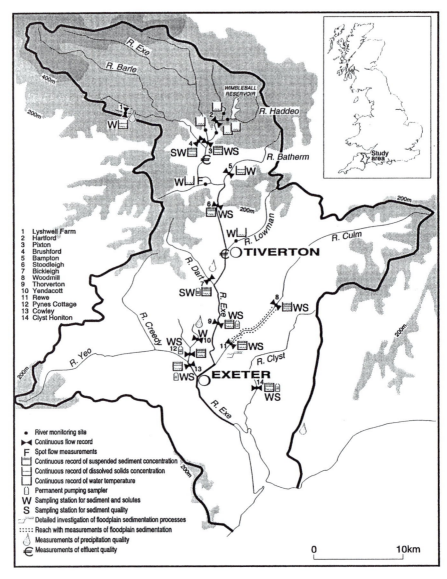

Figure 5.2 A long-term monitoring framework used for a range of process investigations. Measurements have been made in the Exe basin since the 1970s, at a variety of scales, and some of the results obtained are given in Walling (1996a, b). The map was kindly provided by Professor D.E. Walling.

Data on drainage basins, river channels, alluvial fans, and the effects of active tectonics were obtained from the Rainfall Erosion Facility at Colorado State University (Schumm *et al.*, 1987), and exciting experimental results are now being obtained (e.g. Allan and Frostick, 1999).

Historical techniques in physical geography are very useful because experimental data capture is necessarily limited to the period of experimental measurement, which is usually several years at the most, and a maximum of decades even where national recording agencies are used. The potential was reviewed through the cooperation of a geomorphologist and a historical geographer (Hooke and Kain, 1982) who reviewed graphical, written, oral, statistical and non-documentary sources, and the range of archival sources available in the US has been summarized (Trimble and Cooke, 1991) and their potential for historical and environmental studies shown (Cooke, 1992). The methods available are constantly increasing in number and their utility is greater as the complexity of contemporary processes is revealed. One somewhat unconventional data source for reconstructing past event frequency and character is the newspaper. The incidence of eighteenth- and nineteenth-century weather conditions in Scotland was studied from newspaper reports (e.g. Pearson, 1978), maps of hazard distribution in Derbyshire for 1879–1978 were derived from reports in the *Derby Evening Telegraph* (Gregory and Williams, 1981) and the approach has been used to reconstruct global hazards (see Chapter 7, p. 190).

5.3 Process in the branches of physical geography

Contributions made by physical geographers to the study of environmental processes cannot easily be separated from contributions made by practitioners of other disciplines, and the boundaries between disciplines became much less apparent. This was a welcome development because when the Environmental Science Services Administration (ESSA) was instituted in the US in 1965, its concern with the interactions among air, sea and earth and between the upper and lower atmosphere embraced the scope of physical geography, but no physical geographer was involved significantly in the early stages (Hare, 1966). In Britain, with the foundation of the Natural Environment Research Council in 1966, it was notable that physical geography was not mentioned in the charter (Hare, 1966).

In *soil science,* external to physical geography, emphasis had proceeded towards soil processes including soil biology and microbiology (e.g. Russell, 1957), and interest in soil processes and dynamics led to the inception of new ideas including a more process-based vision of the soil profile as based upon additions, subtractions, translocations and transformation of constituents (Simonson, 1959). Greater concern for soil processes was also associated with developments in soil mapping, as the soil series, adopted as the basic unit of field mapping in the 1950s using soils with similar profiles, derived from similar materials under similar conditions of development, acquired a third dimension with the definition of the three-dimensional soil body or pedon (Johnson, 1963). The smallest volumes, usually between 1 and 10 m² on the surface, were the basis for mapping units, or polypedon, which contain more than one pedon. Bridges (1981) argued that soil geography aimed to 'record and explain the development and distribution of soils on the surface of the Earth', but the contribution of physical geographers as primarily a historical or global one was not unanimously accepted in all the books produced by physical geographers in the 1970s. Increased attention on processes was first exemplified by measurement of the variables involved in soil systems, including soil moisture, soil organic matter content and soil pH, which necessitated knowledge of the measurement techniques themselves. Once process variables had been isolated and quantified, then the interaction of dynamic process variables with the spatially distributed variability of soil properties could be investigated (Trudgill, 1983). Although attention to processes at this level could proceed to what some geographers would clearly categorize as soil science, the movement did catalyse the development of a more three-dimensional view of the soil. Thus Runge (1973) interpreted soil properties (s) as functions of organic matter production (o), water for leaching (w) and time (t) in a form $s = f(o,w,t)$ where all parameters are fairly easily measured.

To achieve a more three-dimensional model of soil it was necessary to proceed beyond the catena concept, and developments depended upon a close relationship between soils and the landsurface. One of the most striking developments was the nine-unit landsurface model, proposed (Conacher and Dalrymple, 1977) as an appropriate framework for pedogeomorphic research. This model developed from one suggested earlier in relation to slope processes (Dalrymple *et al.*, 1969) and was important because it introduced the term 'landsurface catena' to refer to a three-dimensional slope extending from interfluve to valley bottom and from the soil/air interface to the base of the soil, with arbitrary lateral dimensions. Each landsurface catena is composed of landsurface units which are identified and defined according to responses to single, or multiple, contemporary geomorphological (in 1969) and also pedological (in 1977) processes. The processes include interactions amongst soil materials, water and gravity, or the mobilization, translocation and redeposition of materials by water flow and mass movements, whilst the responses are identifiable physical and morphological properties of the soil and also of soil morphology. Recognition of aperiodic soil/water/gravity events in a landsurface catenary concept makes this approach appropriate for studying the dynamic interaction between soil properties and landforms. Although Ruellan (1971) distinguished those researchers who attach great importance to the geomorphological processes of erosion and deposition (allochthonists) from those who attribute the major characteristics of soils to pedological processes (autochthonists), Gerrard (1981), in his treatment of soils and geomorphology, argued that soils are the results of the interaction of both sets of processes. Subsequently Gerrard (1993) identified soil geomorphology as '... the integration of pedology and geomorphology' and followed the definition of Olson (1989) that it is 'the study of the landscape and the influence that landscape processes have on the formation of soils'. One aspect previously given relatively little attention by physical geographers was the occurrence of soil erosion. Although there had been some investigations of a functional kind relating soil erosion amount to controlling variables in areas like Zimbabwe (Stocking, 1977), studies of soil loss are potentially very useful (Stocking, 1980) and may be undertaken by detailed process investigations which are usually concerned with parts of the erosion process or with laboratory measurements (e.g. de Ploey, 1983); by empirical investigations which monitor output in relationship to input and use a relation similar to the Universal soil loss equation; and factorial survey methods which Stocking visualized as analysis, and collation of the spatial pattern of all factors which related to soil loss. Although research interest in soil erosion developed in the late 1970s (e.g. Morgan, 1979), it is a paradox that work, for example, developing from the Universal soil loss equation (Wischmeier, 1976), did not materialize, but this was later remedied by research (e.g. Thornes, 1990) that analysed spatial patterns and the relationship to vegetation cover.

In *biogeography* it is not easy to separate natural from cultural biogeography or to disentangle the contributions that originated in ecology from work undertaken by physical geographers. The movement by ecologists towards trophic–dynamic ecology was developed by Lindeman (1942) in a classic paper which treated natural ecosystems on the basis of the capacity of their primary producers (photosynthetic plants) to capture part of the incident solar and atmospheric energy, and to incorporate it into the dry organic matter that would subsequently yield it to the grazing and decay food webs. Focus upon ecosystems in this way also required emphasis upon biogeochemical cycles as the pathways whereby mineral nutrients are cycled through the world systems as in the systems approach (Chapter 4, p. 194). In an ecosystem, identification of trophic levels within the feeding hierarchy in studies of specific areas could then show 'what eats what' and therefore how the trophic structure is built up. In an extremely detailed investigation in the Hubbard Brook forest of New Hampshire, USA, the pathways of chemical elements were studied under natural conditions and then after deforestation and the biogeochemical cycles were investigated in each stage (Likens *et al.*, 1977).

Climatology embraced process investigations in at least three ways, each corresponding to a particular scale. At the *global scale*, numerical modelling was applied to the planetary boundary layer facilitated by computer development in short-range forecasting research, by the construction of general circulation models (GCMs), and thence models of global climate. At the *meso-scale*, models of synoptic and planetary-scale motions constructed for both research and forecasting purposes (Atkinson, 1983) involved general models of meso-scale flows involving the planetary boundary layer (PBL) and models of particular kinds of meso-scale circulation such as sea/land breeze models, urban circulation models, and slope, mountain and valley wind models. Also at this scale, research on precipitation areas showed that the vertical air motion responsible and the areas themselves are organized into a hierarchy (according to the horizontal size). Atkinson (1978) reviewed the small meso-scale precipitation areas (SMPA) and large meso-scale precipitation areas (LMPA) and the way in which these areas were associated with frontal systems. Although the SPMAs last for 3–4 hours and move parallel to and ahead of the front, a theoretical explanation for the size of SPMAs was still required. At the most *detailed scale*, research developments embraced microclimatology (Geiger, 1965) and also applied developments such as agricultural meteorology. In relation to contemporary climatology, Henderson-Sellers (1989b) contended that a quantitative synthesis based upon appropriate description of process and spatial analysis is still urgently needed.

The impact of studies of process was perhaps most substantial and most dramatic in **geomorphology.** At least six significant antecedents existed for the study of geomorphological processes, and the first was undoubtedly the seminal work of Grove Karl Gilbert. In his contributions in the western part of the US not only did he describe physical erosive processes but also derived a system of laws governing progress from initial to adjusted forms (Chorley *et al.*, 1964). In his *Report on the Geology of the Henry Mountains* (Gilbert, 1877) he provided the first major treatment by a geologist of the mechanics of fluvial processes, and in 1914 he published a remarkable investigation into *The Transportation of Debris by Running Water* which included the results of laboratory experiments (Gilbert, 1914). Gilbert is now acknowledged as a brilliant geomorphologist whose contribution anticipated many of the developments half a century later, and whose deductions regarding stream and landscape mechanics 'have given new life to quantitative geomorphology in the twentieth century' (Chorley *et al.*, 1964, p. 572). Sack (1992, p. 251) concluded that:

> Gilbert was an appropriate figurehead for the new paradigm … Identifying Gilbert as the archetypal process geomorphologist helped sanction the process paradigm, distance it from Davisian geomorphology, and ease the paradigm change.

A second antecedent could be found in awareness of work by engineers, and particularly R.A. Bagnold who published his monumental *Physics of Blown Sand and Desert Dunes* in 1941, which stipulated the bases for processes in desert areas. Bagnold subsequently worked on processes involving fluids other than air, contributed to understanding of beach formation by waves based upon wave tank experiments (Bagnold, 1940), to the analysis of fluvial processes (Bagnold, 1960), and in one of his last papers (Bagnold, 1979) he reviewed fluid flow in general. A third source derived from Scandinavia where in 1935, F. Hjulstrøm published results of field and laboratory investigations related to the River Fyris and identified relationships of fundamental significance, later partly amended by Sundborg (1956). Also in Scandinavia an important study of the mass movement processes on the slopes of Karkevagge (Rapp, 1960) was important because it endeavoured to quantify all the processes that affect a slope in a subarctic environment and their relative significance, and concluded that the most effective agent of removal was running water removing material in solution.

Other antecedents all emerged in North America but for slightly different reasons. Research directed by A.N. Strahler at Columbia University (see p. 64) and summarized in Strahler 1992 was the fourth antecedent. In this Columbia school of geomorphology, measurements were made of processes operating on stream channels and slopes of a number of areas (e.g. Schumm, 1956); later the focus turned to coastal processes (Strahler, 1966), and perhaps most significant was Strahler's advocacy of the need for a dynamic basis for geomorphology (Strahler, 1952). This significant paper endeavoured to extend geomorphology from a functional viewpoint towards a more realist view, as indicated by the aim (Strahler, 1952, p. 923):

> ... to outline a system of geomorphology founded in basic principles of mechanics and fluid dynamics, that will enable geomorphic processes to be treated as manifestations of various types of shear stresses, both gravitational and molecular, acting upon any type of earth material to produce the varieties of strain, or failure, which we recognize as the manifold processes of weathering, erosion, transportation and deposition.

A fifth contribution came from the ideas of dynamic equilibrium advanced by J.T. Hack (1960), who argued that the concept of dynamic equilibrium provided a more reasonable basis for the interpretation of topographic forms in an erosionally graded landscape, that every slope and stream channel in an erosional system is adjusted to every other, and that when the topography is in equilibrium and erosional energy remains the same, all elements of the topography are downwasting at the same rate. In this view the accordant summits in areas like the ridge and valley province of the US were interpreted as the result of dynamic equilibrium rather than as remnants of earlier erosion cycles (Hack, 1960, p. 81), but the two approaches were partly reconciled by Schumm and Lichty (1965) as shown in Chapter 6 (p. 153).

A final and sixth antecedent was the investigation of magnitude and frequency of geomorphological processes (Wolman and Miller, 1960) and subsequently of fluvial processes in general (Leopold *et al.*, 1964). The analogy was drawn between the efficacy of geomorphological processes and that of a giant, a man and a dwarf attempting to cut down a forest. The dwarf maintains an assault on the trees for long periods and achieves little (the equivalent of frequent, low-magnitude events); the giant sleeps most of the time but occasionally wakes and causes great destruction (a catastrophic event); whereas the man works regular hours and systematically achieves the greatest effects (events that occur once or twice each year). It was concluded (Wolman and Miller, 1960, p. 54) that:

Closer observation of many geomorphic processes is required before the relative importance of different processes and of events of differing magnitude and frequency in the formation of given features of the landscape can be adequately evaluated.

The publication in 1964 of *Fluvial Processes in Geomorphology* (Leopold *et al.*, 1964) ushered in a new era of process investigation. This book was the first to emphasize contemporary processes and underlying physical principles, and it focused upon river channels, drainage systems, and slopes with reference to climatic-inspired systems. It contended that process involves mechanics which requires understanding of the inner workings of a process through the application of physical and chemical principles. After 1964 greater emphasis upon processes and interaction with other disciplines saw international multidisciplinary conferences such as those on gravel bed rivers, which produced important volumes (Hey *et al.*, 1982; Thorne *et al.*, 1987; Billi *et al.*, 1992) and were extremely significant for furthering the study of fluvial processes and sediment transport and for applications to river management. Processes were used as the theme for Graf's (1988) important book on *Fluvial Processes in Dryland Rivers* (see p. 135) and also for a seminal volume on floodplains (Anderson *et al.*, 1996). Other branches also showed awareness of the need for studies of processes, however, and the need had always been appreciated in limestone areas (Ford and Williams, 1989) which were the focus for many of the earliest, and simplest, process technique developments and for the derivation of modified models of landscape development as proposed by Smith and Newson (1974) for the Mendips. In coastal geomorphology the process foundation was well established by the achievements of engineers and of research institutes, using theoretical and laboratory investigations as well as empirical ones. The coastal geomorphologist therefore focused upon the processes that control shoreline equilibrium and the significance of longshore currents and sediment transport, of wave activity in relation to the swash, nearshore and offshore zones, and to the less-studied influence of tides and of impulsive events such as tsunamis. Research investigations were exemplified by the work of C.A.M. King and other staff from the University of Nottingham who investigated changes in the coast of south Lincolnshire at Gibraltar Point from 1951 to 1979. Processes were reflected in the content of major texts (King, 1972) and one (Davies, 1973) included a map of wave environments of the world. Physical geographers avoided involvement in oceanography except in multidisciplinary programmes, although in consideration of the northwest European shelf, Hardisty (1990) concluded that the future lies in the development and utilization of advanced numerical models of the region.

Each major branch of geomorphology had at least one new textbook which served to incorporate processes and to serve as a baseline from which new research and teaching could develop. In relation to deserts, Cooke and Warren (1973; subsequently Cooke *et al.*, 1993) provided a standard

Geomorphology in Deserts primarily concerned with desert landforms, the materials that compose them, the processes of debris preparation, erosion and transport that modify them, and the environmental factors influencing all these phenomena. Other books focused upon *Aeolian Sand and Sand Dunes* (Pye and Tsoar, 1990) and *Aeolian Geomorphology* (Nickling, 1986), and processes were the theme used for the *Introduction to Aeolian Geomorphology* (Livingstone and Warren, 1996) and constitute a major section in *Arid Zone Geomorphology* (Thomas, 1997). In the fields of glacial and periglacial geomorphology, the first modern textbook (Embleton and King, 1968) was soon followed by others (Price, 1973), and later work utilized a systems framework (Andrews, 1975; Sugden and John, 1976). It was suggested that the study of glacial processes and forms had been left out in the cold and poorly understood because a gulf had arisen between those who study glaciology and those who study glacial landscapes, so that (Sugden and John, 1976, p. 1):

> Perhaps there is a need for a more glaciological type of geomorphology and a more geomorphological type of glaciology. There is now a strong case for a radical dialogue between those studying glacier dynamics and those studying forms. Until this occurs, there can be few spectacular advances such as those recently achieved in fluvial and slope geomorphology.

Glacial processes were studied by investigations in areas including Iceland, Baffin Island, Antarctica, Alaska, Scandinavia and the European Alps, and it was from such research that the understanding and interpretation of past glacial systems was enhanced, as in Scotland by Sissons (1967). Thus glacial geomorphology was defined (Gjessing, 1978) as concerned with bedrock forms and superficial deposits produced by glacial and fluvioglacial processes in areas of present glaciers as well as in areas covered by glaciers during the Quaternary. The influence of theoretical research in glaciology was substantial, and the physicist J.F. Nye (1952) derived equations for glacier flow assuming that ice is a perfectly plastic substance, that it flows down a valley of constant slope, and that the conditions of temperature, accumulation and ablation are simple and uniform; his model could be compared with field observations. Processes were the theme for *Glacial Geologic Processes* (Drewry, 1986) and field investigations of subglacial processes led to recognition of the importance of subglacial till deformation (Hart and Boulton, 1991; Knight, 1993). Geographers were well represented in glaciological research and in 1990 accounted for 11% of the affiliation of the International Glaciological Society, 15% of the authors in the journal, and 26% of the participants at the UK meetings (Knight, 1993).

In periglacial geomorphology, knowledge of contemporary processes was achieved by investigations of permafrost in Siberia, Alaska and Canada, and of cryonival processes in areas such as Spitsbergen. Work such as that on thermokarst in Siberia in relation to the development of lowland relief

(Czudek and Demek, 1970) had a significant influence on research in areas of Quaternary periglacial morphogenesis. Textbooks in periglacial (e.g. Davies, 1969; Pewe, 1969; Washburn, 1973) as well as glacial geomorphology acknowledged the contributions made from other disciplines, from research institutes and from international conferences. Thus the Institute of Arctic and Alpine Research (INSTAR) at Colorado, US, founded in 1951, the Scott Polar Research Institute in Cambridge, UK (1920), and the V.A. Obruchev Institute of Permafrost Science in the USSR (1930) are examples of research institutes that contributed significantly to developments in the science of cold regions. Process investigations were an important ingredient in the new journal *Permafrost and Periglacial Processes* (1990–).

Slopes were the focus for revived interest after 1950 (Strahler, 1950b), the subject of quantitative description and analysis (Bakker and Le Heux, 1952), and then for measurements which could lead to empirical research in specific areas using, for example, the Young Pit (Young, 1960) and providing many indications of rates of erosion (Young, 1974). Slope process investigations first used single empirical measurements and later continuous recording, and then theoretical approaches were admirably exemplified in the book *Hillslope Form and Process* (Carson and Kirkby, 1972); and by stability analyses which utilized the factor-of-safety approach and related approaches used by the civil engineer. These strands complemented the well-established qualitative models of Davis and Penck, and the blend of approaches adopted in textbooks varied according to author and date of writing, with an emphasis on factors and measurement in some (e.g. Young, 1972), a more theoretical foundation stressed in others (Carson and Kirkby, 1972), and both were later combined (e.g. Selby, 1993). In the foreword to a major two-volume compilation on hillslope processes (Anderson and Brooks, 1996), Chorley commented:

> In 1964 I identified 'the assumed minor importance of studies of present processes' as a major impediment to slope research. A main message of the current volume is that the study of reasonably contemporaneous processes lies close to the heart of modern concern.

Spectacular advances had occurred (Sugden and John, 1976) in slope and in fluvial geomorphology, and some 27.7% of British research could be categorized as fluvial including both processes and landform development in 1975 (K.J. Gregory, 1978b) and this was maintained until 1996 (see p. 283), but the pattern was complicated because of the interaction with hydrology. Horton's 1945 paper was fundamental and Chorley (1995, p. 534) assessed its impact:

> Horton's freedom from conventional geomorphic thinking allowed Horton, like Pip in *Great Expectations*, to throw open the windows of an antique house and to cause a quantitative, dynamic wind to blow through the misty corridors of denudation chronology. Like Miss Haversham's house, the old geomorphology was soon to be destroyed.

The impact of the great increase in geomorphological research investigations following Horton's paper and the research of the Columbia School was ascribed to seven contributing themes (Gregory, 1976), namely network morphometry, drainage basin characteristics, hydraulic geometry, river channel patterns, theoretical approaches, dynamic contributing areas, and the palaeohydrology–river metamorphosis approach. The latter owed much to research by Schumm and his students (see Chapter 6, p. 161) whereas the other themes reflected increasing knowledge of, and dependence on, hydrology. In addition there were many research contributions in the field of fluvial morphology which were reflected in the books that succeeded Leopold *et al.* (1964) and were devoted to streams (Morisawa, 1968), to drainage basin form and process (Gregory and Walling, 1973), to water and environmental planning (Dunne and Leopold, 1978) and to alluvial river channels (Richards, 1982; Knighton, 1984, 1998). There were also books produced by geographers dealing with hydrology (Ward, 1967) and with aspects of hydrology such as floods (Ward, 1978; Smith and Ward, 1998). Indeed, although physical geographers perceived hydrology to be a separate field of scientific enquiry and, for example, *Progress in Physical Geography* maintains progress reports on fluvial geomorphology as well as on hydrology, other physical geographers advocated the notion of geographical hydrology. Ward (1979) used the term 'geographical hydrology' to connote the hydrology approach of geographers and noted that:

> ... the engineer resorts to empiricism and coefficients, to simplification and generalization of systems and processes. The geographer on the other hand, is primarily interested in how the landscape works, and in man's interactions with it, and recognizes that water is but one of the terrestrial phenomena in the total complex interacting ecosystem in which he is really interested.

Two decades later there was less separation of physical geography from hydrology and no clear distinction in methods or objectives. In addition to the seven types of fluvial geomorphology investigation specified above, the major focus of research on small instrumented areas (p. 111) developed towards assessment of sediment and solute yields (e.g. Walling, 1983a) and to the assessment of human impact including that on urban areas (e.g. Hollis, 1979). The contribution made by physical geographers to international cooperation and research is exemplified in the work of D.E. Walling, first as Secretary and then as President of the Commission on Continental Erosion of the International Association of Hydrological Sciences. Substantial contributions made by physical geographers to hydrology broadly defined have arisen not only from contributions concerned with the drainage basin and runoff generation but also from hydrometeorology.

Branches of geomorphology and of physical geography did not all gravitate towards process investigations at the same time, but the edited volumes produced, often the outcome of conferences, summarized significant advances. Increasing fragmentation of the branches of geomorphology occurred as they became more closely associated with other disciplines, and Marcus (1979) noted that in addition to maintaining some contact with human geography, 'most physical geographers today keep one foot in the AAG and the other in a cognate scientific society. It is a reality of our professional lives'. This situation has continued and Agnew and Spencer (1999, p. 5) commented that:

> Physical geographers have become increasingly specialized, perhaps working more as environmental scientists than as geographers. One result is that physical geography has fragmented into its component specialisms (climatology, biogeography, geomorphology and hydrology), and the 'split' by climatologists from the UK geographical community some years ago is threatened in other areas of the discipline.

Although such fragmentation was a feature beginning in the 1960s and 1970s, two trends fostered integration, namely the use of common techniques and the adoption of a more realist approach. Techniques were initially specific to particular branches of geomorphology but increasingly became less discipline-based, and this was relaxed further by developments in remote sensing and by opportunities arising from developments in micro-electronics. An early book on techniques in geomorphology (King, 1966) was complemented by specific manuals such as the US Geological Survey *Techniques of Water Resources Investigations*, and these were succeeded by an edited manual of geomorphological techniques (Goudie, 1981a, 1990c) and a volume on field techniques (Dackcombe and Gardiner, 1983). For techniques teaching in physical geography, a self-paced course was developed (Clark and Gregory, 1982). Technical Bulletins of the British Geomorphological Research Group showed the importance accorded to techniques and were succeeded by the Technical and Software Bulletin (Hardisty, 1998) with a computer disk subsequently available through the Wiley website (www.interscience.wiley.com).

Some physical geographers (e.g. Thomas, 1980) argued that the study of the energetics of the landsurface 'has perhaps robbed the subject of some of its scope and depth' but it can be argued that investigations of process energetics extended towards illuminating recent temporal change (p. 168). Inevitably process studies may lead to very specialized investigations, so that a potential danger identified by Conacher (1988, p. 161) is that:

> A number of geomorphologists are becoming increasingly concerned by the pursuit of apparent trivia by some of their process-orientated colleagues. It is argued that the original intention of process research – to explain landforms – has been forgotten.

In fact the original intention had not been forgotten but could not be addressed without adequate modelling (Section 5.6) to relate to temporal change (Chapter 6, p. 138).

5.4 Patterns of process

Whereas measurements and analysis of processes are often undertaken for a specific location – the climatological station, the soil pit, erosion plot, drainage basin or vegetation quadrat – process data and analysis are often required at much greater spatial and temporal scales. Analysis at regional and global scales has been possible because, firstly, sufficient process data have now been collected to allow national and world patterns to be discerned; secondly, because international cooperation remedied past deficiencies of data coverage, quality and conformity; and thirdly, because remote sensing provided further clues to the world picture (see also Chapter 9, p. 236). Spatial analysis becoming increasingly critical in environmental risk assessment was a conclusion reached by Phillips (1988) from an investigation of non-point source pollution.

World spatial systems are an enormous topic, but a dominant trend has been towards the differentiation of the Earth's surface on a more realist dynamic basis to replace the rather static functional treatment previously used. Although exemplified in all branches of physical geography, this trend is probably most obvious in the data-rich field of *climatology* where the impact of remote sensing was profound (Henderson-Sellers, 1990). Maps showing mean values of climatic elements such as precipitation or temperature had long been employed, and numerous attempts made to classify climates upon a world basis. Whereas many such attempts embodied a somewhat static view of the pressure distribution, this was succeeded by what Barry and Perry (1973) termed a kinematic view of the weather map in which the synoptic weather map was viewed in terms of airflow and the movement of pressure systems. This included what Court (1957) described as pressure field climatology and led to a range of approaches which included weather patterns and air mass climatology, all of which provided a more effective reflection of processes. A more dynamic approach was possible with the advent of greater computing power which has allowed the analysis of the necessary amounts of synoptic data. Utilization of satellite data ushered in a new era (Barrett, 1974); such satellites provide observation systems of the Earth and atmosphere, function as highly convenient data-collection platforms, and as connection links between widely spaced ground stations between which large daily exchanges of weather data must take place. Atmospheric classifications were placed in three groups by El-Kadi and Smithson (1992) who concluded that the Kirchhofer classification scheme has advantages over others.

In the field of *biogeography*, world maps of plant formations used for many years had been adopted by attempting to fit some climatic classifications to the pattern of plant distribution. However, the static quality of such maps could be replaced when research on ecological energetics, or nutrient cycling, and on population dynamics led to greater use of net primary productivity (NPP) which is the material actually available for harvest by animals and for decomposition by the soil fauna and flora or their aquatic equivalents. The rate of accumulation of biomass or NPP is expressed as the weight of living matter/unit area/time, and work by the International Biological Programme (IBP) provided more accurate estimates of NPP on a world scale for both continents and oceans, making it possible to rank biome types according to NPP or present processes. This offered an additional dimension to biogeography (Simmons, 1979a), and because it is infinitely renewable but subject to substantial modification by man, the NPP of an area can be viewed as what Eyre (1978) characterized as 'the real wealth of nations'.

Soil geography had a particularly well-established legacy of soil maps which related the one extreme of detailed soil survey to the other of a world distribution. Emphasis had been more towards the local scale because of the demand for soil maps in relation to agricultural and other land-use purposes. Emphasis at this level had been upon soil evolution rather than upon soil dynamics, which had been treated in relation to land capability. However, the increasingly similar basis underlying national soil maps allowed correlation to take place more easily so that a soil map of the world (Bridges, 1978a) could be the basis for soil geography in future years (Gerrard, 1981; Ellis and Mellor, 1995).

In *geomorphology*, as in soil geography, it was not easy to achieve representations of global spatial patterns to complement the long-established static morphological or structured landform regions. However, coastal environments were classified according to wave energy by Davies (1973) to assist a more meaningful correlation between wave type and coastal morphology. Perhaps the most dramatic progress was made in relation to erosion rates, and in addition to world patterns of discharge, Walling and Webb (1983) reviewed earlier attempts to portray world patterns of sediment yield, some of which conflicted quite significantly, and provided a revised map of global sediment yields based upon data from nearly 1500 measuring stations and pertinent to sediment yields from a basis of 1000 to 10,000 km^2 in area. A review of global erosion and sediment yield showed how the lack of long-term records can be remedied by evidence from sources including lake sediments, catchment experiments and space–time substitution, by Walling (1996a) who noted:

> The increasing rate of degradation of the global soil resource and changes in the flux of materials between the land and the oceans are clearly important implications of this aspect of global change which would also benefit from an improved assessment of the magnitude of the changes involved.

Such progress exemplifies the way in which a physical geographer makes a distinctive contribution by deriving world patterns using the most recently refined methods for assessing sediment yield, and presenting data in a form pertinent to a wide range of disciplines. It is not easy to collate data on rates of geomorphological processes, and the problems were cited to be (Goudie, 1995):

- different units or forms being used by various authors (some utilize Bubnoff units, for example);
- the tendency to undertake work in the most dynamic areas;
- a problem of scale arising because one cannot calculate variations in space by multiplying results from one scale to another;
- variations occurring over time reflecting episodic development;
- the influence of human activity.

One might add the need to include consideration of tectonics in areas such as New Zealand (Williams, 1991).

Results obtained from specific areas need to be brought together and, for example, Trudgill (1986, pp. 13–14) concluded that two substantial challenges remain in solute geomorphology, namely the bringing together of fieldwork and theoretical work and the evaluation of the spatial distribution of rates of erosion in order to predict differential landform evolution. These two challenges were subsequently reflected in the progress made in *Solute Modelling in Catchment Systems* (Trudgill, 1995). To achieve greater understanding of spatial variations, closer cooperation between scientists in several disciplines can be of considerable value. For example, Walling and Webb (1983, p. 95) pointed to the need for closer cooperation between limnologists and fluvial geomorphologists which could enable greater understanding of sediment yield.

A more integrated approach was attempted by climatic geomorphology, and an approach by Peltier (1950) endeavoured to relate climate to geomorphological processes on a semi-quantitative basis using mean annual temperature and mean annual precipitation. Peltier later (1975, p. 129) contended that climatic geomorphology is one part of a broader structure of regional geomorphology which includes both tectonic geomorphology and the stratigraphy of continental deposits. He produced a general landform equation in which landform (LF) could be viewed as a function of geological material (m); rate of change of geological material, structural factor (dm/dt); rate of erosion (de/dt); rate of uplift (du/dt); and the total duration of the process (t) in the form LF = f(m, dm/dt, de/dt, du/dt; t). He was then able to propose expressions for an erosion factor (de/dt). Relationships between geomorphological processes and climatic parameters were also approached from a climatological perspective, and Hare (1973) cited a quotation from Budyko and Gerasimov (1961):

... the heat and water balance of the Earth's surface is, as a rule, the main mechanism that determines the intensity and character of all the other forms of exchange of energy and matter between ... the climatic, hydrologic, self forming, biologic and other phenomena occurring on the Earth's surface.

Hare (1973, p. 171) avowed that 'synthesis is easy to announce, but hard to pull off', but a focus upon energy balance, the energy flux and energetics (Chapter 4, p. 94) could underpin analysis of the links between climate and soil, plant and animal life. A considerable amount of research on energy-based, and energy-budget, climatology had been undertaken since 1956 when Budyko (translated 1958) attempted a heat balance of the Earth's surface. Referring to Budyko's classic paper, Lockwood (1997, p. 342) commented:

> On reading again . . . I was struck by how much of it is still highly relevant to existing problems in modelling exchanges at the land and ocean interfaces.

Budyko attempted a physico-geographical zonation of the Earth, and his approach was subsequently developed by systems in which energy and moisture regimes are related to vegetation types (e.g. Grigoryev, 1961), to genetic soil types (e.g. Gerasimov, 1961) and later to geomorphic zonality (Ye Grishankov, 1973). Geomorphologists were the last to explore the potentially fruitful links with energy-based climatology identified by Hare, who commented (1973, p. 188) that '. . . geomorphologists have not put their discipline on an energy balance basis to nearly the same extent, and probably with good reason'. The reason for the later development of geomorphology in this direction was adduced (Hare, 1973) to be due to the facts that fluvial processes tend to be dominated by extreme events rather than balance relationships, so that stochastic methods and extreme value theory are closer to the reality of geomorphic processes than is energy-balance climatology; and that the geomorphological time-scale is longer than that utilized by the climatologist. However, the prospect of an energy-based integrated physical geography including geomorphological processes was visualized by Hare (1973, p. 189) as a direction espoused by Budyko and Gerasimov:

> . . . a common, quantitative, theoretically-based language . . . in my judgement, is what physical geography needs. They lead us, also in the direction that all science aims at, towards prediction.

Further attention focused on spatial variations with the development of climatic geomorphology and by the identification of process domains. *Climatic geomorphology* emerged from attempts to relate process to climate and to emphasize inter-relations between morphological, pedological, vegetational and climatic characteristics of the Earth's surface. Peltier (1950) identified nine different possible morphogenetic regimes, each of which could be distinguished by a characteristic assemblage of geomorphological processes. Peltier (1950, p. 222) acknowledged the importance of the earlier ideas of Penck, Davis and Troll, and especially of Kirk Bryan, whose emphasis on climatic morphology was accredited with the formulation that Peltier offered. However, Peltier proceeded to identify a periglacial cycle which went beyond the relationship between climate and process and attracted criticisms of the kind previously levelled at the Davisian normal cycle. Climatic geomorphology had developed in France somewhat independently under the influence of J. Tricart and A. Cailleux. Morphoclimatic zones recognized by the French school (Tricart, 1957) were related to climates and to processes and also to soils and to vegetation. Indeed, the approach in some ways resembled that of the early Russian school of soil science and involved the recognition of *zonal phenomena* as the direct results of latitudinal climatic belts; of *azonal phenomena* arising from non-climatic control including endogenetic effects; *extrazonal phenomena* which occurred beyond their normal range of occurrence, such as sand dunes on coasts; and *polyzonal phenomena* including those that operate in all regions of the globe subject to the same physical laws. This school of geomorphology produced an introduction to climatic geomorphology (Tricart and Cailleux, 1965, 1972) and a series of volumes each dealing with specific groups of morphoclimatic zones. In Germany, Julius Budel proposed an approach of *klima-genetische Geomorphologie* (Budel, 1963) which became widely known in the English literature after a paper by Holzner and Weaver (1965). Budel distinguished three generations of geomorphology, namely:

(1) *dynamic*, which concerns the study of particular processes;
(2) *climatic*, which considers the total complex of present processes in their climatic framework;
(3) *climatogenetic geomorphology*, which involves the analysis of the entire relief including features adjusted to the contemporary climate and also those produced by former climates.

This model was supported by five climato-morphogenetic zones (Budel, 1963), later expanded to seven (Budel, 1969), and then to eight (Budel, 1977). Each zone was characterized by particular landscape-forming processes and by relief features, and so could also be the basis for understanding past landscape development. Thus the extra-tropical zone of former pronounced valley formation was dominated by relict landscape features both glacial and periglacial.

Climatic geomorphology subsequently attracted both great support and disenchantment, and was the subject of at least two groups of interpretation. One view was that climate governs the character and distribution of landforms, whereas the other view was that processes are related to climate through inter-relationships between the morphological, pedological, vegetational and climatic characteristics of the Earth's surface. It is perhaps the lack of equal familiarity with the achievements of both groups that led to a range of viewpoints and reactions. Thus Stoddart (1968) concluded that although it may be possible to derive a satisfactory methodological basis for climatic geomorphology, in 1968 it was viewed as not new, not well-established and premature. Butzer (1976) contended that:

Climatic geomorphology attempts to cope with the excessive complexity of natural parameters by holding variables such as structure, lithology and man constant. In much the same way implicit or explicit models for describing the evolution of stream channels or drainage basins commonly are used to make simplifying assumptions that eliminate considerations of time, history and sometimes even progressive change.

In relation to palaeoclimatology and the geomorphological record, Barry (1997) provided a summary of approaches used and identified climatic, palaeoclimatic and geomorphological topics that need resolution. Of the latter he suggested the determination of the duration of specific climate regimes which are considered to be relevant to the imprinting of geomorphological and soil characteristics that survive to the present. In their critique of climatic geomorphology, Twidale and Lageat (1994) considered the landform analysis approach in particular, and concluded that climatic impacts are not denied but have been overestimated and that (Twidale and Lageat, 1994, p. 330):

> Climatic factors are important in inducing the operation of processes that find clear expression in landform assemblages the world over, but together they constitute but one of several factors that determine the shape of the Earth's surface at regional and local scales. Climate is certainly not an over-riding consideration in the interpretation of landscape.

Process domains were identified by Thornes (1979, 1983b) as the spatial distribution of work done by several processes according to some particular environmental parameter, such as rainfall intensity, which could separate areas dominated by throughflow from those in which overland flow does most geomorphological work. Thornes (1983b, p. 227) suggested that the idea could be extended to two or more dimensions and that some process domains can overlap, some processes are contained within the domains of others, some are in competition, while some are spatially exclusive and disjunct. Domains therefore represent equilibrium relationships between processes according to the controlling parameters such as climate, infiltration rate or cover density. This idea is analogous to the niche concept in ecology and in equilibrium the environment is partitioned into a set of spatially organized process 'niches' separated by relatively sharp or diffuse boundaries, so that Thornes (1983b, p. 227) described process geomorphology as concerned mainly with the behaviour that determines the character and configuration of the domains. Whereas process geomorphology is concerned mainly with the way in which domains are determined, evolutionary geomorphology is concerned with the initiation and development of the structure giving rise to the domains. Changes can be envisaged in the relative importance of different processes and a particular process can change its domain and interact with that of another process. To analyse development of this kind may require consideration of unstable behaviour as an alternative to the steady-state model (Thornes, 1983b), which is pertinent to environmental change (Chapter 6, p. 163) and consistent with the approach of Graf (1988).

5.5 Processes in time

It is impossible to study processes in physical environment independently of time, although timeless and timebound approaches were distinguished by Chorley (1966). For analysis purposes, time sequences can be thought of as of four types (Thornes and Brunsden, 1977):

(1) *continuous time*, which means that observation is unceasing, such as a continuous record of river discharge;
(2) *quantized time*, when imaginary sections are used to subdivide time, for example when precipitation amounts are measured daily or weekly;
(3) *discrete time*, when interest focuses upon time duration and frequency of events per unit of time;
(4) *sampled time*, when observations can only be made at particular periods, such as weekly measurements of plant growth.

Within any one of these types it is necessary to appreciate that events may occur of a magnitude that would not normally be expected within a particular research programme. The significance of the magnitude and frequency of geomorphological processes was highlighted (Wolman and Miller, 1960) and subsequently the significance of rare events has been scrutinized. On coasts of coral atolls the effects of hurricanes are shown to be significant not only to the reef morphology but also to the organisms and the total environment (Stoddart, 1962), and in drainage basins in general, and along river channels in particular, the impact of rare floods and the time necessary to recover from their effects has been investigated. In a review of previous studies, Gupta (1983) indicated that steam channels are affected by low-frequency, high-magnitude events and the persistence of such events is greater in arid and semi-arid than in temperate areas (Wolman and Gerson, 1978). The significance of flood events in relation to river channels and the way in which relationships change over time is shown in *Flood Geomorphology* (Baker *et al.*, 1988) in which exciting questions are enumerated by Baker (1988, p. 6) as:

> Under what conditions do floods dominate as agents of fluvial landscape change? What are the detailed processes of flood erosion, transport, and deposition? Can the long-lasting effects of extraordinary floods be used to calculate their past magnitudes? What contribution can geomorphological flood studies make to flood-control manage-

ment? These are some of the questions addressed in the scientific discipline of flood geomorphology.

Research showed how glaciers are affected by periodic surges in which ice may be transmitted down the glacier at speeds 10–100 times greater than normal, and this has been identified from small glaciers and also from some ice sheets on a continental or subcontinental scale (Sugden and John, 1976).

Infrequent but substantial and significant events promoted two particular types of study: neocatastrophism and earth hazards. **Neocatastrophism** was revived because it acknowledges the significance, and in some cases the dominance, of events of greater magnitude and low frequency. Originally developed to explain sudden and massive extinctions of life forms in palaeontology, a review of the implications for parts of physical geography was provided by Dury (1980a). Growing interest in storm-generated deposits in sedimentology, and in geomorphology studies of events of great magnitude and low frequency, indicated that stream channels and interfluves must be visualized separately, and that there are differences between arid and humid areas. Therefore when considering environmental processes the significance of large events must be considered, the modifications to a strict uniformitarian view must be assimilated, and step functions and catastrophe theory can be helpful. The significance of large events was explored generally by Huggett (1988, 1989a, b, 1990) who showed (Huggett, 1994) that truly catastrophic floods have occurred in the past and that neodiluvialism claims that superfloods play a cardinal role in the development of some landscapes. He suggested that there is a mounting body of evidence to indicate that superwaves, originated either by submarine landslides or by the impact of asteroids or comets in the ocean, have flooded continental lowlands, and that on continents there is evidence for the catastrophic release of impounded water, possibly resulting from impact events which led to great deluges. Huggett (1994, p. 341) therefore contended:

> After 150 years of gradualistic explanations of flood deposits it seems fair, with the general acceptance of the bombardment hypothesis, to look again at catastrophic explanations. Interestingly, evidence for a diluvial origin of the landscapes and sediments comes not so much from the finding of previously unnoticed phenomena as from the reinterpretation of well-known landscape and sedimentary features.

The channelled scabland of Washington was suggested (Bretz, 1923) to be the result of catastrophic flooding after the drainage of Late Pleistocene proglacial lakes, which filled the pre-existing valleys to overflowing. However, features interpreted, including giant current ripples up to 15 m high, were not accepted by the majority of workers in the 1920s, and Bretz encountered great opposition to his interpretation. Although the opposition (e.g. Alden, 1927) was gradually overcome, it was not until recent work by Baker (1978a, b, 1981; Baker and Bunker, 1985) that the scale of the landscape development was appreciated, occasioned by maximum discharges which may have been as great as 21.3×10^6 m^3 s^{-1}.

Secondly, there have been studies of *earth hazards* which can be thought of as single, such as heat waves, frost, blizzards or fires, or compound, including droughts and tsunami. Certain very dynamic environments are influenced by very large events, and this has encouraged statements about the geography of natural perils in the case of Australia (Blong, 1997) and summaries in the case of New Zealand, which demonstrate the need for training about the dynamic environment (O'Loughlin and Owens, 1987). However, as White (1974b, p. 3) noted:

> Extreme natural events illuminate one aspect of the complex process by which people interact with biological and physical systems ... By definition no natural hazard exists apart from human adjustment to it ...

and so natural hazards are included in Chapters 7 and 8 (p. 187, p. 209).

5.6 The position of process studies

Greater attention to environmental processes was a necessary concomitant of the evolution of physical geography in the second half of the twentieth century, and Sack (1991, p. 29) concluded that: 'The dominant approach used to study landforms in the latter part of the twentieth century is essentially Gilbert's approach'. Gilbert, who worked as a geologist but considered himself a geographer (Gilbert, 1909a), was described by Tinkler (1985) as the 'intellectual patron saint' of American geomorphology, and his views were retrieved to provide a more respectable ancestry (Tinkler, 1985, p. 229) for the new paradigm. This was regarded as placing new wine in old bottles by Sack, who considered (Sack, 1992, p. 251):

> Gilbert was an appropriate figurehead for the new paradigm ... Identifying Gilbert as the archetypal process geomorphologist helped sanction the process paradigm, distance it from Davisian geomorphology, and ease the paradigm change.

However, the focus on processes was not, at least initially, directed towards the study of landforms, and so there have been times when some physical geographers resented the emphasis placed upon processes, in particular the direction in which some process investigations were directed. However, whereas Gilbert stimulated studies of process, there is no doubt that the recent revolution in process techniques (e.g. Table 5.2) enabled a return to the major questions which have to be addressed by physical geography. Five major conclusions may be reached from recent investigations of process. Firstly, there have been advances in the techniques used for process measurement (e.g. Quine *et al.*, 1997) and the major categories are illustrated in

Table 5.2 Examples of techniques that have advanced research on environmental processes

Problem	Technique	Example
Site characterization and location	Electronic distance measurement (EDM)	Construct digital elevation model (DEM)
	Close-range digital workstation (CDW)	
	Global positioning system (GPS)	Repeated detailed topographic surveys
	Digital elevation models	Short-term spatial changes
	Digital camera, digital mapping	Coastal landform change
	Terrestrial photogrammetry	River bank erosion
	Ground penetrating radar	Variations in sediments
	Airborne radar and radio echo-sounding	Basal ice conditions, water volume in lakes in ice sheets
Process measurement		
Remote measurements	Digital loggers	Water quality monitoring, turbidity monitoring
Repeated measurements	Continuous monitoring	Sand traps for aeolian events Bedload samplers
Spatial monitoring	Acoustic Doppler velocimetry	3D velocities in rivers
Tracing	Magnetic techniques Magnetic resonance imaging O^{18}, deuterium	Beach sediment sources Infiltration into soils, pollutant transport in soils Hydrograph separation
Laboratory analytical techniques	Automated analysis	Expanded number and type of samples can be processed, and greater range of properties analysed
	Scanning electron microscope	Characteristics of grains in deposits to indicate transport
Simulation Plot measurements	Erosion measurement	Assessments of different land-use cover
Dating techniques Cs-137 and Pb-210	Dating of sediment accretion	Soil erosion rates Floodplain sedimentation
Sediment analysis	Multisizer	Small samples, e.g. aeolian
Modelling	Generalized linear modelling	Glacier surging, landslide susceptibility

Sources: Hardisty (1998) and previous annual issues; Thrift and Walling (2000).

Table 5.2. For example, infiltration into soils is not an easy process to document, and a quantitative, visible and non-destructive technique for monitoring spatial and temporal water distribution is required. This can be provided by magnetic resonance imaging (Amin *et al.*, 1998), a technique which may also be appropriate for studies of solute/pollutant transport in soils, the migration of fine particles through the macropores of soils, and the infiltration of fine particles into the gravel beds of rivers. Hydrograph separation provided a challenge for many years, and environmental isotopes including O^{18} and deuterium are increasingly used to separate stormflow into its event and pre-event components and so to elucidate sources, pathways and residence times of water in the drainage basin (Buttle, 1994).

Secondly, to ensure that all aspects of environmental processes are investigated (e.g. Brown, 1995; Harvey and Sala, 1988), more parameters have been measured, for example water temperature in the case of rivers, and measurements have been made at a variety of scales. The grizzly bear can be a significant variable in soil erosion in Glacier National Park, Montana, and in digging for food it is estimated that each grizzly can displace 6.8 m^3 of sediment annually (Butler, 1993). More comprehensive data is one reason why, thirdly, results have been related to spatial scales at local, regional, continental and global levels, and it has been increasingly feasible to place results in a temporal context as well (Robinson and Williams, 1994). Fourthly, knowledge and understanding derived from process investigations has enabled closer relationships with modelling investigations. This is illustrated in the case of gullying where Bull and Kirkby (1997, p. 354) concluded:

> ... much detailed research has been carried out on badlands, but the long-term rates of gully development are not well understood ... In the short term theoretical modelling may provide the way forward and a direction for more holistic investigations.

This greater use of modelling may further elucidate landscape change and therefore address the concerns that Thomas (1980) and Ollier (1979) (p. 143) expressed. In a stimulating book on *Fluvial Processes in Dryland Rivers*, Graf (1988, p. vii) contended:

> If geomorphology is to progress to the stage of a mature and useful science that can successfully explain water-related processes in drylands, it must combine the field experience of perception, classification, description, and measurement with effective theory building.

Graf (1988, p. 230) suggested that the landscape that faces the analyst in dryland settings has three components:

(1) a landscape of energy or force derived from the inputs of water from the atmosphere, which is highly variable in space and time and susceptible to change under natural circumstances and to manipulation by human activities;

(2) a landscape of resistance derived from the structure and materials of the Earth, also variable from one place to another;
(3) a suite of geomorphological environments which arises from the combination of energy or force with resistance in earth surface processes, and forms which have specific geographical patterns and specific histories of development.

Such an approach exemplifies, fifthly, the way in which studies of process have led naturally towards temporal change, and in the context of hillslope instability in the tropics an improved knowledge of the processes of slope hydrology is significant (Anderson and Kemp, 1991). More than 20 years of field investigation of land exposed in historical times, since the glacier maximum of the Little Ice Age in the Jotunheimen region of Norway, allowed Mathews (1992) to present a geoecological model. Many of the books quoted in this chapter, such as the volume on floodplain processes (Anderson *et al.*, 1996), also exemplify ways in which process studies are being related to temporal landscape change (e.g. Sugden *et al.*, 1997), which is addressed in the next chapter.

Further reading

It is revealing to look at the work of G.K. Gilbert to appreciate how clearly he saw the need for the study of processes:
> GILBERT, G.K. 1914: *The transportation of debris by running water*. US Geological Survey Professional Paper **86**.

Approaches to processes are illustrated by:
> WILLIAMS, P.J. 1982: *The surface of the Earth: an introduction to geotechnical science*. London: Longman.

To appreciate how processes can be investigated over a number of years, a productive set of field monitoring experiments have been achieved in the Exe Basin and these are reflected in:
> WALLING, D.E. 1987: Hydrological and fluvial processes: revolution and evolution. In M.J. Clark, K.J. Gregory and A.M. Gurnell (eds) *Horizons in physical geography*. Basingstoke: Macmillan, 106–20.

> WALLING, D.E. 1996b: Suspended sediment transport by rivers: a geomorphological and hydrological perspective. *Advances in Limnology* **47**, *Suspended particulate matter in rivers and estuaries* (Special Issue, Archiv für Hydrobiologie), 1–27.

Topics for consideration

(1) As a realist approach becomes more reductionist, concerned with the movement of individual particles, how do we retain an overall landscape perspective?

(2) In relation to a specific location or process domain, are the four main methods of obtaining information on processes (p. 110) sufficient, and are there any other ways of obtaining data?

(3) Difficulties in extrapolating from specific process measurements to the spatial pattern.

(4) Amplify the categories and content of Table 5.2, which gives some techniques that have dramatically advanced process measurements.

|6|

Environmental change

Environmental change, landscape evolution and landscape chronology have been amongst the longest studied subjects in physical geography; they are themes associated with several distinct approaches, and have involved innovative methods and increasingly timely research. Early approaches relied upon established models of landscape development and upon studies of Quaternary geography, both of which at first appeared to be seriously challenged by the emphasis some physical geographers placed upon environmental systems, environmental processes or human activity. However, developments originating from these three approaches, as well as from the evolution of Quaternary science, contributed to the enhanced understanding of, and exciting approaches to, research on environmental change. This chapter outlines the early models (6.1) and the foundation of Quaternary science (6.2), followed by the way in which Quaternary studies were advanced by more sophisticated techniques including those for dating (6.3), and were later reinforced by contributions from studies of environmental systems, environmental processes (6.4) and from studies of human impact. Finally, links to investigations of recent phases of environmental change and multidisciplinary approaches are stressed (6.5).

Reconstruction of the past evolution of the physical landscape and establishing the chronology of environmental change depends upon what can be obtained from 'windows' of limited and varied opacity and size (Lewin, 1980). It is through the 'windows' remaining that we glean the necessary evidence to permit environmental reconstruction (Fig. 6.1); the amount of evidence surviving in a particular area will depend upon the age of the landscape and the subsequent changes to which the landscape has been subjected. Evidence obtained through any window is of four major kinds. First is evidence relating to the morphology of the environment: a portion of a river

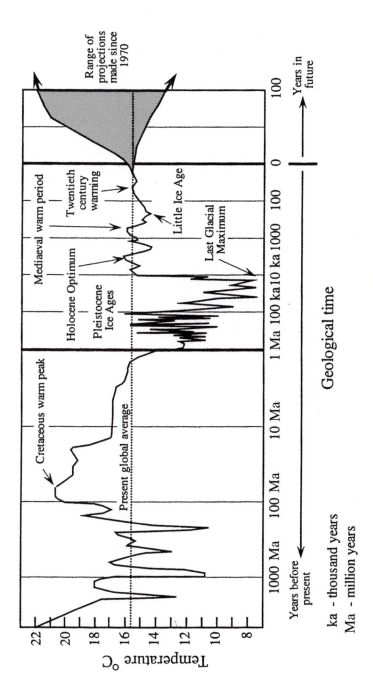

Figure 6.1 Temperature record of the Earth over geological time (after Bryant, 1997, with permission from Cambridge University Press). A logarithmic scale for time has breaks at present and 1 million years BP (compare with Figs 3.3 and 5.1).

terrace could have been a fragment of a much more extensive valley floor in the past. However, a researcher may only see, through available windows, that evidence which is understood and is consistent with the models in current use. Kimball (1948) proposed that historical geology could be divided into two parts: stratigraphy, which deals with what is there, and denudation chronology, which is concerned with what isn't! Secondly, evidence from sediments can be used to make inferences about the mode of deposition and the physical environment at the time of formation, such as the material comprising a river terrace, for example. Thirdly, knowledge of the processes operating in the landscape may be gleaned from historical records or by analogy with situations elsewhere. Linking short-term geomorphological processes to landscape evolution is now more feasible (Sugden *et al.*, 1997). Fourthly, fragments of evidence allow relative and absolute dating to be undertaken by an increasingly varied range of techniques. Evidence collected from such windows depends not only upon current models and academic training, which condition what is seen and how it is interpreted, but also upon the techniques available for dating and environmental reconstruction. Analysis of the four types of evidence has become progressively more sophisticated and has contributed to exciting developments. Whereas some studies of landscape change originated within geomorphology, others derived from investigations of ways in which soils and sediments relate to past systems of vegetation and of climate, so that in this field, possibly more than in any other, it is impossible for one individual to do justice in a single chapter to the breadth of environmental change investigations.

Interpreting past, and predicting future, geomorphological processes and forms on the evidence of present landforms and processes encounters seven problems (Schumm, 1985), extended to ten by Schumm (1991) in *To Interpret the Earth: Ten Ways to be Wrong*:

Problems of scale and place:
- **Time** – the period of time for data collection is invariably too short and we deal with physical systems that operate over varying time spans.
- **Space** – two aspects are *scale*, which involves the resolution at which the object is viewed, and *size*; these are affected by the resolution possible in any investigation.
- **Location** – involves differences between places and perspectives obtained by workers according to their training.

Problems of cause and process:
- **Convergence** – may be referred to as equifinality: different processes and causes can produce similar effects.
- **Divergence** – similar causes and processes produce different effects.
- **Efficiency** – the ratio of the work done to the energy expended, referring to the impact of an event or series of events on a system.
- **Multiplicity** – multiple causes act simultaneously and in combination to produce a phenomenon, so that a multiple explanation approach should

be applied to problems and multiple working hypotheses used as appropriate.

Problems of systems response:
- **Singularity** – the condition, trait or characteristic that makes one thing different from others, and is really the randomness or unexplained variation in a data-set, called indeterminacy by Leopold and Langbein (1963, p. 190).
- **Sensitivity** – refers to the propensity of a system to a minor external change and embraces the proximity of a system to a threshold: if it is near, and sensitive, it will respond to an external influence.
- **Complexity** – the complex response that occurs when the system is perturbed because a complex system, when interfered with or modified, is unable to adjust in a progressive and systematic fashion.

Schumm (1991, p. 94) contended that if studies are detailed and if the data are sufficient then these problems may not exist, but they should be considered by researchers, although three of the ten (singularity, efficiency and multiplicity) may be inherently accommodated by most investigators. Schumm was addressing young earth and environmental scientists rather than philosophers, advocating multiple approaches and drawing the analogy between multiple working hypotheses and differential diagnosis in medicine. He quoted the causes of diarrhoea and identified seven classes of acute and six classes of chronic among a total of 86 specific cases, but noted that the physician will probably give each patient just one diagnosis.

6.1 Models of landscape change

Denudation chronology, developed in the first half of the twentieth century, was essentially founded upon the Davisian approach and model of landscape evolution (p. 38). Research concentrating upon the sequence of evolution of particular areas came to have an unduly significant influence upon the way in which other areas were interpreted. Southeast England was the subject for a monograph of considerable significance (Wooldridge and Linton, 1939). The sequence of landscape evolution deduced for that area, with its emphasis upon drainage evolution and upon erosion or planation surfaces, relied upon interpretations based upon morphological evidence and information from fragmentary deposits to produce a model sequence of early Tertiary planation, mid-Tertiary uplift, late Tertiary planation, the early Quaternary Calabrian shoreline, and subsequent valley development during successively lower Quaternary sea-levels. Other areas in the UK were investigated in the light of development of this model, including Wales (Brown, 1960). In North America, studies in Appalachia tended to dominate in the way that those in southeast England were influential in Britain (Brown, 1961), relying heavily upon the work done by Fenneman (1931,

1938). Such approaches continued until 1970 and were essentially extensions of, and developments from, the Davisian model, whereas a later regional geomorphology of the US (Graf, 1987) adopted a limited number of research themes in each geomorphological region and encouraged theoretical developments from the survey of each region.

Other studies demonstrated how contemporary denudation rates could give clues about rates in geological time (Schumm, 1968), so that with the appearance of grasses in the Cenozoic, the relationships between climate, vegetation, erosion and runoff became much as today, except for the later influence of man. It became appreciated that there were alternatives to a school of denudation chronology based upon the Davisian cycle of erosion or peneplanation model. Developing his early research in South Africa, the geologist Lester C. King proposed continent-wide bevelled surfaces produced by pediplanation as the basis for understanding the geomorphology (see p. 44) of the world's plainlands (King, 1950).

The model of pediplanation extended knowledge of the Earth's surface, involved correlation of surfaces in Australia, Africa and South America with those in the northern hemisphere (L.C. King, 1962), embraced earth movement in the form of cymatogenic arching, as well as exogenetic processes as an integral part of landscape development, and provided a framework that could be used for the interpretation of vegetation patterns for the savannas of the southern hemisphere (Cole, 1963). In an assessment of King's 'Canons of landscape evolution', Ollier (1995, p. 376) concluded that:

> 'Canons of landscape evolution' was a significant article that opened new eyes, but was rapidly overtaken by events. In the 1960s, plate tectonics appeared, a ruling paradigm that seemed to have all the answers. Active margins became the centres of research, but much of King's work (Africa, Brazil, Australia) was on passive margins. Tectonic geomorphology had a big relapse (outside Russia and the eastern block countries) and geomorphologists tended to take their tectonics second hand from geophysicists. In North America and Britain process studies took over. In continental Europe, especially in Germany and France, climatic geomorphology was dominant. So King was an isolated figure, not the leader of a new thrust in landscape studies.

Also conceived in the tropics was a model of landscape development generated against the background of the seasonally humid tropics rather than the semi-arid, which emphasized weathering profiles and igneous rocks and reflected the experience of soil scientists together with the benefit of inputs from climatic geomorphology (p. 128). This model was developed with the realization that deep chemical weathering could be the norm in many tropical areas, and that many temperate areas included tor-like residuals which were remnants of times when climatic conditions were warmer and wetter, involving a double surface of levelling (Budel, 1957). A lower basal weathering front marked the position at which chemical weathering was attacking sound unweathered rock, and on the landsurface, or upper surface of levelling, exogenous processes were eroding, transporting and depositing sediment across a landsurface composed of chemically weathered rock with occasional protrusions of unweathered residuals. The relative functional behaviour of these two levels, reflecting contemporary and past erosion systems, could vary from humid tropics to semi-arid and arid landscapes. This model has been adopted in subsequent research (e.g. Thomas, 1978, 1994), was incorporated in the climato-genetic geomorphology developed from climatic geomorphology by Budel (1969, 1977), provided a way of visualizing the pattern of morphogenetic systems over the Earth's surface during the Cenozoic, and afforded the means of distinguishing contemporary world zones from those of the past when ice sheets were non-existent or much less extensive.

Studies of denudation chronology dominated much of geomorphology and played a major part in physical geography in the 1950s and 1960s. However, alternative approaches focused on processes or human activity, for example, were not appropriate for all areas of the world because landscapes dominated by the remnants of formerly extensive planation surfaces characterize parts of Australia, so Ollier (1979, p. 534) concluded that cyclic theories do not fit well in the Australian scene. Evolutionary geomorphology was not a cyclic approach with a sequence of successional stages but, according to Ollier (1981), the Earth's landscapes as a whole are evolving through time analogous to the concept of an evolving Earth as used in some geology books (e.g. Windley, 1977). He concluded (Ollier, 1981):

> To play a part in the problems of geology over the next few decades geomorphologists must forget their trivial catchments and see mega forests instead of trees ... Dynamic equilibrium, climatic geomorphology and process studies have all been shown to have limited application to wherever geomorphic history is measured in hundreds of millions of years. If we reject cyclic ideas and even uniformitarianism, what have we left? The answer is evolutionary geomorphology.

Landform inheritance (e.g. Pain, 1978) was a concept emphasized by some researchers in environments of this kind. Denudation chronology continued to be used as an approach in areas for which it was most appropriate, and general principles were proposed which included the seven principles enunciated by Brown (1980, p. 11), the 50 canons of landscape evolution (King, 1953, p. 747–50), the seven points of general agreement concerning planation surfaces (Adams, 1975, p. 449), and the tablets of stone, towards the ten commandments of geomorphology (Brunsden, 1990).

6.2 Quaternary science

A consequence of progress over recent decades, especially the development of techniques (Section 6.3), is that physical geographers researching on

Quaternary environmental change have become part of an international multidisciplinary community. The International Union for Quaternary Research (INQUA) was established in 1928 and holds an international congress every four years; when it met in Durban in 1999 the delegates gave 80 different types of affiliation, and geographers represented some 11% of the delegates, exceeded only by geologists who made up nearly 13%. The other 78 groups included many earth and environmental scientists including archaeologists, biologists and oceanographers. The proliferation of disciplines and subdisciplines, as well as their range, reflects the way in which Quaternary science has developed. A consequence of the technical developments has been the way in which the duration of the Quaternary increased from 1.8 to 2.3 million years, and the pace of advances and the relevance to global environmental change encouraged physical geographers, sometimes in collaboration with other scientists, to write important texts including those by Lowe and Walker (1997), Bell and Walker (1992), and Williams *et al.* (1993). Books have also been produced dealing with the Holocene (Roberts, 1989) and with environmental change in general (Huggett, 1997; Mannion, 1999). Investigations of the Quaternary were aided by the development of new journals (Table 5.1) including *Quaternary Research* (1970–), *Boreas* (1972–), *Quaternary Science Reviews* (1982–), *The Holocene* (1991–), *Quaternary International* (1990–) and *Quaternary Geochronology* (1993–).

Studies of Quaternary landscape change were initially founded around the themes of: changes in sea-level; Quaternary landscape change including glacial sequences; glaciology, the Pleistocene and landforms; and studies of other areas including periglacial landscapes and arid landscapes. Research on ***changes of sea-level*** reflected the fact that many areas possessed evidence of stages of Quaternary erosion marked by river terraces and raised shorelines around coastal margins, together with evidence from buried valleys and remnants of former sea-levels which were later submerged. Studies of this kind were first influenced by denudation chronology and then by developments in the study of the Pleistocene where the work of Zeuner (1945, 1958) and various geologists (e.g. Wright, 1937; Flint, 1947; Charlesworth, 1957) had been particularly influential. In the analysis of shorelines and chronological stages, many studies initially followed Baulig (1935) on 'The changing sea level' and other workers including Deperet working on Mediterranean shorelines, and then were influenced by the masterly synthesis achieved by Fairbridge (1961) and by Zeuner's (1958) consolidated view of sea-level change in the Quaternary. The latter provided what was to become a classic sequence of Pleistocene sea-level fluctuations including transgressions associated with interglacials and regressions accompanying the glacial phases; these fluctuations accompanied a gradual sea-level decline during the course of the Pleistocene. This primarily glacio-eustatic explanation depended upon sea-level fluctuation in relation to the amount of water stored in ice caps during glacial and interglacial phases. Other worldwide eustatic causes of sea-level change include sedimentary infilling of ocean basins, which could give a sea-level rise of 4 mm per 100 years (equivalent to 40 m in a million years) (Higgins, 1965); orogenic eustasy whereby orogenic uplift creates ocean basins of different total volume; and geoidal eustasy whereby the ocean surface reflects the variations in the geoid surface due to the Earth's irregular distribution of mass, which can give a difference between lows and highs of as much as 180 m. In addition, a number of local factors can be responsible for sea-level change (Goudie, 1983) including: glacio-isostasy whereby the Earth's crust responds to the development or removal of large ice sheets; and hydro-isostasy when a similar response occurs as a result of large bodies of seawater or lake water on continental shelves and lake basins; orogenic and epeirogenic activity; compaction of sediments; and the increased gravitational attraction associated with large Pleistocene ice sheets. An example of glacio-isostasy was provided by detailed investigation of raised shorelines in Scotland (Sissons, 1976) which revealed a complex pattern of late Devensian and Flandrian raised shorelines. They have been differentially uplifted and are slightly diachronous because the parts nearer to the centre of uplift were lifted clear of the sea earlier than the peripheral remnants (Smith *et al.*, 1980).

The Mediterranean model was superseded by the advent of information from a greater range of sites elsewhere in the world, with correlation other than on simple altitudinal grounds, and with more complete methods of absolute and relative dating. The advent of oxygen isotope dating of deep-sea sediment cores (Section 6.3) was probably most significant and also very important in relation to climatic change and glacial chronology. In addition, use of uranium-series dating of uplifted coral coasts (Chappell, 1974) and of amino-acid dating, especially as applied to the coasts of the US (Wehmiller, 1982), advanced knowledge to allow separation of sea-level and tectonic components on rapidly uplifting coastlines. Particular attention has been devoted to late Quaternary sea-levels, especially those of the Holocene, but it is still not possible to be certain about world sea-levels before the last quarter of a million years. Simple correlation cannot easily be made between different parts of the world in view of plate tectonics and the amount of movement associated with the displacement of particular plates. In an extensive review of work done by the International Geological Correlation Programme Project 61 (IGCP 61) Kidson (1982) concluded that the search for a universal eustatic curve should be abandoned; that regional differences in changes in the geoid mean that eustatic sea-level curves can only have regional validity; and that no part of the Earth's crust can be regarded as wholly stable. Recent research on sea-level changes has tended to range across a variety of time-scales but to concentrate upon the most recent stages (e.g. Tooley, 1978), and it has been suggested (Tooley, 1994, p. 188) that greater attention should be devoted to clarifying the links between rapid rates of sea-level rise and climatic change, and to elaborating dynamic sea-level changes, particularly extreme water levels during storm surges.

This exemplifies the way in which Quaternary research can link to studies of contemporary environmental processes and to implications for future global change (Chapter 9, p. 244).

In areas dominated by Quaternary deposits, research after the late 1950s was directed towards the interpretation of the stages and nature of *Quaternary landscape change*, and this involved new conceptual models of glacial landform development well exemplified by the way in which patterns of deglaciation were developed to involve stagnant as well as active ice (e.g. Sissons, 1958, 1960, 1961, 1976, 1977) and benefiting from the study of contemporary glacier behaviour (Sugden, 1996). Research by physical geographers proceeded sequentially to embrace increasing involvement in chronology, to utilize new techniques, and to relate to investigations of contemporary processes, to progress towards a multidisciplinary focus and to contribute to modelling strategies. Influential textbooks by Quaternary geologists (e.g. Flint, 1947; Charlesworth, 1957) were produced when the disparities between areas such as northern Europe, central Europe and North America were perhaps most apparent. The classical model of Penck and Bruckner (1909) involved four main glaciations and was based upon sediments which represented only a small proportion of the time span; unconformities between each successive terrace probably concealed events lost to the record locally or regionally (Bowen, 1978). The four-fold sequence, named from Alpine rivers, continued to feature, but additional models of the glacial sequence had been proposed for northern Europe based upon sequences on the north European plain and in Scandinavia; for the British Isles; and for central North America (Bowen, 1978). Although physical geographers did not contribute to their development, these model sequences were increasingly reflected in physical geography research, as knowledge of the sequence in any one area and correlation between areas was necessary to proceed towards explanation of local, and then of regional, patterns. By the late 1970s, books by physical geographers adopted structures in which chronology and the classic models appeared at the beginning. Reconstruction of change in particular areas was a major theme benefiting from the greatly expanded range of available techniques, with biogeographers investigating landscape change aided by palynology. The sequence of glaciations established by the evidence from deep-sea cores was confirmed by the loess record from central Europe (Kukla, 1975), the USSR and China (Goudie, 1983).

Research on *glacial landforms* has a very long and distinguished history. Whereas it was initially very deductive and tended to be somewhat isolated from research on chronology and glacial sequences, this changed as the chronological sequence became clarified, with investigations of processes in contemporary areas aiding the interpretation of past sequences. This enabled the production of an integrated landscape model (e.g. Shaw, 1994, Fig. 9), allowed re-examination of long-held beliefs about glacial activity (Harbor, 1993), and led to a suggestion that any reconstruction of past glaciations needs to take into account deforming bed conditions over unconsolidated sediments (Hart, 1995). Glacial geomorphology therefore became associated with investigations of glacial processes (see p. 121).

The more extensive range of methods for dating and environmental reconstruction also benefited the investigation of **Quaternary morphogenesis in areas beyond ice sheets**. Although emphasis was initially upon recognition of the variety of landscape features, sediments and structures that could be developed under periglacial conditions, subsequently a greater knowledge of phases of periglacial landscape development was achieved (e.g. Washburn, 1973). In Poland and other countries in Europe this emphasis was clearly evident in research in the 1960s, reflected in *Periglacial Geomorphology* (Embleton and King, 1975) and *The Periglacial Environment* (French, 1976, 1996). Whereas periglacial geomorphology in Europe was associated with Quaternary research, in North America it was a branch of process geomorphology and regarded as part of geocryology (French, 1987), and the importance of this area, with research undertaken by several disciplines, stimulated the new journal *Permafrost and Periglacial Processes* (1990–). Earlier developments by physical geographers, which proceeded somewhat independently, were located in arid, semi-arid and subtropical areas. These were additional to interpretations of planation surfaces and were particularly concerned with the alternation of pluvial and arid phases in the Quaternary and their possible relationship to those of temperate latitudes; and with the sequence of valley development particularly as revealed by the chronology of alluvial deposits. Researchers in some areas studied the significance of human activity extending over many centuries. This was exemplified by the research of K.W. Butzer in the Middle East and the Mediterranean where he reconstructed stages of landscape change as related to Quaternary stratigraphy and climates. These studies were founded upon analysis of fluvial, lake, aeolian and cave sediments together with inputs from related disciplines including palaeobotany, palynology, palaeoclimatology and archaeology (Butzer, 1964). Other work in the Mediterranean also investigated the association between valley evolution, human activity and climates (Vita-Finzi, 1969), whereas in hot desert areas investigations of fossil dunes in India, Africa and Australia together with changes in the extent of pluvial lakes in Africa, the Middle East and North America (Goudie, 1983) provided further insight into the chronology of Quaternary environments.

6.3 Technique developments including dating

Contributions by geographers in the second part of the twentieth century ranged from studies of particular sites, often landform-related, to contributions embracing local and regional chronology, and thence to research which involved collaboration with, and achieved respect from, scientists in other disciplines, including analysis of models of ice sheet behaviour and landscape development. This was somewhat analogous to the development of studies of process (Chapter 5) because, from initial concern with small-scale deposits or sequences or with process investigations, the move was to larger-scale regional interpretations and models. However, Quaternary research was revolutionized by new techniques and interpretations developed particularly after detailed analysis of deep-sea cores.

Greater involvement with chronology became possible as new techniques complemented long-established ones, leading to refinement of dating and the detailed interpretation of past environmental conditions. Long-established techniques included analysis of varves in lake deposits indicating an annual or seasonal rhythm, used since the pioneer work of de Geer in Sweden in 1912 (see Boygle, 1993); analysis of tree rings by dendrochronology linked very convincingly to the climate record; relative dating of landforms; and palynology and other types of microfossil analysis including non-marine mollusca. Some of these techniques were extended, including the use of the wing cases of fossil beetles (Coleoptera) to provide information about the climatic characteristics of palaeoenvironments, because coleoptera are very sensitive to moisture conditions. Palynology was used by many biogeographers, and pollen analysis became the basis for reconstructing the detailed vegetation history of many specific areas, although emphasis subsequently evolved from interpretation of individual sites to the reconstruction of patterns of change in Britain (e.g. Barber, 1976) and the tropics (e.g. Flenley, 1979). Such studies meant that chronology became more central to research and other techniques facilitated reconstructions of local sequences of environmental change. Physical geographers, together with biologists, limnologists and other researchers, used palynological and chronometric dating techniques to investigate a large number of cores and were able to establish the nature of the flora and fauna in the late Quaternary and at the transition, during the early Flandrian, from tundra to closed deciduous woodland. Other techniques used included mineral magnetic properties, valuable because they are preserved for long periods, are environmentally diagnostic in many situations, and have parameters which are easy to measure. Initially applied in the context of lakes and their drainage basins, which can be used as units of sediment-based ecological study (Oldfield, 1977), the technique was subsequently shown to have widespread application to correlation of lake sediments, differentiation of weathering and pedogenesis, identification of sediment sources (and possibly long sediment sequences from major lake basins in non-glaciated regions), near-shore marine sediments in morphogenetically dynamic areas, and also to cave sediments, alluvial fills, river terrace sequences and loess successions (Oldfield, 1983a). The achievements of Holocene palynology were reviewed (Macdonald and Edwards, 1991) to show the questions which need to be resolved, and to indicate potential future advances in elucidating human activity and vegetation change (Edwards and Macdonald,

1991). The broader range of palaeoecological evidence for environmental change (Mannion, 1989a) and geochemical data from peat deposits, lake sediments, ice cores and tree rings (Mannion, 1989b) can play a major role in documenting ecosystem change over the last 200 years.

The historical emphasis, described as Quaternary ecology (Simmons, 1980), has prevailed in biogeography since the 1960s. Progress in documenting Quaternary vegetation sequences linked to changes of climate was achieved for a variety of world areas and contributed to the reconstruction of world patterns. IGCP Project 158B was devoted to the study of lakes and mires in the temperate zone (Berglund, 1983) and to collation of data that enabled generalizations to be made about trends in vegetation change, and hence in climate, during the last 15,000 years. This international project organized from a Department of Quaternary Geology involved inputs from physical geographers as well as from biologists, geologists and archaeologists. Quality control of palynological data was achieved by calibration with contemporary studies of pollen rain, and analysis can give detailed information on former environmental conditions as illustrated by the interpretation of the recent history of a floodplain woodland in southwest Ireland (A.G. Brown, 1999).

Isotopic methods provided geochronometric dating, complementing the use of varves, dendrochronology or palaeomagnetic evidence which had provided a chronology relative to a floating scale in stratigraphic investigations. Radiocarbon or ^{14}C dating was first applied in 1949 and, together with other evidence, initially provided the chronology for the last 50,000 years and was later extended to 75,000 years. Radiocarbon dating was used on wood, charcoal, peat, organic mud and calcium carbonate in molluscs, foraminifera and bones. Uranium nuclides ^{238}U and ^{235}U decay to stable lead, and this decay is the basis for a radiometric method applied to molluscs, coral and deep-sea sediments, which usefully complements ^{14}C dating because it can be used for materials up to 350,000 years old. Potassium–argon dating can be applied to volcanic rocks such as lavas and tuffs and, although it is difficult to measure the decay, it attains its maximum usefulness in the middle and early Pleistocene and can be used for a range greater than 20,000 years. The range of techniques has been reviewed by Bowen (1978) and Goudie (1981a; 1990c). Deep-sea cores provide a complete stratigraphical record whereas the continental record is necessarily less perfect. With the development of appropriate piston corers in the late 1940s it was possible to collect columns of sediment 10–30 m in length which contained material that could be dated by radiometric and other means, and could also provide environmental information, for example by analysing the frequency of sensitive foraminifera. Abundant information was yielded by cores: one extracted from the Pacific near the equator from a water depth of 3120 m gave a sequence of 900,000 years including eight complete glacial cycles and nine terminations with rapid glaciations. The new hydraulic piston corers (Shackleton and Hall, 1983) provided better samples as the basis for more detailed analysis, to allow a greater degree of isotopic structure, and enabled the climatic variability of the early Pleistocene to be analysed in similar detail to the late Pleistocene.

Dating and the greater range of available techniques revolutionized the study of the Quaternary, and Boulton (1986) characterized a paradigm shift in glaciology. Not only was the range of available techniques extended, but could be undertaken more rapidly, employed smaller samples, and often gave better resolution of results. Radiocarbon dating undertaken by a new method in the late 1970s using smaller samples and giving results more rapidly meant that not only was a bottleneck of Quaternary research reduced, but in addition, new methods such as uranium series disequilibrium dating methods and stable isotope analyses were extended. Oxygen isotope analysis of foraminifera in deep-sea sediments yielded a time series of globally time-parallel geological events. Thermoluminescence dating (TL) was initially developed in the early 1960s for archaeological samples, but later extended to sediments which had been exposed to daylight at the time of deposition, an exciting development applied to samples in the range 100–1000 years old, with the ability to date individual grains from a sample (Duller, 1996). Closely related to thermoluminescence dating is optically stimulated luminescence dating (OSL) (Goudie, 1990c), which has wider potential application to alluvial deposits. As new techniques continued to be provided, including a new method of tephrochronometry using thermoluminescence giving accurate dating for 400 ka, W.C. Mahaney (1995) assessed the position:

> The application of new dating methods to problems in geomorphology, geology, climatology and archaeology continues to widen, and new refinements are taking place at a faster pace, making for more diverse and broader applications. From new dates on the palaeomagnetic timescale, and higher resolution in U-series dating of corals, to advances in thermoluminescence and aminostratigraphy, the dating-methods discipline is finding itself filled with a host of old and new practitioners, making important advances in a dynamic area of the earth sciences.

Dating methods have featured as regular progress reports by W.C. Mahaney in *Progress in Physical Geography* since 1986. As dating has developed into a substantial field it is not possible to describe all techniques here, but they are summarized in Section V of *Geomorphological Techniques* (Goudie, 1990c), and are the subject for reviews (e.g. Mahaney, 1984) and specific treatments such as that on *Absolute Age Determination* (Geyh and Schleicher, 1990) which considers principles and applications of methods for the geosciences, and *Radiocarbon Dating* (Lowe, 1991) which outlines recent applications and future potential. In addition there are volumes that are more directly user-orientated, such as *Quaternary Dating Methods* (Smart and Francis, 1991), and some that are specifically directed to field teaching such as *Glacial Analysis: an Interactive Introduction* (Hart and Martinez, 1997) which aims to teach the user a variety of techniques in order to be able to reconstruct the glacial sedimentology of a site, and so to reassemble the processes that operated to produce the resultant landforms and sediments.

Successful application of this greater range of techniques to material from a diverse range of environments and locations demonstrated the analysis of deep-sea sediments and ice cores to be fundamental. Although data from Greenland and Antarctica ice cores correlate very well from one core to another, the transfer function was needed to link the results to climate variations and to correlate the ice core data with deep-sea chronologies over the surface of the globe. The Vostok core represents 400,000 years and embraces four or five complete cycles, while Heinrich events (e.g. Adams *et al.*, 1999) have proved to be useful in correlating ocean sequences with those from Greenland ice cores. The global sequence continues to be refined, but cycles of c. 19–23 ka have been identified prior to 2.8 million years ago, and perhaps going back as far as 7.8 million years; cycles of 41 ka feature in the period from 2.8 to 1 million years ago; and in the last 1 million years, 100 ka cycles have dominated. As a result of these developments a new time-scale of chronozones has been developed: the beginning of cooling began between 3.3 and 2.8 million years ago, the stages of the Pleistocene may have begun 2.8 million years ago, and the Holocene started 10,000 years ago, although some authorities think that it could have been 13,500 or 6500 depending upon location. The length of the Quaternary has now been extended so that when INQUA was founded in 1928 it was concerned with environmental change only in the last 1 Ma; this has now extended to 2.8 Ma, and papers submitted to its meetings can deal with periods in the last 5 Ma. Data collected from all areas of the world are now readily available and the World Data Center – A for Palaeoclimatology exists to archive and distribute palaeoclimatic data, including information about past environments generated from computer model simulations and geophysical calculations. It can be accessed on the World Wide Web at http://www.ngdc.gov/paleo/paleo.html or by e-mail at paleo@ngdc.noaa.gov.

Such advances mean that Quaternary science has been transformed, is a very exciting, specialized field, and necessarily multidisciplinary. Not only does it embrace practitioners from the earth and environmental sciences, but also from the physical sciences; as hominids appeared on the scene some 2.8 million years ago, it also includes archaeologists and geoarchaeologists.

Developments such as the advent of scanning electron microscopy also assisted in the elucidation of environmental change. Since the 1960s it has been possible to examine the surface of sand grains in detail (Krinsley and Doornkamp, 1973) and subsequently other materials studied under the electron microscope included organic materials, tills and soils (Whalley, 1978). Subsequently cathodoluminescence and high voltage electron microscopy furnished detail that even normal electron microscopes could not provide (Bull, 1981). The size, shape and sculpturing of diatoms are taxonomically diagnostic and because of their siliceous composition are often preserved in stratigraphic deposits where they can be used to infer past environmental conditions (Moser *et al.*, 1996). In some cases geographers adapted techniques, as in the use of soil micro-organisms as an indicator of the relationship between land-use change and soil characteristics (Maltby, 1975); developed techniques with a great range of applications based upon magnetic susceptibility (e.g. Oldfield, 1983a); and dating using ^{210}Pb which gives great accuracy for recent sediments (Appleby and Oldfield, 1978) as noted above. The range of techniques available to interpret and date Quaternary sediments is summarized in Parts 3 and 5 of *Geomorphological Techniques* (Goudie, 1990c).

6.4 Conceptual approaches to temporal change

Advances in techniques and the growth of multidisciplinary studies of the Quaternary led in one of two related directions: either towards greater emphasis upon chronology and reconstruction of past environmental conditions; or towards more process- or model-based investigations subsequently linked to studies of global change (p. 243). In relation to landscape evolution, Dorn *et al.* (1991, p. 302) noted three broad trends:

(1) new surface exposure dating methods;
(2) enhanced understanding of geomorphological process linked with improved techniques for laboratory and sediment analysis;
(3) advances in GIS, digital elevation data and fractal mathematics.

The revolution in techniques for investigation of environmental change (p. 148) coincided with that for environmental processes (p. 133), enabling the two to converge to stimulate new conceptual perspectives on environmental change.

One paper that stands out, not just in geomorphology but in physical geography as a whole, was the seminal paper by Schumm and Lichty (1965) entitled 'Time, space and causality in geomorphology', which reconciled previously separate views. Subsequently reviewed as a classic (Kennedy, 1997, p. 420), it had at least two original ingredients that account for its impact. First was the distinction of three separate *time-scales*:

(1) cyclic or geological time – which encompasses millions of years as required to complete an erosion cycle;
(2) graded time, which may be hundreds of years, during which a graded condition or dynamic equilibrium exists;
(3) steady-state time, typically of the order of a year or less, when a true steady-state situation may exist.

Second was the accompanying explanation of the status of geomorphological variables according to the time-scale being investigated. A variable that
is dependent at one time-scale may be independent at another, and this
aspect of status was illustrated by a well-cited table (Table 6.1) showing the
application to drainage basin variables. This paper has been considered as
providing, with the magnitude/frequency concept, the main conceptual
basis of thinking up to 1985 (Tinkler, 1985). It is not always possible to separate different time- and space-scales so conveniently, as events occurring at
different time- and space-scales may have a net effect upon the system,
partly due to sensitivity to initial conditions. Lane and Richards (1997,
p. 258) concluded that the geomorphologist should consider links between
the wealth of 'traditional' conceptual ideas of landform behaviour (e.g.
Brunsden and Thornes, 1979) and new ideas of non-linear response.
Udvardy (1981) offered a very similar scheme of three time-scales which
were:

(1) the secular scale with spatial dimensions of about 100 km and time
    dimensions of about 100 years;
(2) the millennial scale which covers at least post-Pleistocene time and spa-
    tial scales of up to 1000 km where climate and sea-level change are the
    major factors operating;
(3) evolutionary time, or the phylogenetic scale, where the time-scale may
    be up to 500 million years and the spatial extent may reach 40,000 km,
    so that continental displacement may be important.

This tripartite scheme could reconcile separate views in biogeography in the
way that Schumm and Lichty (1965) facilitated reconciliation of different
viewpoints in geomorphology. Researchers in different disciplines use
several time-scales when considering environmental change; Driver and
Chapman (1996) differentiated between:

(1) 'now time';
(2) generational time-scale – 10–100 years for sustainable development;
(3) century time-scale – 100–1000 years;
(4) late Quaternary time – 1000–10,000 years.

Such conceptual approaches to time-scales are a necessary basis for the
development of new models. Other conceptual ideas, of which a mere selection can be given here, have developed particularly within fluvial geomorphology, an emphasis noted by several authors (e.g. Rhoads and Thorn,
1996) but perhaps also reflecting the predilection of the author. The way
change over time is viewed by physical geographers has necessarily reflected
the geological tradition which was basic to much early physical geography,
especially as some physical geography is located in departments of geology
in the US. The distinction between *timeless* and *timebound changes* was
highlighted by Chorley and Kennedy (1971, p. 251) by illustrating how
cause and effect passed from *becoming* at one level of integrative organiza-

Table 6.1 (A) The status of river variables during time spans of decreasing duration (after Schumm and Lichty, 1965)

		Status of variables during designated time spans		
River variables		Geological	Modern	Present
1	Time	Independent	Not relevant	Not relevant
2	Geology	Independent	Independent	Independent
3	Climate	Independent	Independent	Independent
4	Vegetation (type and density)	Dependent	Independent	Independent
5	Relief	Dependent	Independent	Independent
6	Palaeohydrology (long-term discharge of water and sediment)	Dependent	Independent	Independent
7	Valley dimensions (width, depth, slope)	Dependent	Independent	Independent
8	Mean discharge of water and sediment	Indeterminate	Independent	Independent
9	Channel morphology (width, depth, slope, shape, pattern)	Indeterminate	Dependent	Independent
10	Observed discharge of water and sediment	Indeterminate	Indeterminate	Dependent
11	Observed flow characteristics (depth, velocity, turbulence, etc.)	Indeterminate	Indeterminate	Dependent

(B) The status of drainage basin variables during time spans of decreasing duration

		Status of variables during designated time spans		
Drainage basin variables		Cyclic	Graded	Steady
1	Time	Independent	Not relevant	Not relevant
2	Initial relief	Independent	Not relevant	Not relevant
3	Geology	Independent	Independent	Independent
4	Climate	Independent	Independent	Independent
5	Vegetation (type and density)	Dependent	Independent	Independent
6	Relief or volume or system above base level	Dependent	Independent	Independent
7	Hydrology (runoff and sediment yield per unit area within the system)	Dependent	Independent	Independent
8	Drainage network morphology	Dependent	Dependent	Independent
9	Hillslope morphology	Dependent	Dependent	Independent
10	Hydrology (discharge of water and sediment from system)	Dependent	Dependent	Dependent

tion, to *being* when adopting a characteristic structure or morphology at a higher level, to *behaving* by adopting a characteristic morphology or structure at a still higher level (Fig. 6.2).

Change over time implicitly involves some notion of *equilibrium/disequilibrium* conditions, long referred to in geomorphology (Kennedy, 1992; Thorn and Welford, 1994), but which depends initially upon methods of analysis for temporal data. Three types of serial dependence identified by Thornes and Brunsden (1977) include:

- trend, which is the long-term pattern with short-term variations excluded by a filtering technique;
- periodicity, because of regular periodic or cyclic fluctuations, which may be controlled by climate, for example;
- persistence, either due to the presence of physical factors which produce the output series or as a consequence of the collection and manipulation of the data-set.

When data are compared (see also Fig. 5.1), time series statistics can establish the autocorrelation between two or more series covering the same time

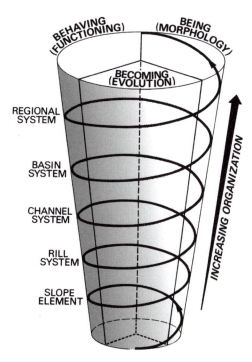

Figure 6.2 Systems architecture in time, developed from a diagram used by Chorley and Kennedy (1971, Fig. 7.1) based upon Gerard (1964).

period. This allows advances to be made beyond the process-response systems often analysed using correlation and regression analysis. In physical hydrology this enabled insight to be gained into the structure of comparatively long-term responses and occasionally of a whole basin system (Anderson, 1975). Equilibrium (Chapter 2, p. 31) has required attention throughout physical geography and it is imperative to establish how, at a particular time, there is equilibrium or stability in a system. In relation to ecosystems, Hill (1987) provided a review of the development of the concept from the 1970s and suggested definitions for the terms. Although definitions of equilibrium condition or environmental stability (Moss, 1999) are needed to differentiate when change occurs, the notions should not be taken too far, as shown by Kennedy (1994) in a paper entitled 'Requiem for a dead concept'.

Environmental change occurs when there is an alteration of equilibrium or stable position; an intriguing approach to the sequence of disruption was provided by Graf (1977) in his application of the *rate law* in geomorphology. Graf (1977, 1988) proposed that a rate law in the form of a negative exponential function, similar to that used to describe the relaxation times of radioactive materials and chemical mixtures, could provide a useful model for relaxation times in geomorphological systems. Graphical representation of the disrupted system (Fig. 6.3) involved the change from steady state or equilibrium A to steady state D passing through reaction time B, the time needed for the system to absorb the impact of the disruption, and the relaxation time C, the time period during which the system adjusts to new conditions. In addition to the magnitude of disruption, D–A, it is important to know the duration of the reaction time and the relaxation time. Graf (1977) applied the rate law to the development of a small fluvial system in the

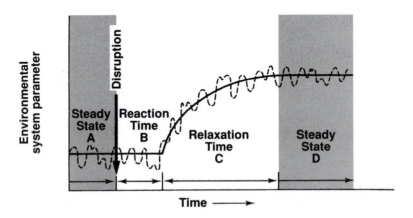

Figure 6.3 Response of a geomorphological system to disruption (after Graf, 1977, 1988).

Denver area of the US. Temporal environmental change (A to D in Fig. 6.3) can be autogenic inherent system dynamics, or it can be allogenic, whereby it is superimposed on the system as a result of climatic forcing or human impact.

Critical limits, boundary conditions and yield points beyond which change may occur are important in other disciplines, but formed a comparatively small part of the geomorphological literature, so that Coates and Vitek (1980) suggested that the two doctrines of catastrophism and uniformitarianism should be supplemented by a third base concerned with *thresholds*, defined (Fairbridge, 1980, p. 48) as:

> ... a turning point or boundary condition that separates two distinct phases of interconnected processes, a dynamic system that is powered by the same energy source.

Thresholds may be applied to studies of tectonics, hydrological processes, glaciology, eustasy and sedimentology. Glacier surging, characteristic of the flow of some glaciers, exemplifies the threshold situation because once the threshold is exceeded the surge occurs with velocities often several orders of magnitude greater than normal – the glacier then becomes quiescent, with the lower reaches raised in elevation and the upper reaches lowered. Subsequently instability is built up again as the upper part rises and the lower part ablates until another surge occurs. Thresholds were categorized by Schumm (1979) to include:

- extrinsic, the levels at which a system responds to an external influence such as climatic change;
- intrinsic, which are crossed, for example, when an internal variable changes, in the way that long-term weathering reduces the strength of slope materials until slope failure occurs;
- geomorphic, which are inherent in the manner of landform change and are thresholds of landform stability exceeded either by intrinsic change of the landform itself or by a progressive change of an external variable.

Analysis of thresholds led away from ideas of progressive erosion and progressive response to altered conditions, towards the notion of complex response whereby a fluvial system seeks a new equilibrium which could vary from one area to another. In the understanding of temporal change, thresholds should specify the process boundary conditions when change, which may be expressed in morphological adjustments, will occur. An enthusiastic search for thresholds included studies of the Central City district of Colorado where mining developed rapidly in the nineteenth century and subsequently declined just as rapidly (Graf, 1979b). The distinction was made between gullied and ungullied valley floors in the Piceance Creek area of Colorado, where Patton and Schumm (1975) plotted a linear relationship between valley slope and drainage area as the threshold separating the gullied from the ungullied valley floors. Threshold conditions are very difficult to specify quantitatively, and if the relation between an input series (X_t), or forcing function as it is sometimes called, and the output series (Y_t) is envisaged as a transfer function $Y_t + gX_t$, where g is the impulse response function, then it has often been assumed that the threshold relation will be linear. However, several complications may occur (Church, 1980) because the history of the landscape may dictate that a specific situation is supply-limited, and that the output sequence may result from the combined effect of several conditions rather than a single one. Therefore the outcome of this pattern of complex response is that geomorphological event sequences may be more intermittent than the forcing sequence (Church, 1980, p. 18). Specific studies have included investigations of large rare events to establish their significance (Starkel, 1976; Costa and O'Connor, 1995), and there may be a difference in the significance of such events between temperate and arid areas; Wolman and Gerson (1978) proposed a time-scale for effectiveness that relates recurrence intervals of an event to the time needed for a landform to recover to the form existing prior to the event.

A more general approach to thresholds emerged by considering the extent to which parameters involved in disturbed environmental systems correspond in the timing of their disruption, and then in analysing the temporal sequence of disruption. Such timing was considered by Knox (1972, 1995) in relation to valley alluviation in southwest Wisconsin where he found a disunity between hillslope processes and channel processes. Thus during shifts from humid to arid climatic regions (Fig. 6.4) there may be time lags in the rate at which the relative vegetation cover (B) responds and

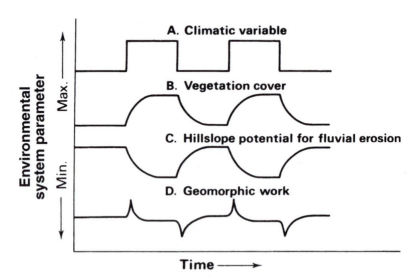

Figure 6.4 Phasing of environmental parameters in response to disruption (after Knox, 1972).

the way in which hillslope potential for fluvial erosion (C) and geomorphological work change. Douglas (1980) suggested that such lags have to be removed when contemporary process-response systems are quantified and that the possible existence of lags and their magnitude in past environments have to be investigated. More recent research (Knox, 1999) showed how human influences also need to be taken into account.

Threshold relationships have often been regarded as linear, whereas the incidence of change may be rather more complicated. This is particularly because of the Hurst phenomenon, whereby there is a tendency for a non-periodic grouping of similar values over long periods of time, and 'a short term realization of the event sequence in nature will not sample all scales of variability in the process and ... in consequence, distribution statistics derived from the sample realization will be biased' (Church, 1980). In investigating the spatial variation of fluvial processes in semi-arid areas, Graf (1982) investigated the way in which *energy* varies spatially in channel networks, and, in a subsequent paper on the Henry Mountains, Utah, he suggested the way in which stream *power* varied along the channel in 1883, which contrasted with the downstream increase in power obtained in 1909 and 1980 (Graf, 1983). Graf (1979b, p. 266) argued that 'the tradeoff between force and resistance lies at the heart of explanation in geomorphology'. By establishing a relationship between tractive force in dynes, calculated for a ten-year recurrence interval, and biomass on the valley floor in kg m^{-2}, he established a threshold relationship differentiating the entrenched and the uncut valley floors. This stimulating paper provided a means of expressing the relationship between force and resistance in process terms and identified a threshold situation that could be employed to interpret spatial distribution of valley floors entrenched since the mining activities of the 1830s; such a relationship can be employed to indicate areas of valley floor instability at present and in the future. Other concepts such as complex response have been used, together with a systems framework; and equilibrium utilizing the notion of critical stream power (Fig. 6.5) was used in an intriguing way by Bull (1991), illustrating how the basis for modelling can be refined.

Landscapes vary in their *sensitivity* to change, and Brunsden and Thornes (1979) derived a series of four fundamental propositions of landform genesis generated by process–form studies, which were:

(1) *Constant process–characteristic form.* For any given set of environmental conditions through the operation of a constant set of processes, there will be a tendency over time to produce a characteristic set of landforms.

(2) *Transient behaviour.* Geomorphological systems are continually subject to perturbations, which may arise from changes in the environmental conditions of the system or from structural instabilities within the system.

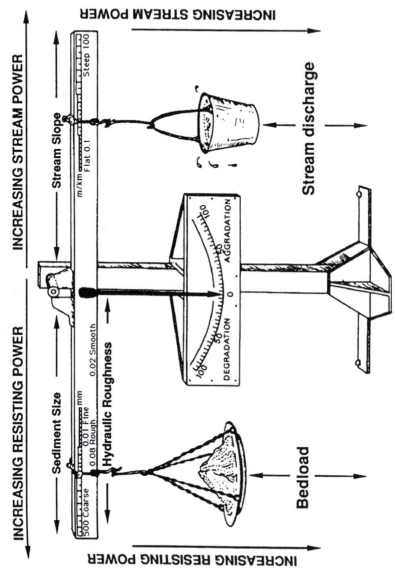

Figure 6.5 Application of power to stream aggradation and degradation (Bull, 1991, with permission from Oxford University Press, originally from the notes of E.W. Lane and modified from Chorley *et al.*, 1984). Zero is the threshold of critical power. Increases or decreases of one or more of the important variables may cause the mode of stream operation to depart markedly from the threshold condition.

(3) *Complex response*. The response to perturbing displacement away from equilibrium is likely to be temporally and spatially complex and may lead to a considerable diversity of landforms.

(4) *Sensitivity to change*. Landscape stability is a function of the temporal and spatial distributions of the resisting and disturbing forces and may be described by the landscape change safety factor, considered to be the ratio of the magnitude of barriers of change to the magnitude of the disturbing forces.

They also suggested a transient-form ratio (TFr), which can express the sensitivity of landform to both internally and externally generated changes, in the form:

$$\text{TFr} = \frac{\text{Mean relaxation time}}{\text{Mean recurrence time of events}} \qquad (6.1)$$

Sensitivity has been used in several ways, and Downs and Gregory (1993, 1995) showed the benefit of clarifying definitions that have been used:

- as the ratio of disturbing to resisting forces;
- as the relationship of disturbing forces to specific threshold conditions;
- as the permanence of morphological changes according to their time-dependent ability to recover from a disturbance;
- by analogy with sensitivity analysis, in which the linear and nonlinear response of geomorphological systems is assessed in relation to alterations in multiple controlling variables.

New models were required to develop from established ones (p. 100) and developments were stimulated by empirical results. Whereas Linton (1957) had taken data on contemporary rates of erosion to indicate that the production of a planation surface by subaerial erosion could take of the order of 10 to 110 million years, Schumm (1963a) combined knowledge of rates of denudation with information on rates of tectonic uplift to compare rates of denudation with rates of orogeny. A very significant paper (Schumm, 1965) used relationships between mean annual precipitation and mean annual runoff, mean annual sediment yield, and mean annual sediment concentration, all for different mean annual temperatures, to provide a basis for Quaternary palaeohydrology from which changes could be interpreted. Subsequently, river metamorphosis (Schumm, 1969) was proposed to describe changes of river channels which could be instigated as a result of changes in discharge and sediment load, and such changes were visualized against the background of a classification of alluvial channels (Schumm, 1963b). Other important approaches to the fluvial system included work by Dury (1964a, b, 1965) on underfit streams which explored the implications of stream shrinkage due to climatic change. Investigation of relationships between changes of river channel shape and planform in response to changes in discharge and sediment transport were summarized in a book (Schumm, 1977) which included a structure for the idealized fluvial system, embracing the zone of production (zone 1), the zone (2) of transfer, and the zone (3) of deposition. Wolman (1967a) explored river channel adjustments downstream of reservoirs and urban areas (Wolman, 1967b). Subsequent investigations of river channel change identified effects of flow regulation due to reservoirs, urbanization, or other land-use or channel changes (Gregory, 1977, 1987c; Petts, 1984; Brookes, 1988).

Progress in other branches of geomorphology embraced thresholds in relation to process dynamics and in arid areas by collating information on contemporary processes. Goudie (1983) collected details of the environmental consequences of dust storms, and then proceeded to indicate the way in which they have varied over time. The ergodic hypothesis was employed by Czudek and Demek (1970) who analysed the distribution of alas character and produced a six-stage model showing the progress of a surface with syngenetic ice wedges to a thermokarst valley. Notable developments in the field of glacial geomorphology included a reconstruction of the Laurentide ice sheet (Sugden, 1978) which concluded that landscapes of glacial erosion are equilibrium forms fashioned largely when ice sheets are at a maximum and reflect periods of steady-state conditions of the order of 100,000 years. A steady-state model of the late Devensian ice sheet which covered much of Britain (Boulton *et al.*, 1977) summarized patterns of glacial erosion and deposition over Britain and suggested that the central parts of the ice sheet were relatively inactive, that relatively high erosional intensities in marginal areas were produced by high marginal velocities, which in turn precluded thick lodgement till and drumlin formation near the margin and concentrated them in internal zones.

The *ergodic hypothesis* has been used, particularly in geomorphology, to investigate change over time because under certain circumstances space–time transformations may be permissible as a working tool (Chorley *et al.*, 1984). This approach was used to suggest how drainage networks may develop using till sheets of different ages, and developed for tilted surfaces in California (Talling and Sowter, 1999); and also using the profiles of coastal cliffs where the cliffs have been protected from basal marine erosion, as in the Laugharne Burrows Spit in South Wales (Savigear, 1952). Craig (1982, quoted by Tinkler, 1985, p. 224) concluded that the ergodic hypothesis enabled useful models of landform behaviour to be tested, although space–time substitution does have limitations (Paine, 1985). The possibility of using models of contemporary processes was anticipated by Thornes (1983a, p. 327) who concluded that:

> ... real progress will only be achieved by establishing models of contemporary processes and driving these models by independently assessed climatic or other input parameters for past times.

In addition to space–time substitution, other approaches to investigation of erosional responses to environmental change are provided by inferences from deposits, repeat observations, and radioisotope measurements (Brown and Quine, 1999b). Physical models have continued, and the University of Colorado's 9 × 15 m rainfall erosion facility was used to investigate drainage network change and channel adjustments (Schumm, 1977) which subsequently formed the major topic for a book on *Experimental Fluvial Geomorphology* (Schumm *et al.*, 1987) (see Chapter 5, p. 114). However, throughout the branches of physical geography it is necessary to be aware of 'the geographer's insatiable appetite for new methods of analysis irrespective of the constraints of data' (Alexander, 1979), but new emphases often using partial models have been developed.

Whereas many analyses in physical geography concentrated upon linear methods, thresholds automatically involve a different form of distribution. Linear relationships, usually established by regression, have been the subject of improper use in the earth sciences as shown by Williams (1983), and so other models have been used. Distributions such as step functions provide a method of resolving series into distinct sequences (Dury, 1980b, 1982). Non-linear models and catastrophe theory, since proposed by Thom (1975) for the description and prediction of a number of discontinuous processes, have also been employed (Graf, 1979a), relating changes which are catastrophic transitions or flips of the system state from one domain of operation into another, perhaps irreversibly. Seven elementary catastrophes were provided (Thom, 1975), and of these the cusp catastrophe may be utilized for modelling temporal changes in environmental systems.

Many developments emanated from geomorphology and hydrology. Approaching evolutionary geomorphology, Thornes (1983b) concluded that the emphasis was shifting from the observation of equilibrium states *per se* to the recognition of the existence of multiple stable and unstable equilibria, which may be the basis for new models of geomorphological evolution. In an unstable system there is no return to equilibrium and it may not be possible to predict successive behaviour. In reviewing progress in temperate palaeohydrology, Thornes and Gregory (1991) considered ways in which stability and multiple stable states can be approached and showed that, by knowing the structure of the stability–instability relationship, it should be possible to predict the transitions. In dynamic systems analysis, shifts in a single parameter may produce a sequence of changes involving a transition from a single stable equilibrium, to a bifurcation with two stable states, and ultimately to further bifurcations with many stable states (Thornes and Gregory, 1991, p. 535). Non-linear methods therefore hold considerable promise for future modelling of environmental change (Phillips, 1995; Lane and Richards, 1997). Concepts of self-organization in landscapes, including many of those noted above, are grouped by Phillips (1999a) into two: firstly those concerned with evolution of order and regularity; and secondly the differentiation of landscapes into more diverse spatial units. A theory of spatially divergent self-organization related to the latter has been developed (Phillips, 1999b), and Phillips notes (1999a, p. 466):

The extent to which field testable hypotheses are generated, or explanations provided based on process mechanics or landscape history, will ultimately determine the utility of self-organization concepts and methods in physical geography.

Developments in soil geography included those related to geomorphology (e.g. Gerrard, 1981) and palaeosols (Boardman, 1985). Criteria for the identification of palaeosols in relation to the distinction of soil, regolith and weathering profiles were refined, and the use of laboratory techniques, such as using the forms of phosphorus present or the level of amino-acid nitrogen to indicate the biological activity, was significant (Gerrard, 1981). Decomposition and distribution of organic matter, or humon, allowed the original population of plants and animals and their associated environments to be interpreted, and the infra-red absorption spectra of humic acids could be utilized to recognize palaeosols and to specify the type of former vegetation environment that existed. Further developments included the use of weathering indices, radiocarbon dating of profiles, analysis of soil pollen and of opal phytoliths, which are the remnants in the soil profile of silica originally absorbed by plants and precipitated in plant cells which remain in the soil when the plant decays. Conceptual advances in soil development included hypothetical curves for soil development over time under fluctuating and constant climates (Birkeland, 1974). In Australia the formulation of the concept of K-cycles (Butler, 1959) gave a useful methodology to elucidate the stratigraphy of soil development. This included the concept of the groundsurface to represent the development of a soil mantle and the recognition of K-cycles, each of which is composed of an unstable phase (K_u) and a stable phase (K_s). In each K-cycle, erosion and/or deposition in the unstable phase is succeeded by soil profile development in the succeeding phase and, in a particular landscape, there may be evidence of up to eight K-cycles preserved in the soil landscape. In addition to the K-cycle approach, Huggett (1982) referred to the review provided by Daniels *et al.* (1971) and to the temporal nature of soil–landscape relationships proposed by Vreeken (1973, 1975) to connote the overall topographical influence on soil properties within a drainage basin. When datable land surfaces are found it should be possible to develop chronofunctions which show how soil properties change with time and, although Birkeland (1974) and Gerrard (1981) reviewed the general relationships of soils, weathering and geomorphology, there is still further need for the development of mathematical models. A model for a soil catena or two-dimensional landscape slice, and for a soil landscape or three-dimensional landscape segment, were built by Huggett (1976a) to embrace the redistribution of mobile soil constituents by throughflow. Huggett (1982) concluded that:

> Perhaps the biggest untapped area of research in soil geography is the building of mathematical models which link soil and vegetation processes to geomorphological processes on slopes and in landscapes.

Other advances include the use of fractals (Nortcliff, 1984), which refers to temporal or spatial phenomena that are continuous but not differentiable and exhibit partial correlation over many scales; soil properties are visualized as fractals (Burrough, 1983) because increasing the scale of mapping reveals more and more detail.

In biogeography, investigation of palaeoenvironmental change was influenced by contributions in other disciplines, and *Quaternary Palaeoecology* (Birks and Birks, 1980; Berglund *et al.*, 1996) and *Biology and Quaternary Environments* (Walker and Guppy, 1978) were very significant for physical geographers. Much work in biogeography and ecological theory was founded in the temperate and boreal regions of the world, so the tendency was for the simple and the geologically young to become accepted as normal. It was therefore argued that the tropics should be viewed as the norm for developing theory and basic principles (Whitmore *et al.*, 1982) rather than relying too heavily upon regions where soils and vegetation are generally less than 10,000 years old, because 'the Pleistocene, rather than the Holocene, is the norm in ecology' (Flenley, 1982, in Whitmore *et al.*, 1982). In the context of human impact on the environment, Oldfield (1983a) argued that physical geographers emphasize models of ecosystem change which are deterministic, progressive and evolutionary in character. The succession-climax model was still very evident in biogeography, despite its weakness in the light of recent empirical evidence and the difficulties encountered when trying to reconcile it with systems theory, while models with a cyclic or harmonic element had been developed. Oldfield (1983b) therefore argued for a model of ecological change that differs in principle from both the evolutionary and the cyclic models and is more analogous to the steady-state models developed in other branches of science. He proposed that lakes, together with their sediments and drainage basins, provided an appropriate focus to utilize the new methods available (Oldfield *et al.*, 1983). The steady-state model proposed (Oldfield, 1983b) is in many ways analogous to models employed in geomorphology (e.g. Fig. 6.3) and extensive shifts between steady states have often been triggered by processes of soil depletion and erosion with the associated consequences for soil structure, water content and nutrients. In biogeography generally, Stoddart (1983) noted how a number of recent books had not identified 'the swirling controversies which have dominated the research frontier of the subject for a decade'. For example, a new approach (e.g. Nelson and Platnick, 1981) was founded upon Hennig's phylogenetic systems usually termed 'cladistics', in contrast to approaches based upon conventional evolutionary taxonomy. Cladistics as a method of classification used graphs of relative affinity, termed 'cladograms', which do not require any *a priori* assumptions about the nature of the relationships, including evolutionary relationships, involved. Such a cladistic approach led to the rejection of methods employed by evolutionary taxonomists and then to different degrees of 'denunciation' of the Darwinian contribution (Stoddart, 1983). When considering the processes responsible for distributional patterns in space and time, it was suggested (Croizat, 1978) that these arise from either chance dispersal or vicariance. Dispersal came to be associated with Darwinian biogeography, whereas vicariance biogeography (Nelson and Rosen, 1981) is concerned with biotic distributions which may be congruent with plate-tectonic reconstructions, although there are other explanations for vicariant distributions including Pleistocene climatic changes and sea-level changes (Stoddart, 1981). The literature applying cladistic methods to a number of biogeographical problems was reviewed by Stoddart (1978, 1983), who suggested that both vicariance biogeography and dispersal are appropriate tools at particular levels of enquiry, and that only Udvardy (1981) had attempted to specify the scales which might be appropriate as a background for research investigations (see p. 153). Whereas the approach used in historical biogeography and vicariance biogeography can be characterized as static and showing its inadequacy (Hengeveld, 1993), the dynamic approach, which analyses spatial patterns as a result of processes operating over longer or shorter time periods, is used in ecological biogeography. A more dynamic approach and closer links between previously separated branches, such as the development of biogeomorphology (e.g. Viles, 1988), facilitated advances in the analysis of environmental change (see p. 169).

As climate is a fundamental control upon environmental systems, the study of climatic trends is vital for the investigation of change; the climatic record of the post-glacial and also of the Quaternary as a whole is of great significance to several branches of physical geography, as is the new journal *Climatic Change* established in 1978. Geographer climatologists necessarily consider climate change (Lockwood, 1979a), focusing on the energy balance, general circulation and statistical dynamical models (Lockwood, 1983a) that afford the main types of approach used. These include, firstly, the causes of climatic fluctuations (Lockwood, 1979b) and the advances in climate theory (Barry, 1979) where energy budget models are promising but require adequately distributed world data to refine their performance. Secondly, increased use of models enhanced the reconstructions of climates for geological periods. Particularly for the Quaternary it has been established that there is a connection between the Earth's orbit and ice ages, but the exact nature of a connection between the long-term periodic variations in the Earth's orbital parameters, as embraced by the Milankovitch theory, and the occurrence of glacial–interglacial transitions requires elucidation (Lockwood, 1980). Thirdly, short-term fluctuations in climate have attracted considerable interest from physical geographers. This extends from the last 75–100 years, when the pattern of global trends is still difficult to isolate because of major gaps in the world date network especially over the oceans (Barry, 1979), to consideration of shorter periods, because on a global scale the variability of weather is more pertinent to food production than are climatic trends. During the historic time-scale, changes of climate have been investigated using a range of historical techniques, including diaries and records together with sedimentary evidence and information from faunal remains, archaeology, tree rings and ocean deposits illuminating climatic fluctuations in western Europe (e.g. Manley, 1952). On a relatively short time-scale, investigations of the influence of vegetation on climate (Lockwood, 1983b) and the effects of the southern oscillation, which is a variation in pressure between the eastern and western Pacific over a period of years, and the anomalously warm sea surface called El Niño off the coast of Peru (Lockwood, 1984; Waylen, 1995) have attracted attention. Fourthly, attention to modelling of climate *per se* has been especially related to global change, which is a further area where process has been associated with change (e.g. Bryant, 1997) and where past variation is an important key to understanding future climate (Bradley and Jones, 1992). Research by physical geographers on climate change has been facilitated by involvement in multidisciplinary research and also by research focused on global change (Chapter 9). Whereas weather forecasting is a long-established objective for the meteorologist, climate forecasting has now become a priority, and the first objective of the Global Atmospheric Research Program (GARP) was to achieve greater understanding of short-term weather processes and thence to improve forecasting. However, not all modelling is directed toward future climates because the refinement of numerical models and their application to time-bound problems, such as the characteristics of glacial and interglacial climates, offers a link between studies of contemporary climates, quantitative modelling approaches and environmental reconstruction that the physical geographer cannot afford to ignore.

6.5 The multidisciplinary trend

Concern that greater specialization inexorably leads towards fissiparist tendencies, increasing the divisions between branches of physical geography, has not been substantiated by recent trends for two types of reason. Firstly, similar models and conceptual approaches have become widely used, so that connections have been forged internally between branches of physical geography, and secondly, greater links have been developed externally with other disciplines.

The first trend, of links within physical geography, has been evidenced by examples in the last few pages. The way in which biogeographical studies have progressed towards the lake–drainage basin system necessarily fosters links between biogeography, geomorphology and hydrology, and the ecosystem watershed concept was advocated (O'Sullivan, 1979). In environmental systems there is a recurrent interest in domains from the geomorphologist which is reminiscent of the niche concept used by the ecologist, and the need to progress towards a vision of time-scales is as fundamental in geomorphology (Schumm and Lichty, 1965) as it is in ecology (Udvardy, 1981). A perspective was provided by Driver and Chapman

(1996) who suggested that their four time-scales (p. 153) could be visualized in relation to four different worlds, namely the physical/chemical, that of human artefacts, the biological, and that of conscious human thought. A specific example linking the present landscape and environmental change and bridging the branches of physical geography is provided by Roberts (1996, Fig. 2.2) who explains changing dambo hydrology under four different scenarios of climate and human impact. Following his review of the stability of the East Antarctic ice sheet, which he suggests has been stable for at least 14 million years, Sugden (1996) proposes the development of a hermeneutic geomorphology to build up the interpretative and historical aspects of the science to complement the analytical approach of the experimental sciences. In this and other instances, the physical geographer can make a very significant contribution. Particularly stimulating development is the way in which research links investigation of environmental processes to study of environmental change, progressing beyond consideration of the problems (e.g. p. 136) and of the broad strategies (e.g. Sugden *et al.*, 1997) to undertake specific research investigations which actively develop the link (Brown and Quine, 1999a). This enables greater exploration of non-linear models and consideration of alternative types of response. Thus in considering the relationship between sedimentary/erosional and hydrological responses, Brown and Quine (1999b, Fig. 1.5) distinguished hyper-sensitivity, with super-abundant sediment, low vegetation cover, unstable slopes and unarmoured beds, and contrasted under-sensitivity states. Recognition of such states, and of the switches from one to another, can be particularly important in studies of the specific effects of global change (see p. 249).

   A second trend of greater linkage with the several disciplines concerned with environmental change has arisen because themes on the research agendas of branches of physical geography have been prominent in, and sometimes imported from, other earth sciences. Thus thresholds achieved prominence in ecology (May, 1977) and there is potential for further application of fractals in relation to environmental data as exemplified by spatial analysis of soil properties (Nortcliff, 1984; Rodriguez Iturbe and Rinaldo, 1998). Research on postglacial climatic change was shown by Prentice (1983, p. 278) to benefit from contributions from several disciplines because 'Cross checks among palaeoclimatic estimates obtained in different ways will be extremely useful, but no one method or discipline is privileged'. Desertification is one problem which relates to past environmental change but is also inextricably linked with the future (see p. 185). Research cooperation has been fostered by the development of multidisciplinary programmes, including the International Geological Correlation Programme (IGCP), which embraces a number of interdisciplinary and international research investigations that involve physical geography research. One multidisciplinary programme of research on palaeohydrology was focused initially on the temperate zone as an IGCP project (Berglund, 1983; Starkel,

1981, 1983; Gregory, 1983; Starkel *et al.*, 1991) and has subsequently focused upon global continental palaeohydrology (GLOCOPH) as a Commission of INQUA involving more than 200 scientists in over 35 countries. This is just one international project that benefits from multidisciplinary collaboration (Gregory *et al.*, 1995; Branson *et al.*, 1996; Benito *et al.*, 1998) and in one of the contributions the range of global programmes relating to human activities was summarized (Gregory, 1995a). Multidisciplinary research is aided by national organizations for Quaternary scientists including AMQUA in America, DEUQA in Germany, NORDQUA in Scandinavia and the QRA (Quaternary Research Association) in Britain; and by the International Union for Quaternary Research (INQUA) with many commissions to coordinate research in specific fields. Closer links between branches of physical geography, such as between geomorphology and biogeography, have encouraged research links with other disciplines including biologists, geologists, archaeologists and palaeoclimatologists, all increasingly aware of the importance of global climate change. One important outcome is fields of research at the interface of disciplines (e.g. Davidson and Shackley, 1976), such as the recent development of geoarchaeology (Brown, 1997). In *Reconstructing Quaternary Environments* (Lowe and Walker, 1997), recent developments in Quaternary research are reviewed in their first chapter where the emphasis on multidisciplinary and interdisciplinary approaches in the study of palaeoenvironmental changes is noted. These developments are also important for human activity, as described in the next chapter.

Further reading

One of the current texts on environmental change outlines the present situation:

LOWE, J.J. and WALKER, M.J.C. 1997: *Reconstructing Quaternary environments* (2nd edn). Harlow: Addison Wesley Longman.

To cover the Holocene and link to the present:

ROBERTS, N. 1994: *The Holocene. An environmental history.* Oxford: Blackwell.

A complementary view linked to systems and providing a foundation for global change:

HUGGETT, R.J. 1997: *Environmental change. The evolving ecosphere.* London: Routledge.

One approach providing background is:

BRUNSDEN, D. 1990: Tablets of stone: towards the ten commandments of geomorphology. *Zeitschrift für Geomorphologie* Supplementband **79**, 1–37.

Further developed by:

BAKER, V.R. 1996: Tablets of stone: ten commandments or a golden rule? In S.B. McCann and D.C. Ford (eds) *Géomorphology sans Frontières*. Chichester: John Wiley, 59–67.

Topics for consideration

(1) How have recent approaches to environmental change developed in different countries?

(2) Equilibrium and non-equilibrium views can be applied to environmental change and each has its merits.

(3) Ways in which interpretation of sequences of environmental change have been aided by knowledge of contemporary environments are illustrated by Brown (1991), interpreting the mantle of superficial deposits in relation to palaeohydrological development, and by Walling *et al.* (1998), demonstrating the amount of fine-grained sediment stored on the floodplains and channel bed of the main channels of the Ouse and its tributaries. Are there similar studies in other branches of physical geography?

|7|

Human activity: an increasingly dominant theme

Although the effects of human activity were increasing throughout the twentieth century, they did not attract significant attention from physical geographers until the 1950s and 1960s. Instead it was environmental change before human activity, or the study of processes largely unaffected by humans, that claimed prior attention as indicated in Chapters 5 and 6. This situation prevailed despite the fact that the theme of human–environment relations has never been far from the heart of geographical research, and for many had often been the overriding theme. Antecedents for the investigation of human activity are outlined (7.1), prior to a review of the ways in which research has established the magnitude of human influence (7.2). This proceeds to the way in which the study of Earth hazards (7.3) and urban physical geography (7.4) provided foci for geographical research, leading to the implications which arise for the scope, content and focus of physical geography (7.5).

Although not all geographers accepted the growth of interest in human activity, the investigation of human impact together with hazard research offered the prospect of closer links between human and physical geography (e.g. Johnston, 1983c). When Hewitt and Hare (1973, p. 23, 31) assessed the position they

> ... accepted the time-honoured division of physical geography into climatology, soil science, geomorphology, and biogeography, holding that these were concerned with the explanatory description of large, tangible and plainly visible parts of the natural landscape, or the directly measurable crude properties of the atmosphere. When, however, we began to teach a course of generalized man–environment relations, in the crisis atmosphere of the early 1970s the physical geography of our background appeared totally inadequate. It simply

did not, and does not, offer a framework on which to hang a convincing story ... the geographer, when he analyses the material properties of the man–environment systems, must base himself on the central functions of that system, rather than on the traditional divisions of physical geography.

The central functions that Hewitt and Hare identified were partly addressed by the advent of the systems approach (Chapter 4), but also required much greater attention to human activity. Study of the physical environment, in which the significance of human activity was very prominent, came about for several reasons which are used as subdivisions for this chapter. Few studies in physical geography focus exclusively upon human impact, but there are many that analyse human activity as a dominant influence.

7.1 Antecedents for the investigation of human impact in physical geography

Despite frequent allusions to human–environment relationships by geographers, the signposts that were evident from the mid-nineteenth century onwards were largely ignored, and physical geography developed until the mid-twentieth century largely independently from studies of the impact of human activity. On reflection this is surprising in view of the clear pointers waiting to be recognized. In the introduction to a book on *Man and Environmental Processes* (Gregory and Walling, 1979), the pointers available were resolved into a century of milestones (to 1960), a decade of papers in the 1960s, and a decade of readings published in the 1970s. When *Human Activity and Environmental Processes* was produced (Gregory and Walling, 1987), these three headings were supplemented by 'an emphasis upon processes' which had characterized developments in the 1980s.

The first milestone was undoubtedly *Man and Nature* published in 1864. Its author, George Perkins Marsh, conceived the book as a

> ... little volume showing that whereas others think the earth made man, man in fact made the earth ... The object of the present volume is: to indicate the character and approximately, the extent of the changes produced by human action in the present condition of the globe we inhabit; to point out the dangers of imprudence and the necessity of caution in all operations which, on a large scale, interfere with the spontaneous arrangements of the organic or the inorganic world.

To demonstrate the magnitude of changes wrought by humans upon nature, Marsh devoted chapters to vegetable and animal species, woods, waters, sands, and to the projected or possible geographical changes by mankind. His book (p. 31) provided a foundation for the conservation movement (Mumford, 1931), it proved to have a great influence upon the way in which the land was visualized and used (Lowenthal, 1965), and its full title *Man and Nature or Physical Geography as Modified by Human Action* clearly indicated the direction in which it was pointing. The clearest reflection of the way in which the full implications of Marsh's book were not appreciated in physical geography is shown by the fact that he is not referred to in several other important books (Chorley *et al.*, 1964; Harvey, 1969; Johnston, 1979, 1983a, 1997; Holt-Jensen, 1981), although Marsh was cited in the third edition of *Geography: Its History and Concepts* (Holt-Jensen, 1999). Such omissions were not universal, because Stoddart (1986) referred to the substantial contribution of Marsh, but they are consistent with the trend in physical geography for nearly a century after *Man and Nature* was published. Other occasional nineteenth-century pointers included Charles Kingsley's book on *Town Geology* published in 1872, and the objection of Kropotkin (1893, p. 350) to the trend to exclude man from physiography (see also Chapter 10, p. 258).

A second milestone was *Man as a Geological Agent* (Sherlock, 1922) and a related article (Sherlock, 1923) in which the focus was geological with particular attention devoted to denudation by excavation and attrition, to subsidence, accumulation, alterations of the coast, the circulation of water, and to climate and scenery, and in addition biological aspects were mentioned. By focusing upon the impact of humans, Sherlock emphasized the contrasts between natural and human denudation and concluded that in a densely populated country such as England (Sherlock, 1922, p. 333) 'Man is many more times more powerful, as an agent of denudation, than all the atmospheric denuding forces combined'. In 1922, the year that Sherlock's book was published, H.H. Barrows's presidential address (Barrows, 1923) to the Association of American Geographers (see Chapter 10, p. 258) and proposal to separate specialized branches, including geomorphology, climatology and biogeography, from the parent subject was not adopted. In the US much physical geography did develop in geology departments, but exactly 50 years later in a paper with the same title as Barrows's, Chorley (1973) concluded that the control system could be an appropriate focus, that it would clearly incorporate human activity and focus upon the links between human and physical environment, and that:

> It is clear, however, that social man is, for better or worse, seizing control of his terrestrial environment and any geographical methodology which does not acknowledge this fact is doomed to in-built obsolescence. (Chorley, 1973, p. 167)

In the 1930s and 1940s, although the impact of human activity was becoming increasingly evident from the effects of the dust bowl, the creation of the Tennessee Valley Authority (TVA), and the much greater use of fertilizers, this situation was not fully acknowledged in physical geography and this is perhaps particularly startling given the appearance of books on soil erosion (e.g. Jacks and Whyte, 1939). However a further major and third milestone was to come with the publication of *Man's Role in Changing the Face of the Earth* (Thomas, 1956), a book of 1193 pages, based upon an interdisciplinary symposium with international participation which was organized by the Wenner-Gren Foundation for Anthropological Research and held at Princeton, NJ, in 1955. This monumental achievement acknowledged the work of Marsh and also that of the Russian geographer A.I. Woeikof (1842–1914), who had developed a utilitarian approach to the study of the Earth's surface, acknowledging the impact of human activities. The 52 chapters of the work (Thomas, 1956) were organized in three parts: the first retrospective, elaborating the way in which humans have changed the face of the Earth; the second reviewing the many ways in which processes had been modified; and the third concerned with the prospect raised by limits on the role of mankind.

This important volume, subsequently republished in parts, did not immediately have the effect that might have been expected. However, its significance became more fully appreciated in the 1960s as books and papers reflected increasing awareness of human effects in at least four ways. Firstly, a series of review papers, sometimes based on inaugural or presidential addresses, included reviews of humans and the natural environment (Wilkinson, 1963), advocacy of the need for study of anthropogeomorphology (Fels, 1965) and a revival of the title used by Sherlock when Jennings (1966) stressed that 'Man as a geological agent' is significant because studies of contemporary processes are nearly always heavily biased by anthropogenic effects. A very significant paper by E.H. Brown (1970) characterized mankind as both a geomorphological process in relation to its direct, purposeful modifications of landforms, and also as indirectly effective through the human influence upon geomorphological processes. Gradually, therefore, human activity began to achieve greater significance in physical geography, although few textbooks before 1970 allocated much space to the effects of human activity. Certainly the notion of the noosphere (Trusov, 1969) as a new geological epoch initiated as a consequence of human activity, although capable of wider use in geography (Bird, 1963, 1989), did not find unequivocal acceptance. However, Svoboda (1999) has explained the Homosphere as the biosphere modified by *Homo sapiens* and suggested that this sphere of human influence and presence is one from which the noosphere as a new 'thinking layer' emanates.

A second strand of research in the 1960s was a consequence of the increasing focus upon processes which often highlighted human activity, and led to the inauguration of research specifically designed to measure the magnitude of human influence by comparing modified and unmodified areas or by measuring one area before, during and after the human effects. Thus it was possible to investigate processes in a small drainage basin before, during and after the effects of building activity (Walling and Gregory, 1970). Equally significant was research on landscape evolution, which now focused on time-scales intermediate between those used for earlier chronological studies and those adopted in process investigations. Such research was sometimes located in areas where the effects of human activity had been sustained for many centuries. Thus in the Mediterranean basin the evolution of valleys (Vita-Finzi, 1969) could be understood only by reference to human activity. More biogeographically, the significance of cultural biogeography was exemplified by the work of Professor D.R. Harris, initially in the field of historical ecology but then embracing the domestication of plants and animals (e.g. Harris, 1968), and later inter-action with archaeology upon his appointment as Professor of Archaeology in the University of London in 1979.

As a result of research of this kind, a further third strand could be detected which devolved upon the investigation of Earth hazards and of physical geography from a socioeconomic viewpoint. Instrumental in launching the study of hazards was Professor Gilbert White's *Changes in Urban Occupance of Flood Plains in the United States* (White, 1958) which stimulated much subsequent research (see p. 188). An excellent illustration of an integrated approach was provided by *Water, Earth and Man* (Chorley, 1969a), an edited volume reviewing aspects of the physical and socioeco-nomic environment that relate to water, in which Chorley and Kates (1969, p. 2) affirmed that: 'it is clear without some dialogue between man and the physical environment within a spatial context geography will cease to exist, as a discipline'. To proceed towards such a dialogue, Chorley (1969a) believed that using an integrated body of techniques for physical and human geography, and emphasizing resources of the physical world such as water, could be achieved, in a book in which (Chorley and Kates, 1969, p. 3):

... Geographers, freed from the traditional distinction between human and physical geography and with their special sensitivity towards water, earth and man, have in these both opportunity and challenge.

Studies of this kind exemplified some shift in the attitude to physical envi-ronment, as environmental perception had a significant influence on the research undertaken by physical geographers.

A fourth and final strand evident in the 1960s was the atmosphere created by the inception of international research programmes and of increasing environmental concern. Programmes dependent upon interdisci-plinary as well as international participation included the International Biological Programme (1964–74), the International Hydrological Decade (1965–74) which embraced human influence as one of its major themes, and the Man and the Biosphere programme. Environmental concern mush-roomed during the 1960s, stimulated by warnings about the impact of human action and by debates about the extent to which earth resources were finite, presenting the basis for a pessimistic or optimistic future for Spaceship Earth. Simmons (1978) noted that geographers had taken little notice of the wave of concern for the environment which peaked about

1972, so that he proceeded to argue in favour of a humanistic biogeography (see Chapter 10, p. 260).

At the beginning of the 1970s, in an exploration of the role and relations of physical geography, Chorley (1971) had proposed that control systems offered an approach whereby human activity acts as a regulator in natural systems; this was explored (Chorley and Kennedy, 1971) and perhaps initiated a more consolidated attitude to the significance of human activity (see Chapter 4, p. 88), when the available literature on human action was very limited. The deficiency was remedied first by a series of collected readings (Coates, 1972, 1973), in which (Coates, 1973, p. 3) it was noted:

> A twentieth century innovation has been man's growing awareness of his impact on the environment. With few exceptions, such as the case of George Perkins Marsh, man before 1900 had a hostile view of the Earth; his need was to conquer and to subdue nature.

Similarly Detwyler (1971) collected previously published papers to produce *Man's Impact on Environment;* a collection of essays devoted to research on environmental problems was organized by the Association of American Geographers' Commission on College Geography (Manners and Mikesell, 1974); and an edited collection reviewing *Man and Environmental Processes* (Gregory and Walling, 1979) was later revised as *Human Activity and Environmental Processes* (Gregory and Walling, 1987). In 1981 A.S. Goudie published *The Human Impact, Man's Role in Environmental Change* (Goudie, 1981b), with subsequent editions in 1986, 1990 and 1993. This book (Goudie, 1986a, p.1):

> ... seeks to find out whether, and to what degree, humans have during their long tenure of the Earth changed it from its hypothetical pristine condition. For as Yi fu Tuan put it: 'The fact of diminishing nature and of human ubiquity is now obvious'.

Goudie successfully approached the task by dealing with human impact on the major components of the environment, namely vegetation, soil, waters, geomorphology, climate and the atmosphere. A concluding chapter directs attention towards forward-looking problems, including the significance of human influence on nature, with the conclusion (Goudie, 1986a, p. 294) that:

> ... in many cases of environmental change it is impossible to state without risk of contradiction, that people rather than nature are responsible. Most systems are complex and human agency is but one component of them, so that many human actions can lead to end-products which are intrinsically similar to those that may be produced by natural forces.

Since the fourth edition of his book (Goudie, 1993b), Goudie and Viles (1997a) have published *The Earth Transformed: An Introduction to*

Human Impacts on the Environment, which explored 'the many ways in which humans have transformed the face of the Earth'. It placed the transformations into a historical context, seeing how humans have changed through time by focusing upon the biosphere, atmosphere, waters, land surface, oceans, seas and coasts, with a conclusion directed towards a sustainable future. This focus upon the magnitude of human impact rather than upon research investigations showed that human activity was established as a core theme in research and was perhaps therefore appropriate for the pre-higher education market. As indicated at the beginning of this section, the major milestones in the recognition of the impact of human activity could be supplemented by an *emphasis upon processes* which was the theme for the book edited by Gregory and Walling (1987) and is a central theme for Section 7.2 below.

The foregoing review demonstrates what now seems almost unbelievable: that physical geography for so long avoided the significance of human activity and thence the potential which associated studies could afford. It is indisputable that physical geographers entered the field later than they should have done, so that, at least in the 1960s and 1970s, reference often had to be made to studies of human impact undertaken by other disciplines rather than physical geography. This is one reason why Hare (1969) noted 'sometimes I think that Geography as a science deliberately stays out of phase with the climate of the times'. Fortunately a decade later geographical endeavour was more in tune with the climate of the times when Hare (1980) directed geographical attention to whether the planetary environment was fragile or sturdy. The seminal work on *Man's Role in Changing the Face of the Earth* (Thomas, 1956) was succeeded by *The Earth as Transformed by Human Action* (Turner *et al.*, 1990), affectionately known by the six editors as ET, a volume of 720 pages with four main sets of chapters:

- Chapters dealing with global changes over the past three centuries in major aspects of human activity relative to environment transformation, namely population, technology, institutions and social organization, trade, urbanization and awareness of human impact.
- Chapters concerned with the long-term assessment of natural change in the biosphere, involving 18 chapters in five sets dealing with the last 300 years of impacts on major states and flows of the globe, including land transformation, water flows, oceans, climate and atmosphere, fauna and flora, flows of carbon, sulphur, and so on.
- Studies of historical and contemporary human impact on the environment in 12 regions of the world.
- Three chapters addressing the contribution that different perspectives in social science could make to human-induced environmental transformation.

This significant contribution was assessed by Malone (1994) who concluded that:

... the community of geographers have once again demonstrated a unique, scholarly capability for synthesis. The world urgently needs and anxiously awaits their next contribution to this increasingly important task ... Alternatively a new grouping of disciplines, that would include engineers and scholars in ethics and values, might reach new heights by standing on the shoulders of the nearly 100 scholarly giants who produced ET – a book that takes its respected place among its 1864 and 1955 predecessors.

The approach adopted exemplifies geography as human ecology, and although its structure is more complex than the elegantly simple structure used by Goudie (1993b) and Goudie and Viles (1997a), it covers the main aspects of research contributions and raises questions of the appropriateness of disciplinary boundaries. One year earlier Simmons (1989) produced *Changing the Face of the Earth* as a history of human impact upon the natural environment of the Earth, focusing primarily upon the ecological, and intended as a tribute to the influential book on *Man's Role* edited by Thomas in 1956. The ecological thrust proceeds through primitive man and his surroundings, advanced hunters, agriculture and its impact, industrialists and the nuclear age. The stimulating conclusions (Simmons, 1989, pp. 378–96) deserve wide reading, and range from the regional distribution of environmental alteration, and types of the future, to a focus on energy, entropy and culture, thus echoing the emphasis upon energetics that emerged from the studies of systems (Chapter 4, p. 92). Human activity has now been completely absorbed into the physical geography research agenda with important implications for the way in which applied research is undertaken (Chapter 8), for the way in which cultural approaches are viewed (Chapter 10), and with greater awareness of the global environment and the potential implications of global environmental change (Chapter 9).

7.2 Assessing the magnitude of human activity

Physical geography assimilated the importance of human activity in altering environment and environmental processes, which together are the focus of study for physical geography, so that specific research investigations were devised to assess human impact. Such assessments were achieved by:

- studies that relied upon the already-documented impact of human action and provided reviews;
- investigations designed to establish the impact of human activity on the physical environment and the consequences of those impacts;
- research directed towards investigation of human impact on environmental processes;
- participation in collaborative research – particularly necessary for global-scale investigation.

Management implications have arisen from each of these four strands (Chapter 8), but approaches to research in the branches of physical geography focus on the methods adopted.

Human activity is an obvious central theme in ***biogeography*** and this is reflected in the approach taken by Simmons (1989). Two particular trends, albeit related, proceeded in this direction. Firstly, studies emphasized the historical sequence of human impact beginning with the ways in which prehistoric groups caused environmental changes (Goudie, 1989). Goudie argued that fire, hunting, agriculture and settlement together with the exploitation of minerals all had substantial effects, and that:

> although much of the concern expressed about the undesirable effects of humans has tended to focus on the role played by sophisticated industrial societies, this should not blind us to the fact that many highly significant environmental changes were and are being achieved by non-industrial societies. Even though the population densities of human hunter and gatherer predators are necessarily low . . . they have achieved substantial modifications of some major vegetation types through the use of fire. (Goudie, 1989, p. 17)

Research of the kind referred to was accomplished by studies of Quaternary ecology which successfully established the major stages of vegetational change and also included the influence of human activity, effective through the modification of biogeographical systems. This naturally fostered links with archaeology (Simmons and Tooley, 1981). In Britain the impact of prehistoric man had been considered as a particularly appropriate task for geographical research (Curtis and Simmons, 1976, p. 257), and the significance of the Bronze and Iron Ages in reducing the forest cover of the British landscape, particularly in upland Britain, was demonstrated, particularly where blanket peat growth may have been influenced by deforestation rather than simply by climatic change (Simmons, 1980).

Analysis of mineral magnetic properties (p. 152) prompted Oldfield (1983b) to propose a steady-state model of ecosystem change related to man's impact on environment as an additional alternative to more familiar successional and cyclic models, relevant to other time-bound developments (Chapter 6, p. 141). Whereas research of this kind was initially within the confines of biogeography, it subsequently involved other branches of physical geography, and indeed the boundaries between the branches were relaxed particularly with the enlarged range of techniques available for environmental reconstruction (see p. 134). Particularly fruitful was the development of links with archaeology; Brown (1997) provided *Alluvial Geoarchaeology* to give archaeologists the necessary background in environmental science and also to provide a source of multidisciplinary information on recent and contemporary alluvial environments.

This linked to the second theme evident in biogeography, which was the progress made towards cultural biogeography and historical ecology. This approach, very evident in the US and applied to the most recently settled areas, involved reconstruction of changes in biogeographical systems in recent decades and centuries, incorporating the influence of human activity. Thus Murton (1968) reconstructed the immediate pre-European vegetation on the east coast of the North Island of New Zealand. However, particularly in long-settled countries, it is difficult to differentiate such biogeographical studies from those by historical geographers or economic historians; it was a historical geographer (Darby, 1956) who mapped and compared distribution of forest in AD 900 with that in AD 1900. Progress in this second research theme was also reflected in the presentation of biogeography in textbooks. Although some sought to focus upon the original vegetation, and therefore to relegate human activity to a chapter or afterthought, this was becoming increasingly invalid as the effects of human activity were recognized to be so substantial. I.G. Simmons devised a title *Biogeography: Natural and Cultural* to encapsulate the approach (Simmons, 1979a). Having suggested that, within geography, biogeography teaching has lacked a crystallizing focus, he contended that there had been something of a 'supermarket approach' whereby a number of packages had been bought off the shelves and put in the course trolley, so that Simmons (1979a, p. 1) offered a stronger conceptual framework. Acknowledging the concepts of Dansereau (1957) that humans create new genotypes and new ecosystems, he considered nature without humans as natural biogeography in Part 1 of his book, and then used this as a datum against which to set the larger Part II in which cultural biogeography is concerned with the human effects of changing the genetic make-up of plants and animals, in redistributing them over the Earth's surface, and in altering the structure of many ecosystems. Although such an approach could encompass only part of the traditions of study in geography, it allowed the human impact on individual taxa and ecosystems, on biotic resources, their conservation and protection, to find a place in a coherent whole (Simmons, 1980, p. 148), although he expected that methodological pluralism would continue in biogeography. This approach led towards consideration of natural resources (Simmons, 1974) and conservation (Simmons, 1979b), and *The Ecology of Natural Resources* (Simmons, 1974) was succeeded by different approaches in *Earth, Air and Water* (Simmons, 1991) and *Environmental History* (Simmons, 1993b), which in some ways followed the theme of *Changing the Face of the Earth* (Simmons, 1989). This is returned to in Chapter 10 (p. 260).

In the study of *soil geography*, two formative strands similar to those in biogeography could be discerned, one arising from studies of soil profiles and their evolution, and one more concerned with recent historical changes. Associated with late Quaternary vegetation changes were changes of pedogenesis and of soil systems. In Britain, for example, podzolization succeeded brown-earth formation, and, because human action was involved in the vegetation change, there was also an effect upon soil profiles (Bridges, 1978b).

Recent historical changes of soils were more numerous than the degree of research attention accorded by physical geographers might suggest, but the effects of salinity, drainage, fertilizers, pollution and land-use change all have considerable significance for soil pedons and soil processes. Bearing in mind the magnitude of soil erosion which has occurred in sensitive areas, including the US dust bowl and parts of China and Africa, it is surprising that more attention was not directed to this aspect of human impact. It has been argued that special attention should be focused upon the resilience and potential for recovery of the soil profile in view of the inputs induced by man (Trudgill, 1977, Chapter 8), and the importance of the problem was underlined by Toy (1982) in a review of accelerated erosion, when he concluded that such erosion can be considered to be the pre-eminent environmental problem in the US by virtue of its widespread occurrence and comparative cost. A review by one geographer who contributed to the development and application of indices of soil erodibility (Bryan, 1979, p. 207) includes the following aphorism:

> It has been said that if each soil conservationist stopped the movement of one grain of soil for each word he has written on the topic, the problem of soil erosion would disappear.

Physical geographers were not perhaps as prominent as they could have been in the investigation of the human impact upon soil systems, although subsequent work on processes has tended to remedy this.

Whereas the long-term significance of human activity upon plants and animals, soil characteristics and their distributions provided clear research foci, in *climatology* the emphasis was perhaps more evident at several spatial scales, with the impact first appreciated at the local, subsequently extended to the meso-, and thence to the global scale. Within this progression the atmosphere was increasingly viewed as a resource (Chandler, 1970) and research interests concentrated upon (Chandler, 1970, p. 3):

> ... those aspects of the atmospheric sciences, which ... geographers with a knowledge of physics and mathematics are especially well equipped to study, namely the particular conditions of the planetary boundary layer ... In draining the marsh, clearing the wood, cultivating the fields, and flooding the valleys, man has inadvertently changed the thermal, hydrological and roughness parameters of the Earth's surface and the chemical composition of the air ... There are indeed very few 'natural' boundary layer climates remaining ...

Human activity was included as a final chapter in some textbooks (e.g. Lockwood, 1979a) whereas in others its significance was accorded a more central position. Thus Oke (1987) emphasized the physical basis of physical, topo-, local, meso- and regional climates in the boundary-layer climates of the lowest kilometre of the atmosphere. He dealt with natural climates of non-vegetated and vegetated surfaces, topographically affected climates and the climates of animals, finally progressing to man-modified environments embracing conscious and unconscious modification of boundary-layer climates, particularly those including urban climates. Subjects to which physical geographers have contributed include changes in the radiation balance, changes in precipitation including conscious and inadvertent ones, manmade climates, and pollution. Perhaps in the area of man-made climates the physical geographer climatologist contribution was most significant. The work of Chandler (1965) on the *Climate of London* stands as an exemplary model in this field, and subsequent research on thunderstorm activity (Atkinson, 1979) clearly demonstrated the kind of contribution that can be made to document the inadvertent effects of man. The contribution by geographers to the investigation of atmospheric pollution by carbon dioxide, aerosols, fluorocarbons, thermal pollution, and other pollutants and the effects which they have, was less evident. However, Chandler, a member of the Royal Commission on Environmental Pollution established in the UK in 1970 (Chandler, 1976), remarked:

> ... atmospheric pollution has been with us since the creation; it became much worse ... with the Garden of Eden and infinitely worse following the mechanical ingenuity of James Watt. (Chandler, 1970, p. 10)

Interest in these problems increased as the reality of global change, and particularly of global warming, gradually became accepted (Chapter 9, p. 243).

In *geomorphology* and *hydrology*, assessments of human impact were focused upon four themes. Firstly, research focused upon chronology of landscape change in which it was demonstrated (Vita-Finzi, 1969; Butzer, 1974) how human activity was a pertinent factor influencing the course of alluvial chronology; in Greece, Davidson (1980a) demonstrated that erosion was underway during the Bronze Age but that spatial variations were substantial. In a commentary on the contribution by Vita-Finzi (1969), Grove (1997) concluded:

> In *The Mediterranean Valleys*, Vita-Finzi provided an invaluable launching pad for further exploration of the inter-relations of human history and environmental change in Mediterranean landscapes and, I believe, pointed us in the right general direction; the mission has yet to be successfully completed.

Subsequent research involved a resurgence of links with archaeology (Brown, 1997) and with other disciplines. The benefits of interdisciplinary cooperation are exemplified in palaeohydrology where the significance of human activity was summarized (Gregory, 1995a) and the human dimensions of palaeohydrological change (A.G. Brown, 1996) prompted three areas for applied palaeohydrological research:

- estimation of past frequency distributions and their relationship to climate;

- past and future representation of risks;
- use of analogies in palaeohydrology where the medieval climatic optimum could provide a possible analogy for global warming in NW Europe.

Secondly, there were many investigations of changes due to the impact of human activity during decades rather than millennia, for example, upon river channels (Gregory, 1981; Park, 1981). A particularly important viewpoint offered by Strahler in 1956 provided fundamental guidelines which acknowledged earlier work by engineers, hydrologists, soil scientists and geologists, and formulated an improved understanding of fluvially produced landscapes. This included a diagram of drainage basin transformation from low to high drainage density, which related severe gullying that could occur on slopes to aggradation along the main valley axis (Fig. 7.1A).

Thirdly, studies of geomorphological processes affected by human activity were exemplified by Wolman (1967a) who suggested how sediment yield varied at the present time between urban and non-urban areas, to provide a model of change of sediment yield in the northeast of the US since 1700 (Fig. 7.1B). In his review of mankind, vegetation and the sediment yields of rivers, Douglas (1969) suggested that although little account had previously been taken of human influence, his studies of rates of erosion in a wide range of climatic conditions in eastern Australia showed that present sediment yields are far in excess of those which may have prevailed in the geological past. Methods of calculation of such denudation rates were greatly refined (e.g. Walling and Webb, 1983; Walling, 1996b), and Walling (1979a, 1987, 1996a) reviewed the way in which hydrological systems are modified by humans, with results from catchment experiments documenting the effects of land clearance, land drainage, recreational pressure, strip mining, conservation measures and urbanization. In the eastern part of the US the significance of conservation measures has been demonstrated (Fig. 7.1C). The magnitude of human activity was also reviewed for coasts where sea walls, breakwaters, dredging and dumping, sediment supply and vegetation effects can all be significantly affected by human activity (Bird, 1979, 1987) and the beach sediment budget can be profoundly changed (Clayton, 1980b). Whereas effects upon coastal processes have long been considered, at least in qualitative terms, more recently appreciated was the significance of human activity in affecting permafrost as illustrated in Canada (R.J.E. Brown, 1970); in influencing endogenetic processes by loading effects from dams and reservoirs, water injections, irrigation and structures; by withdrawal effects due to groundwater, oil, gas and mineral extraction; and by surface excavation effects (Coates, 1980). Important inputs to environmental management are afforded by studies of human influence upon permafrost and upon endogenetic processes. Studies of impacts in relation to particular environmental processes have enabled, for example in the case of

(A)

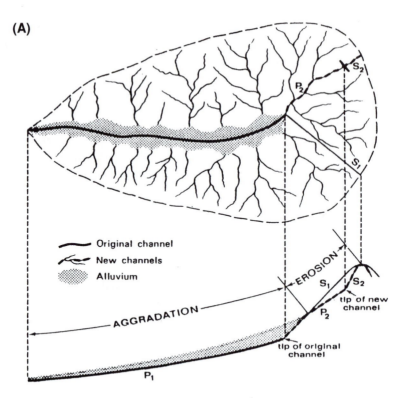

Original channel
New channels
Alluvium

EROSION

AGGRADATION

S_1
S_2
tip of new channel

P_2
tip of original channel

P_1

(B)

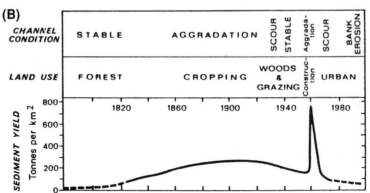

CHANNEL CONDITION	STABLE	AGGRADATION		SCOUR STABLE	Aggradation	SCOUR	BANK EROSION
LAND USE	FOREST	CROPPING		WOODS & GRAZING	Construction	URBAN	

SEDIMENT YIELD Tonnes per km²

800
600
400
200
0

1820   1860   1900   1940   1980

(C)

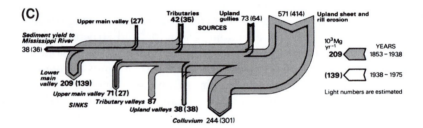

Upper main valley (27)
Tributaries 42 (35)
Upland gullies 73 (64)
571 (414)
Upland sheet and rill erosion

SOURCES

Sediment yield to Mississippi River
38 (36)

10^3 Mg yr^{-1}
209   YEARS 1853 – 1938
(139)   1938 – 1975

Light numbers are estimated

Lower main valley 209 (139)
Upper main valley 71 (27)
SINKS
Tributary valleys 87
Upland valleys 38 (38)
Colluvium 244 (301)

the impact of dams (Petts, 1984), detailed investigations of geomorphological and hydrological processes which have been innovatively extended to link with ecology, as exemplified by studies of fluvial hydrosystems (Petts and Amoros, 1996).

A fourth theme in geomorphology was to demonstrate the world impact of human activity. Although no geomorphology texts were structured explicitly to emphasize human impact, as in the case of Simmons (1979a) in biogeography and Oke (1978) in climatology, some progress towards global assessment was made. Starkel (1987) produced a world distribution of human impact and detailed research investigations have been extrapolated in attempts to refine progressively the estimates of world erosion rates (Chapter 5, p. 126) including those affected by mankind. A world map of the status of human-induced soil degradation was produced (cited in Ellis and Mellor, 1995), and results from a wide variety of areas were reviewed by Gregory and Walling (1973, pp. 342–58) and by Goudie (1995). Demek (1973) concluded that the effects of human society on the development of the Earth's relief already exceeded the effects of natural geomorphological processes and that 55% of the dryland surface of the Earth was already intensively used by man, 30% partly modified, and the remaining 15% is either unmodified or slightly modified. Hooke (1999) showed how the role of human activity was now greater than that of any other geomorphic agent in certain areas of the US where he compared rates of movement by humans and by rivers (Fig. 7. 2).

Studies were extended to regional or global scales and this is particularly exemplified by studies of what Cooke (1976) characterized as an 'empty quarter'. This quarter, the deserts of the world, demonstrates a sensitivity to human impact, although one that is not easily separated from the effects of fluctuations in climate, as studies of arroyo development clearly showed (Cooke and Reeves, 1976). Desertification was taken by Grove (1977, p. 299) to mean:

> the spread of desert conditions for whatever reason, desert being land with sparse vegetation and very low productivity associated with aridity, the degradation being persistent or in extreme cases irreversible.

Figure 7.1 Human activity and river channels. From Gregory and Walling (1987). Copyright John Wiley & Sons Limited. Reproduced with permission. (A) A change of land use can induce channel extension by gullying and downstream aggradation marked by floodplain growth (after Strahler, 1956). (B) This depicts how changes of sediment yield since 1700 in the northeast US were associated with human-induced alterations of land use, and produced changes in river channel stability (after Wolman, 1967a). (C) This represents the magnitude of sediment production from different sources in Cook Creek before and after conservation measures (after Trimble, 1983).

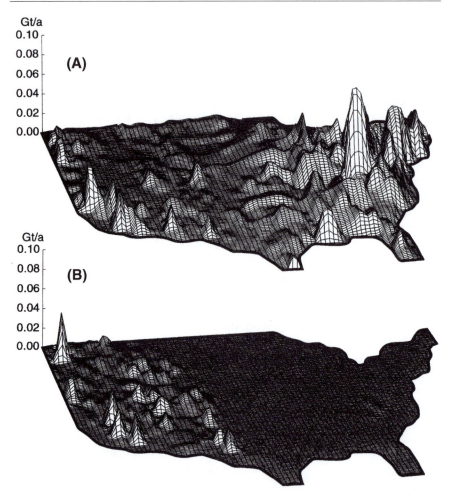

Figure 7.2 Variations in earth moved by humans (A) and rivers (B) (after Hooke, 1999). Copyright John Wiley & Sons Limited. Reproduced with permission. Variations in peak height represent movement in gigatonnes per annum in a grid cell measuring 1 degree in both latitude and longitude.

Desertification was also defined by Williams and Balling (1996) as:

> land degradation in arid, semi-arid and dry sub-humid areas resulting from various factors, including climatic variations and human activities.

Desertification arising from climatic change and from unwise use of land by human action was the subject for international collaboration, including that through the auspices of the International Geographical Union (Mabbutt,

1976), and interdisciplinary cooperation through a number of UNESCO programmes, and the MAB (Man and the Biosphere) programme, which have directed attention to the impact of human activities and land-use practices on arid and semi-arid regions. The *World Atlas of Desertification* (Middleton and Thomas, 1992, 1997), when revised in 1997, included an assessment of the status of human-induced soil degradation in south and southeast Asia (ASSOD) as a regional successor to the global assessment of human-induced soil degradation (GLASOD) used in the first edition in 1992. Such achievements are relevant to the global approach to physical geography (Chapter 9, p. 237).

7.3 Earth hazards

Desertification and some other global-scale changes are in part instances of earth hazards. As studies of the magnitude of human impact encouraged physical geographers to move towards applied problems, it was necessary to alter attitudes to the physical environment and this occurred as a result of three tendencies. The first was the tendency to focus upon extreme events because these are the ones that may occasion damage and costs; they are the ones for which landscape management strategies must be designed; and this was a departure from the earlier emphasis upon the average experience. In addition to the familiar atmospheric hazards, it was possible to include less immediately obvious ones associated with volcanoes (Clapperton, 1972; Blong, 1984) and earthquakes (Keller and Pinter, 1996). Endogenetic processes, although not studied as extensively as exogenetic ones until redressed to some extent by books such as *Volcanoes* (Ollier, 1988), gave a focus for study by physical geographers as illustrated by the eruptions of Etna (Clapperton, 1972). Study of volcanoes in space and time was advocated by Clapperton (1972) to build a bridge between volcanology and volcanic geomorphology, which would be analogous to the fields of glacial and fluvial geomorphology. It has thus been possible to identify and study the complete range of earth hazards, including the geomorphological (Slaymaker, 1994).

Secondly, the juxtaposition of investigations of the physical environment with studies of their socioeconomic relevance was exemplified by a focus on water (Chorley, 1969a) and on the *Value of the Weather*, in which Maunder (1970) introduced the range of atmospheric hazards and evaluated their cost in terms of studies of economic impact and of hazard relief. A later study of the hurricane impact (Simpson and Riehl, 1981) also epitomized the way in which a focus can be placed on the nature of the hazard and the socioeconomic cost. A third trend centred upon growing awareness of the difference between the real world and the way in which environment was perceived, because it was such perception that often influenced decision-making and therefore management. It was therefore axiomatic that knowledge of environment was time-dependent and the prevailing environmental perception at the time of decision-taking could be very significant. Early progress in the study of environmental perception arose from socioeconomic geography, such as the attitude of farmers to the drought hazard on the Great Plains (Saarinen, 1966) and the political economy of soil erosion (Blaikie, 1985), but later research extended to the physical environment.

In the context of a volcanic eruption in Papua New Guinea, a book entitled *The Time of Darkness* (Blong, 1982) provided a fascinating evaluation of myth and reality that compared 54 versions of local legends in Papua New Guinea with the scientifically reconstructed Tibito tephra eruption, which involved a thermal energy production of 10^{25} ergs, one of the greatest eruptions of the last 1000 years. Natural hazards provide potential interaction between physical and human geography. For example, a book (Blaikie *et al.*, 1994) published midway through the International Decade of Natural Disaster Reduction (IDNDR) explored two models: the pressure and release model for disaster relief and reconstruction, and the access model which focuses on access to resources and maintenance of livelihoods.

These three ingredients collectively formed the background for natural hazards research which originated in North America, particularly developed from the work of White at the University of Chicago, later at the University of Colorado. He was leader of the Commission on Man and Environment (1968–) of the International Geographical Union, his earlier research being directed to flood plains and, following the Flood Control Act of 1936, his research group posing five questions about the physical hazard, adjustments made, and policy implications. The study classified adjustments as modifying the cause, modifying the loss or distributing the loss, and established that, while flood control expenditures had multiplied, the level of flood damages had risen, and that the natural purpose of reducing the toll of flood losses by building flood control projects had not been achieved (White, 1973). Subsequent research on a spectrum of natural hazards was collected during work of the IGU Commission (White, 1974a) in which a natural hazard was defined as:

> ... an interaction of people and nature governed by the coexistent state of adjustment in the human use system and the state of nature in the natural events system. Extreme events which exceed the normal capacity of the human system to reflect, absorb or buffer them are inherent in hazard. An extreme event was taken to be any event in a geophysical system displaying relatively high variance from the mean. (White, 1974b, p. 4)

White (1973) regretted that in the 1960s, geographers turned away from some environmental problems just as specialists in neighbouring fields discovered them, but a number of books by physical geographers have subsequently summarized this and other work on earth hazards (e.g. Whittow,

1980; Perry, 1981; Alexander, 1993; Blong, 1997; Smith and Ward, 1998). Two studies were especially influential. Hewitt and Burton (1971), analysing the record for southwestern Ontario, found that in a 50-year period there would be one severe drought, two major windstorms, five severe snowstorms, eight severe hurricanes, 10 severe glaze storms, 16 severe floods, 25 severe hailstorms and 39 tornadoes. They therefore defined the hazardousness of a place as the complex of conditions that define the hazardous part of a region's environment. Hazards were taken to be *simple*, which included a single damaging element such as wind, rain, floodwater or earth tremor; *compound*, which involves several elements acting together above their respective damage thresholds, such as the wind, hail and lightning of a severe storm; and *multiple*, when elements of different kinds coincide accidentally or follow one another, as a hurricane may be succeeded by landslides and floods. A further progressive step was made by Burton *et al.* (1978, 1993) in *The Environment as Hazard*. Beginning from a consideration that suggested that the natural environment is becoming more hazardous in a number of complex ways because losses are rising, catastrophe potential is enlarging, with costs falling inequitably amongst the nations of the world, they proceeded from hazard experience to consideration of choice on an individual, collective, national and international level. They concluded (Burton *et al.*, 1978, p. 221) that an increase in disasters might continue in the decade of the 1980s and that reduction of disaster potential cannot easily be achieved; although loss of life will be reduced substantially, loss of property is most likely to occur in rapidly developing countries and, overall, the forces propelling the world toward more and greater disasters will continue to outweigh by a wide margin the forces promoting a wise choice of adjustments to hazard.

Natural hazard research necessarily focuses upon the inter-relation of geophysical events and human activity, and as such is an important feature of recent research, one that some such as Parker and Harding (1979) have proclaimed to be of central and traditional concern to geographers:

> Natural hazard studies provide a novel and imaginative vehicle for teaching aspects of physical and human geography. They draw attention to the dynamic relationship between man and environment, provide an opportunity to introduce the concept of perception and enhance general environmental awareness which is an important element of environmental education.

Hewitt (1983) viewed the outcome of research on natural hazards as the emergence of a dominant view which conforms to the idea of a paradigm embracing three main areas:

(1) An unprecedented commitment to the monitoring and scientific understanding of geophysical processes – geological, hydrological and atmospheric – as the foundation for dealing with their human significance and

impacts. Here the most immediate goal in relation to hazards is that of prediction.

(2) Planning and managerial activities to contain the geophysical processes where possible.

(3) Emergency measures, involving disaster plans and the establishment of organization for disaster relief and rehabilitation.

A series of chapters edited by Hewitt (1983) demonstrated how natural hazard is not uniquely dependent upon geophysical processes, human awareness is not dependent upon geophysical conditions, and reaction to disaster may be dependent upon ongoing social order rather than explained by conditions or a reaction to calamitous events. Hewitt (1983) concluded that most disasters are characteristic rather than accidental features of places and societies where they occur; risk arises from ordinary life rather than rareness; and natural extremes are to be expected more than many of the social developments that pervade everyday life. This provides an alternative perspective but still one in which man and hazards interact. The range of hazards studied and researched has continued to expand and has implications for landscape management. Thus Goudie and Viles (1997b) reviewed *Salt Weathering Hazards* as an environmental hazard which may become increasingly important regionally, with the impact of irrigation, and globally with the consequences of climate change.

As interest in environmental hazards increases, there is inevitably a question about the extent to which such hazards may be increasing in magnitude and in frequency, and this is a research theme that associates naturally with global change. Although increasing awareness of environmental hazards can arise because hazards are indeed becoming more frequent, it is also a fact that more people may be vulnerable and that hazards are now reported more readily and more rapidly than was the case before communications were so developed. Thus the *Annual Register of World Events*, produced since 1757, together with the contents of *The Times*, indexed since 1790, show increasing numbers of events, but the increase revealed (Fig. 7.3) owes much to the increased reporting of hazardous events (Gregory, 1992).

7.4 Urban physical geography

Urban physical geography has not been identified by separate chapters in many texts devoted to human impact (e.g. Goudie, 1981b, 1993b; Gregory and Walling, 1987; Goudie and Viles, 1997b) although some writers have claimed that the urban environment is sufficiently distinctive to warrant attention as a specific milieu. A directive towards urban physical geography was provided by several volumes (Detwyler and Marcus, 1972; Coates, 1973, 1976a), reinforced by research on urban climates (e.g. Chandler, 1965) which stimulated physical geographers to document the magnitude

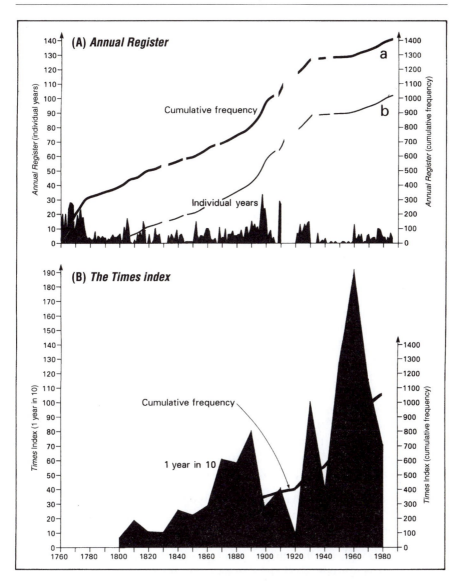

Figure 7.3 World events reported in the *Annual Register* and *The Times* (after Gregory and Rowlands, 1990, with permission from Philip Allan, publishers of *Geography Review*).

and character of urban heat islands, of precipitation modification, of atmospheric pollution and of air movement. Because temperature is continuous in time and space, whereas precipitation is continuous in neither, urban climatologists have had much more success in specifying and explaining urban effects on temperature than on precipitation amount (Lowry, 1998). Knowledge of atmospheric processes in urban areas, such as the Metropolitan Meteorological Experiment (METROMEX), with a field programme up to 1975 based in St Louis (Chagnon *et al.*, 1977), was significantly increased and this is probably the nearest that urban climatologists have come to conducting true field experimentation (Lowry, 1998). Understanding the way in which a city generates its own climate led towards applications so that Chandler (1976) suggested that our cities must be purposefully planned in order to optimize the environment of urban areas and to avoid a series of structural and functional design failures. From a symposium on physical problems of the urban environment, Chandler *et al.* (1976, p. 57) concluded that as the conversion of land to urban uses involves considerable modification of the natural environmental system:

> Physical geography can usefully contribute to the determination of public policies, with respect to the management and development of urbanized areas.

Some of the most specific contributions by physical geographers have been made in hydrology, including: analysis of increased discharges (Walling and Gregory, 1970; Hollis, 1975; Walling, 1979a); investigation of sediment yield (Walling, 1974); subsequently the field of water quality and pollutants related to urban source areas (Ellis, 1979); analysis of river channel changes downstream from urban areas (Leopold, 1973; Gregory, 1981; Gregory and Whitlow, 1989; C.R. Roberts, 1989); with results of scientific investigations on *Man's Impact on the Hydrological Cycle in the United Kingdom* collected by Hollis (1979). In the fields of urban climatology and urban hydrology there is an important research contribution to be made by the physical geographer, who may be in collaboration with scientists in other disciplines. One geographer (Douglas, 1981, p. 315) contended that:

> Understanding of the dynamics of the biophysical components of the city and the way their functioning impinges on people is a vital part of urban studies.

He argued (Douglas, 1981) that an attempt must be made to link the city as a habitat or ecosystem with the city as a social system, and developed an ecosystem approach to the study of cities. This approach, used earlier in relation to disciplines such as architecture (Knowles, 1974), embraces population ecology, system ecology, the city as a habitat, followed by energy and material transfer in cities being used to highlight spatial contrasts within the city and between cities and also to differentiate cities from rural environments. Douglas quotes support for the integrated study of urban areas such as that by Bunge (1973):

> Physical geography is much needed in an urban setting . . . Cities are a karst topography with sewers performing precisely the function of

limestone caves in Yugoslavia, which causes a parched physical environment, especially in city centres.

Perhaps more than in other environments, the analysis of city problems reveals (Douglas, 1981, p. 360) the inadequacy of attempts at understanding through 'blinkered, disciplinary approaches', and so Douglas (1983) elaborated an ecosystem view of the city. He used the background of the economic system as an ecosystem and the city as a dependent system to proceed to the energy balance, water balance, mass balance, geomorphology, biogeography and waste disposal of the city, prior to looking at geographical aspects of urban health and disease, and at management and planning designed to reduce environmental hazards. He emphasized the view that:

> . . . in academic scholarship the divorce between the 'two cultures', the humanities and the sciences, sadly persists, and only at the practitioner level is there some collaboration between the social worker and the public health engineer . . . In examining cities one cannot be simply scientific or simply sociological. (Douglas, 1983, p. 202)

The validity of the physical geography of the city is confirmed by studies undertaken of specific cities. Los Angeles has been a very productive location for a number of case studies (Cooke, 1977) and was used by Whittow (1980) as an example of an environment which experiences an accumulation of hazards. This and other examples are included in *Urban Geomorphology in Drylands* (Cooke *et al.*, 1982), which used a very particular way of developing urban physical geography in an applied direction (see p. 220). In the UK in the late 1990s the Research Councils allocated funding for multidisciplinary research programmes on cities, demonstrating the topical significance.

7.5 Implications for physical geography

When surveying geographical geomorphology in the 1980s, Graf *et al.* (1980) observed that the human factor in geomorphology received insufficient attention until the late 1960s but that researchers were then recognizing the effects of human activity in many processes, ranging from relatively minor disturbances to almost complete control. They concluded that, especially in relation to fluvial processes, explanations that ignore the role of human activities run the risk of eliminating one of the most significant variables, and further that:

> We believe that no other group engaged in geomorphic research is as well qualified to grapple with the human factor as are geographers. Our broad training in both physical and cultural systems and our appreciation of landscape change in the natural and human senses give us perspectives and insight that are rarely found in other disciplines.

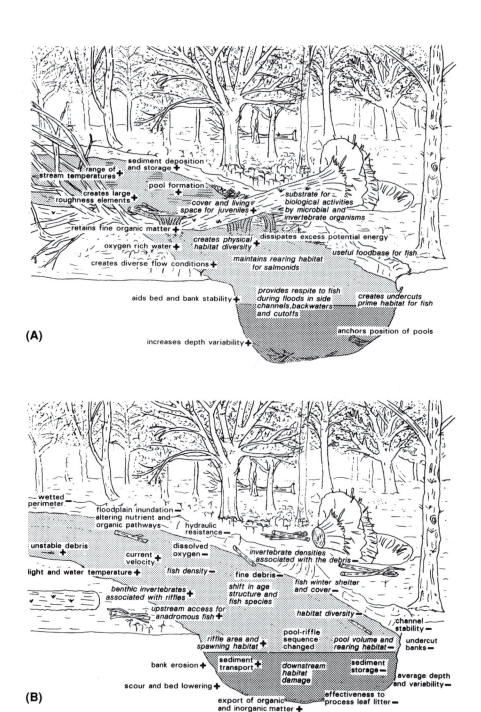

(A)

range of
stream temperatures +

sediment deposition
and storage +

pool formation +

creates large +
roughness elements +

cover and living +
space for juveniles +

substrate for
biological activities
by microbial and
invertebrate organisms

retains fine organic matter +

oxygen rich water +

creates physical +
habitat diversity +

dissipates excess potential energy

useful foodbase for fish

creates diverse flow conditions +

maintains rearing habitat
for salmonids

aids bed and bank stability +

provides respite to fish
during floods in side
channels, backwaters
and cutoffs

creates undercuts
prime habitat for fish

anchors position of pools

increases depth variability +

(B)

wetted
perimeter

floodplain inundation
altering nutrient and
organic pathways

hydraulic
resistance

invertebrate densities
associated with the debris

unstable debris
+

dissolved
oxygen

current
velocity +

light and water temperature +

fish density

fine debris

fish winter shelter
and cover

benthic invertebrates
associated with riffles +

shift in age
structure and
fish species

upstream access for
anadromous fish

habitat diversity

channel
stability

pool-riffle
sequence
changed

pool volume and
rearing habitat

undercut
banks

riffle area and
spawning habitat +

bank erosion +

sediment
transport +

downstream
habitat
damage

sediment
storage

average depth
and variability

scour and bed lowering +

export of organic
and inorganic matter +

effectiveness to
process leaf litter

> Geographers need to work closely with engineers and geologists in order to share with them such wider ranging concepts as spatial analysis and emphasis on the man–land interface. These concepts are endemic to geography, but they may be quite foreign to other workers. (Graf *et al.*, 1980, p. 281)

Since 1980, research in all branches of physical geography has begun to respond to the opportunity outlined by Graf and coworkers; control systems, as an integral part of the environmental systems approach, have been assimilated into research and teaching in physical geography. It is now generally agreed, as Goudie and Viles (1997a) conclude, that the human transformation of nature has been going on for a very long time and has been very pervasive. However, the complexity of the human impact is increasingly appreciated and, for example, Goudie (1986a, Table 8.2, p. 295) contrasts the natural and anthropogenic causes for 12 aspects of the physical environment, including ground collapse which can be the result of natural karstic causes or can arise from anthropogenic reasons such as dewatering by over-pumping. Attention to human impact has not been confined to physical geography, of course, and Gregory and Walling (1987, p. 446) estimated that more than 50 interdisciplinary international programmes existed, focused upon some aspect of human activity. Research results from many of these programmes have involved physical geographers and have provided useful information for further research.

Research investigations by physical geographers have provided more detailed insight into the variety of human impact. For example, Trimble and Mendel (1995) suggested that the role of cattle and other grazing animals, although long appreciated in general, deserves much more detailed investigation by geomorphologists, and that the cow is an important agent of geomorphological change, through heavy grazing on the uplands and grazing and trampling in riparian zones. They concluded that future studies should include both empirical and deterministic modelling to provide greater insights into the effects of grazing. Research investigations have often encouraged multidisciplinary collaboration, as in the case of research on woody debris in river channels and the extent of human impact reflected in its removal and management. This has been the focus of research interest by geomorphologists and by ecologists, initially using rather distinctive approaches but then each benefiting from multidisciplinary research. This illustrates how many studies have progressed in physical geography.

Figure 7.4 Woody debris in river channels (after Gregory and Davis, 1992). Copyright John Wiley & Sons Limited. Reproduced with permission. (A) shows the system with coarse woody debris present, compared with the situation in (B) where the debris has been removed (see p. 196).

Initially, woody debris in river channels was investigated as an integral part of the stream channel process system, but such process studies led to investigations of the significance of human impact because of the extent to which clearing of debris had taken place (Fig. 7.4), often depending upon the perception of land-owners and of forest and river managers. Studies of coarse woody debris progressed towards more applied investigations including a basis for alternative management strategies (Gregory and Davis, 1992); knowledge of the relationship between the floodplain forest and the river channel processes as an essential foundation for the effective management and restoration of catchments and rivers (Gurnell and Gregory, 1995); and a sequence of linked management options (Gurnell *et al.*, 1995) which can benefit commercial forestry. Specific investigations of human impact have led not only to comprehensive identification of details of environmental process–human interaction and applications of the results, but also new monitoring techniques have been devised to give baseline data against which to gauge future changes. In addition a more holistic view has been adopted which not only tends towards the global but is also more multidisciplinary. Perhaps there is also greater awareness of the sensitivity of physical environments and of environmental processes, which is why sensitivity (Downs and Gregory, 1995) can be extended to human impact in general (Gregory, 1995a).

Studies of human impact in physical geography led in at least three directions. Firstly a move towards *applications* of physical geography has occurred, as the nature and magnitude of human impact gives some basis for environmental impact assessment and is germane to notions of sustainability. An investigation of environmental change in the Kalahari demonstrates the benefit of integrated studies compared with solely ecological analysis (Doughill *et al.*, 1999). Secondly, aspects of *global change* have been specified as the global dimensions of human action are more completely understood, with a greater appreciation of 'the blue planet' partly aided by propositions such as the Gaia hypothesis. Thirdly, studies have led towards a more *cultural physical geography* because, as natural regions or biomes no longer exist, wilderness (Simmons, 1993, p. 159) is increasingly restricted, and yet the difference between nature and individual and public preference for particular environments may become increasingly divergent. These three topics provide subjects for the next three chapters.

We should not go overboard, however; *either* to assume that human activity is all about global change scenarios, because many detailed research investigations continue to be necessary to improve understanding of environmental processes and change; *or* to assume that the prospects are all necessarily gloomy. Positive options may arise from detailed research. Michael Frayn's (1968) novel *A Very Private Life* begins with the statement 'Once upon a time there will be a girl called Uncumber'. Uncumber lives in a dome, a closed system, separated from the external polluted world. Uncumber's one ambition is to get out of the dome. We should know enough about the range and nature of human impacts to ensure that the physical environment does not deteriorate so far as to necessitate the option of living in domes, separate from the 'natural' environment.

Further reading

To appreciate the greatest antecedent, which was a pioneer of the ecological approach to environmental problems and an important foundation for the conservation movement:

> MARSH, G.P 1864: *Man and nature or physical geography as modified by human action*. New edition edited by D. Lowenthal, Belknap Press, 1965.

A general, and very readable introduction to human impact is provided by:

> GOUDIE, A.S. and VILES, H. 1997: *The Earth transformed: an introduction to human impacts on the environment*. Oxford: Blackwell.

An excellent example of natural hazards in Australia is given in:

> BLONG, R. 1997: A geography of natural perils. *Australian Geographer* **28**, 7–27.

Topics for consideration

(1) What are the difficulties in separating the effects of human action from natural environmental processes?
(2) Are physical geographers particularly well-trained to investigate, and to demonstrate, the effects of human activity upon the physical environment and environmental processes?
(3) What types of further analysis of the nature, magnitude and consequences of particular types of human impacts are required?

8

Applications of physical geography and environmental management

Whereas physical geographers up to 1980 exhibited considerable reticence about involvement in applications of their research, much greater awareness developed in the last two decades of the twentieth century. This was partly because of the increasing knowledge of potential global change (Chapter 9), and thence of the need to consider how the environment may develop in the future. However, Sherman and Bauer (1993, p. 225) quote the physicist D. Kleppner (1991, p. 10) who cautioned: 'If you should suddenly feel the need for a lesson in humility, try forecasting the future'. An increasing number of possible applications of physical geography have arisen from the research on the themes reviewed in Chapters 4 to 7, and the adjective 'applied' has been placed in front of geomorphology, climatology and pedology, to create a number of book titles; since 1981 the journal *Applied Geography* has published papers on applications of geographical research. Applications of physical geography, and the involvement of physical geographers in environmental management, have come about in two main ways: firstly through research designed to extend investigations characteristically undertaken by physical geographers towards applications; and secondly by direct involvement of physical geographers in management of specific environmental issues and in problem-solving, sometimes in a consultancy role. The development of applied physical geography is considered (8.1), prior to the types of application at present evident in the branches of physical geography (8.2), the role of physical geographers in a consultancy context (8.3), and the potential for further applications (8.4).

8.1 Development of applied physical geography

The number of research and review papers and books directed towards applications of physical geography increased towards the end of the twentieth century, and this was reflected in the content of *Progress in Physical Geography*, where not only were papers increasingly concerned with 'relevant' subjects, but more dealt with specific problems such as: water erosion on British farmers' fields (Evans, 1990); impact of weather and climate on transport in the UK (Thornes, 1992); biotechnology and environmental quality (Mannion, 1995); modelling the spread of wildfire (G.L.W. Perry, 1998); and predicting regional climate change (Mitchell and Hume, 1999).

Increasing awareness of potential applications of physical geography research came about as a result of at least three trends. Firstly, in the traditional subdivisions of physical geography it became appreciated that there is a range of ways in which research may be of use in relation to current environmental problems and needs, and Gregory (1998, Table 2.1, p. 15) suggested that, in addition to the conventional division into pure, applicable and applied research, it is also possible to envisage planning, management and sustainability research. These types were recognized for applications of palaeohydrology but are adapted for physical geography with examples in Table 8.1. The categories in Table 8.1 are not immutable and may change, as Lier (1989, p. 76) observed:

> Logic demands that 'applied geography' be used for geographic knowledge that has been applied to something ... but practice in geography and in other fields has sanctioned 'applied' for uses where 'applicable' would have been grammatically correct.

Table 8.1 Types of research and application

Pure research ('blue skies') Basic research not specifically related to environment problems and not profitable in the current state of knowledge or technological development

Applicable research ('grey skies') Investigations that give results or new facts which may be applicable to environmental problems

Applied research Research where results are related to environmental problems in a specific area

Planning research Elaboration of a strategy for action to be carried out over a specific period of time, involving greater coherence and purpose

Management research Application of appropriate skills and principles in decision-making with critical control exerted over people, activities and resources

Sustainability research Good practice in human exploitation of earth resources whereby resources are capable of being maintained at a certain rate or level

The examples in Table 8.1 include the way in which studies of environmental hazards provide information pertinent to decision-making, illustrated by an atlas portraying the characteristics of drought in Britain in 1975–76 and its physical and economic consequences (Doornkamp *et al.*, 1980). Specific research investigations have been extended, to contribute to applied problems, fulfilling the conclusion of Sherman (1989, p. 127): 'If our science is good, then applications will follow as part and parcel of normal science'.

Secondly, there has been a more direct and conscious intention to direct research and teaching towards applications of the kind shown in Table 8.1. This has been shown in a final chapter devoted to applications in many books; in the increasing popularity of the applied theme as a major ingredient in inaugural lectures (Chandler, 1970; Douglas, 1972; Stephens, 1980; Embleton, 1982); in Presidential Addresses (e.g. Henderson-Sellers, 1989b; Cooke, 1992; Thornes, 1995; Sugden, 1996); and by the use of the applied or resource management theme for series of books. The advantages of applied research have been advocated by many writers. Thus Gerasimov (1968) advocated constructive geography as providing 'The theoretical basis and practical recommendations for man's transformation of the environment for the benefit of society', and he subsequently traced (Gerasimov, 1984) the way in which constructive geography had been used for a variety of studies of the planned transformation of the natural environment, to enable effective use of natural resources. Fifteen issues of a special series *Problems of Constructive Geography* were published from 1975 to 1980 by the Institute of Geography of the USSR Academy of Sciences. During the 1970s, which he characterized as the 'Environmental Decade', Coates (1971) edited a volume on *Environmental Geomorphology*, which was defined as:

> ... the practical use of geomorphology for the solution of problems where man wishes to transform landforms or to use and change surficial processes ... In addition environmental geomorphology includes extraction of surficial material and protection of certain landscapes, such as beaches, which benefit man. The goal for geomorphic environmental studies is to minimize topographic distortions and to understand the inter-related processes necessary in restoration, or maintenance, of the natural balance.

New journals such as *Environmental Management* (1976–), together with more specific ones such as *Regulated Rivers*, provided for the publication of papers from a range of disciplines including physical geography. *Applied Geography* was established in 1981 as a journal

> ... devoted to the publication of research which uses geographical theory and methodology to resolve those human problems that have a geographical dimension. These problems may be related to the assess-

ment, management and allocation of the world's physical and/or human resources.

This mission was reviewed in 1992 when Hansom (1992) commented that although the emphasis in the papers published in the first 12 years had been on geography applied, there is also a need for applied geography that is user- and action-orientated because this

> ... may extend theoretical geography by taking scientific method into the evaluation and implementation phases ... there is a need for a vehicle to disseminate such research and to bring it to the attention of a wider audience ...

The journal continues to state that only through a clear understanding of the relevant physical, behavioural and information systems can we resolve such problems. The comment stimulated Kenzer, who had edited a book on *Applied Geography: Issues, Questions and Concerns* (Kenzer, 1989) to argue that although recent issues of the journal had uncovered a wealth of difference between 'pure' and 'applied' geography, little attention had been paid to what 'applied' actually means. Kenzer (1992) did not see applied geography as a separate field, but Sant (1992) responded and concluded that the enthusiasm of students is

> ... not for geography as an esoteric discipline but for geography asso-ciated with a wide range of environmental and social enquiry dealing with social issues.

This therefore indicates the more explicit or purposeful applied geography that has developed from existing investigations; although it is not easy to categorize the papers appearing in *Applied Geography* between 1989 and 1998, of 210 papers published, 44% dealt with a subject within or of direct relevance to physical geography. The subjects in this decade included a sig-nificant number that extended across the traditional boundaries of physical geography, but the dominant themes of the physical geography papers were water, including hydrology and water management (27%); soil erosion and aspects of soil related to development (19%); biogeography and aspects of land classification (29%); atmospheric topics (7%); and some 7% primarily directed to techniques including GIS, databases and remote sensing. Such investigations, largely emerging from existing branches of physical geogra-phy, often included relationship to global change.

Palm and Brazel (1992) considered applied research to be that which uses geographical theory or techniques to understand and solve specific empiri-cal problems. They saw most applied physical geography as treating human society either as the target of a physical system, as in research on the climatic effects on human comfort, or as a phenomenon that perturbs a physical process in the way that urban heat islands are created (Palm and Brazel, 1992, p. 353). Perhaps it is time to reassert the importance of the physical environment, as suggested by Agnew and Spencer (1999, p. 8), who proposed topics worthy of consideration to include:

- thresholds and constraints of environmental systems, their sensitivity and resilience;
- environmental uncertainty and decision-making: how comfortable are physical geographers with the managerialist agenda?
- environmental degradation. Sustainability places a renewed interest in areas previously tackled by physical geographers such as biogeochemical fluxes, pollution impacts, flows of energy and mass between the ocean, land and atmosphere.

Agenda 21 points to the role of neglected and under-represented social groups, the importance of education and capacity-building, areas largely ignored by physical geographers.

Thirdly, there is direct involvement of physical geographers in applications of physical geography. Such direct participation could be in practice, whereby geographical training and expertise is used in employment, for example with environmental organizations, in the water industry, or in nature conservation; in membership of international and often multidisciplinary groups, for example concerned with desertification (e.g. Professor D. Thomas, see p. 187); on advisory bodies as exemplified by Professor W.L. Graf in the US (see p. 223); on national committees as exemplified by Professor Edmund Penning Rowsell as Chair of the MAFF Research Advisory Committee on River and Coastal Management; or in consultancy, perhaps in conjunction with engineers, as necessary in the case of site engineering projects.

These three kinds of applications of physical geography research exemplify how physical geographers, not always as physical geographers, have contributed to decision-making, although the contributions may have been advisory rather than mandatory. During the First World War the description of terrain characteristics of the Flanders battlefield was an early signpost towards applied physical geography for military purposes (Johnson, 1921). Research to characterize the physical environment was perhaps the first major development of applied physical geography and one close to the heart of the geographical approach. There are many antecedents for the description of the physical environment in a way which is pertinent to its utilization and management, including, for example, the way in which Berg (1950) recognized major physiographical zones of the USSR in a historico-genetic approach (Isachenko, 1977), useful in relation to land utilization. However, many methods of environmental description have been developed outside geography and subsequently been adapted, utilized and embraced within physical geography. Vink (1968, 1983) used the term 'landscape ecology', although, as he pointed out, this was first used by Troll, to connote the interaction between geography (landscape) and biology (ecology) with applications to land development, regional planning and urban planning. Although landscape ecology may be envisaged as an approach that interprets land-scape as supporting inter-related natural and cultural systems (Vink, 1983), it is possible to envisage a major task of landscape ecology as being to describe and characterize landscape according to relationships between the biosphere and the anthroposphere. Landscape ecology was defined by Kupfer (1995, p. 19) as:

> ... the study of how spatial scale and heterogeneity affect ecological processes ... Landscape ecological principles are drawn from a diverse array of disciplines and fields, including physical and human geography, biology, geology, forestry, wildlife management, landscape architecture and planning ... The central focus of landscape ecology is the inter-relationship between landscape structure – the spatial patterning of ecosystems across space – and landscape functioning – the interactions or flows of energy, matter and species within and among component ecosystems.

In the light of application of landscape ecological principles to nature-reserve design and functioning, Kupfer (1995) stressed the way in which landscape ecological theories have integrated existing principles from applied biogeography and population biology and also provided unique insights and a fresh opportunity for biogeographers to be on the cutting edge of theory development. In their study of landscape and plant community boundaries in biogeography, Kent *et al.* (1997, p. 21) advocated continued research into transitional areas using remote sensing, GIS, network analysis and fractal geometry to provide a new focus for spatial analysis in biogeographical research. They also argued that: 'Landscape ecology is now recognized as a distinctive discipline that is of clear relevance to biogeography and spatial ecology'. Although landscape ecology is of interest to a range of disciplines, and 1987 saw the creation of a new journal *Landscape Ecology*, depiction of the spatial patterns pertinent to landscape ecology can be envisaged from the physical geography viewpoint at three related levels: systematic, quantitative and integrated.

At the *systematic* level an inventory can be made, followed by representation of the spatial patterns created by some aspects of geomorphology, climatology, soil geography, hydrology or biogeography. In each branch of physical geography, earlier emphasis upon a static, often morphological, approach was succeeded by a focus upon processes; and surveys by national mapping agencies produced topographical maps, soil maps, geological maps and maps of superficial deposits and also provided climatic and hydrological data. However, it was often necessary to develop the available published information in a way appropriate for environmental use and management. Thus the topographic map conventionally employs contours, but slope maps are more fundamentally significant in relation to, for example, land-use practices. Slope categories can be directly related to angles at which agricultural implements can operate (e.g. Curtis *et al.*,

1965), to the incidence of mass movements (e.g. Cooke, 1977), or to suitability for different types of building construction or for other specified activities (e.g. Cooke *et al.*, 1982, Table III.3). New methods of depicting systematic aspects of the environment emerged, including ways of portraying spatial variations of environmental parameters relating to water on hydrological maps (Gregory and Walling, 1973; UNESCO, 1977). In order to analyse land-surface form, morphological maps were initially proposed (e.g. Savigear, 1965; Gregory and Brown, 1966) and these were succeeded by geomorphological maps. Bakker (1963) proposed that such maps should ideally satisfy five major requirements, which included the principles of morphological characterization, of geomorphological–genetic interpretation, of dating, of characterizing the substrate, and of sedimentology or sedimentology/pedology. It was not feasible to meet fully all five requirements, and the extent to which several of the five criteria of morphology, evolution, dating, lithology and sedimentology are emphasized varies from one mapping approach to another. Geomorphological maps were produced and published in several countries (Demek, 1972; Demek and Embleton, 1978; Embleton, 1988) and in one of the most ambitious schemes in Poland many physical geographers contributed to a national scheme which involved publishing maps at a scale of 1:50,000. Mapping schemes capable of international application were evolved and an approach summarized by Cooke and Doornkamp (1974, 1990) was adapted to conditions in several areas (Cooke *et al.*, 1982). In many specific applied projects, geomorphological mapping was the central technique. Although the approach associated with geomorphological mapping must undoubtedly feature in applications of geomorphology, the map itself has been superseded by the database and GIS, with the growth of remote sensing greatly enhancing the potential to be realized (Townshend and Hancock, 1981; Kent *et al.*, 1993).

Quantitative approaches can now be based on the direct analysis of remotely sensed digital data, but quantitative techniques have been available for much longer. Many attributes of the environment have been expressed quantitatively, as in the case of many drainage basin characteristics (e.g. Gregory and Walling, 1973). At least two other developments have been quantitative in character, including the parametric approach and automated cartography, both later extended with the advent of geographical information systems (GIS). The parametric approach is one in which measurements of environmental parameters are used for the division and classification of land on the basis of selected attribute value, as reviewed by Ollier (1977), and under the heading of general geomorphometry by Evans (1981). Although the actual parameters used vary from one study to another, King (1970) used 12 measurements to furnish four parameters (slope angle, rate of change of slope, contour curvature, and unit catchment area) that could define land elements as a basis for producing a land element map. Land resource information systems as developed by Davidson and Jones (1986) using soil survey records can be the basis for single-factor maps and for land resource assessment. A further quantitative approach is founded in automated cartography, whereas a greater variety of patterns could be derived from an initial survey as exemplified by the case of soil survey data (Rudeforth, 1982).The advent of GIS and the greater availability of more complex models has enabled further developments.

Many of these approaches embrace some degree of integration of several aspects of the physical landscape, but perhaps the most *integrative* approach is the land systems method. This approach has been extensively reported in textbooks (e.g. Mitchell, 1973, 1991; Cooke and Doornkamp, 1974, 1990; Hails, 1977; Dent and Young, 1981; Verstappen, 1983) and extended to *Land Resources* by Young (1998). The Australian Commonwealth Scientific Industrial Research Organization (CSIRO) began extensive resource surveys in 1946 in undeveloped parts of Australia and Papua New Guinea. In subsequent years these surveys acquired information on the geology, climate, geomorphology, soils, vegetation and land use of areas and, by collating the information, designated land systems as areas or groups of areas with recurring patterns of topography, soils and vegetation with a relatively uniform climate. A land system could be divided into small units or land facets which in turn are composed of individual slopes or land elements. The land system approach has been modified for application to the problems of urban and suburban areas (e.g. Grant *et al.*, 1979) and a range of related approaches was described by Vink (1983). An information system has subsequently been developed and can be used to describe extensive regions (Cocks and Walker, 1987). Structured methods of land classification provide a database to give a practical summary of the distribution and management of resources in specific localities (Cooper and Murray, 1992), and a numerical land classification based on 1 km grid squares (Cherrill and Lane, 1995). These quantitative approaches are similar in objective to the land systems approach: an approach which has the advantage of combining many aspects of environmental character and is not use-specific. It was criticized because it does not readily relate to a particular use; because it is not easily related to results obtained from field surveys (Wright, 1972); and because it reflects an emphasis upon a static view, according to evolution of environment rather than upon the dynamics of the environmental system. Moss advocated an alternative approach based upon how the environment works and what it does, rather than merely what it is (Moss, 1968, 1969a, b).

Land evaluation is the estimation of the potential of land for particular kinds of use, and Dent and Young (1981) included productive uses such as arable farming, livestock production and forestry together with uses that provide services or other benefits such as water catchment areas, recreation, tourism and wildlife conservation. Therefore the essence of land evaluation is the comparison of the requirements of land use with the resource potential offered by the environment. Although such evaluation has the advantage that it relates to a particular form of environment use, it has the disadvantage that, because the approach is use-specific, it cannot be utilized for other purposes and so is very labour intensive. Progress towards land evaluation could be realized from more general land system approaches; Isachenko (1973b) showed how three scales of landscape research for planning purposes could be developed to evaluation maps which classify terrain for a particular purpose, prediction maps indicating the modifications likely to arise, and recommendation maps showing the measures which could be used to change the environment, illustrated by evaluation of land with reference to tourism (Isachenko, 1973a).

Extensions of integrated surveys are an admirable way of providing a basis for land evaluation, and it may be possible to extend systematic data sources. National surveys of rock types, superficial deposits and soils have provided the bases for landscape evaluation applications, and in the case of soil survey it is possible to proceed from maps of soil bodies to soil quality maps and soil limitation maps, and subsequently to land classification (Davidson, 1980b) in terms of soil crop response, present use, use capabilities, and recommended use (Vink, 1968, 1983). Soil survey contributed a major component in the derivation of systems of land capability, well illustrated for agricultural uses by Bridges and Davidson (1982b). Assessment of land suitability can be extended to demonstrate yield potential (Ngowi and Stocking, 1989), illustrated by the way in which an ecological land classification approach was used to obtain a first approximation of the distribution of capability to support wild rice in northern Saskatchewan (Weichel and Archibold, 1989). From a study in central Malawi (Young and Goldsmith, 1977), soil survey and aerial photographs were used to identify seven soil landscapes which were each evaluated in terms of six major kinds of land use, so that the requirements of land use could be compared with the land qualities of the mapping units to give the basis for a quantitative economic analysis. This gave an exemplary instance of the way in which land evaluation can lead towards an economic assessment, but Young and Goldsmith (1977, p. 430) cautioned that the trend of translating land evaluation into economic terms was short-lived since it depended upon changes in costs and prices and on somewhat arbitrary assumptions about discount rates and shadow pricing. Landscape evaluation can be developed from land systems or landscape ecology surveys, and in the USSR this has fostered the development of landscape geochemistry, particularly in relation to the extension of soil geography to accommodate the impact of technology (Glazovskaya, 1977). This has been particularly appropriate in the case of forecasts of environmental impact. The vulnerability of the natural geosystems as a result of the impact of the proposed Kansk-Achinsk lignite and electric power project was analysed (Snytko *et al.*, 1981). Other approaches to landscape evaluation, greatly facilitated by application of GIS, have developed within specific branches of physical geography (Section 8.2) or through consultancy applications (Section 8.3).

8.2 Applications in branches of physical geography

Progressive development of applications of physical geography allows recognition of four main types of applications in the branches of geomorphology, climatology, soil geography, biogeography and hydrology, namely:

(1) description, depiction and auditing of the environment in a relevant way;
(2) investigation and analysis of environmental impacts;
(3) evaluation of the environment intended to show how certain characteristics are appropriate for a particular form of utilization;
(4) prediction and design, which are primarily concerned with future uses and are policy-related.

These four stages are set out in Table 8.2 with examples of applications achieved in branches of physical geography. To complement the applied books (e.g. Oliver, 1973; Smith, 1975; Hails, 1977; Hobbs, 1980; Craig and Craft, 1982; Verstappen, 1983; Cooke and Doornkamp, 1990), a significant number of books were produced in the 1990s, either as surveys of the state of the art, or as edited collections which were often the outcome of conferences, but all dedicated to the progress and potential of applications of physical geography. Examples of the 'state of the art reviews' include those directed to the coast in *Coastal Problems: Geomorphology, Ecology and Society at the Coast* (Viles and Spencer, 1996), *Beach Management* (Bird, 1996), and *Atmospheric Pollution – A Global Problem* (Elsom, 1987). Edited volumes have included *Applied Climatology: Principles and Practice* (Thompson and Perry, 1997), *Geomorphology in Environmental Planning* (Hooke, 1988), *Coastal Defence and Earth Science Conservation* (Hooke, 1998) and *Geomorphology and Land Management in a Changing Environment* (McGregor and Thompson, 1995). These are in addition to the many books with a final section on applications of research. For example, the final quarter of *Volcanoes and Society* (Chester, 1993) deals with the societal implications of volcanism, and *Karst Geomorphology and Hydrology* (Ford and Williams, 1989) has a final chapter on karst resources, their exploitation and management. Such volumes illustrate concern with applications throughout physical geography and its branches, and there are also applications associated with global change (Chapter 9).

Description, depiction and auditing of the environment has developed from progress made in landscape ecology (p. 203) and is essential for many specific applications. In fact, it is the essential objective of GIS, which may be employed as a vehicle to underpin analysis of applied problems. Whereas mean annual temperature and mean annual precipitation were the data series used frequently by physical geographers in the past, the number of frost-free days and the length of the growing season are examples of climatic indices which are of greater relevance to agriculture and land use. Many examples of the ways in which climate can be described in a relevant way

Table 8.2 Types of applications in branches of physical geography

Type of application	Geomorphology/ hydrology	Meteorology/ climatology	Soil geography	Biogeography
Description, depiction and auditing of the physical environment in a relevant way	Erosion potential of coast; scenic character of landscape; slope erosion classification	Growing season defined according to specified criteria; spatial pattern of urban climate and pollution	Classification of land according to potential for soil erosion	Ecological land classification; landscape ecology
Investigation and analysis of environmental impact	Extent of coastal floods; effects of dams on flows and channels downstream	Drought impact in particular areas; incidence and impact of acid rain	Effects of crop practices on soils; accumulation of metal contaminants in soils	Impact on wetlands; impact of pressures on national parks, nature reserves
Evaluation of environment to show how certain characteristics are appropriate for a particular form of action	Characterization of avalanche slopes in relation to transport and to settlement location	Weather conditions in relation to transport, e.g. roads and airports; personal comfort indices	Evaluation of soil capability for agricultural and other land uses	Land use – vegetation systems in relation to environmental management
Prediction and design concerned with future and policy-related uses	Specific highway location and design according to slope stability	Recommendations about road treatments for weather hazards	Vegetation and textiles to control soil erosion; retention of soil pollutants	Habitat management for plant and animal conservation; buffer strips along rivers

were given by Barry and Perry (1973), Hobbs (1980), Bryant (1997) and Thompson and Perry (1997). Environmental geology offers a way of emphasizing many aspects of the environment in relation to management problems and has been the subject for a number of books and approaches in the US (e.g. Keller, 1996; see p. 212), and it can be useful to enumerate all the pertinent variables in tabular form as in the context of urban areas (e.g. Chandler *et al.*, 1976). At this descriptive or audit stage it is possible to think of the requirements necessary to identify the variables involved in a particular environmental situation, to propose ways of describing the physical environment or the processes operating in a way relevant to the problem in hand, and to indicate how data may be obtained from existing records or by field or remotely sensed survey, and how they can be analysed. For this purpose the approach of landscape ecology and the methods of GIS are appropriate. Natural hazards, which are naturally occurring geophysical conditions that threaten life or property, have to be described, are listed by Alexander (1999b, p. 423) and shown in Table 8.3 (see also Chapter 5, p. 133).

Environmental impacts have been investigated in some detail, with emphasis placed upon the magnitude of environmental impact so that future estimates can be made. Natural phenomena are hazardous only in relation to patterns of human settlement, land use, and socioeconomic organization which collectively represent vulnerability (Alexander, 1999b). When a natural hazard impacts, the result may be a natural disaster (Alexander, 1993), distinguished from other forms of disaster such as disease, epidemics and wars. Perspectives on environmental geomorphology, emphasizing how new techniques could extend applications, were given by Coates (1990) who stressed that geomorphologists must become better sales personnel for their discipline, with involvement in policies and decision-making. In relation to coastal defences and earth science conservation, Hooke (1998) discussed the importance of education in increasing awareness about benefits of sustainable management plans, and the scientific, educational and recreational value of heritage.

Evaluation of environment and environmental processes is at the stage where research has attempted to show how certain characteristics of the environment are appropriate for particular forms of utilization, and this develops from methods of land evaluation. Coates (1976b) styled engineering as 'the art or even the science of using power and materials most effectively in ways that are valuable and necessary to man', and then proceeded to suggest that the geomorphologist

... must become involved in the tools of engineering because if construction causes irreparable damage to the land–water ecosystems due to lack of geomorphic input the earth scientist cannot be absolved of blame. Thus it is imperative that the geomorphic engineer be involved in the decision-making processes that plan and manage the environment.

Table 8.3 Natural hazards: monitoring techniques and predictability (after Alexander, 1999b, with kind permission from Kluwer Academic Publishers and Professor D.E. Alexander). See also Table 9.1, p. 241.

Hazard	Examples of monitoring methods	General predictability of impacts
Earthquakes	Seismometers, accelerometers, tiltmeters, extensiometers, radon meters, radar scans	High for broad-term, low for immediate-term
Tsunamis	Seismometers, tide gauges, pressure transducers, taut wire buoys	Generally high, although less so in close proximity to point of genesis
Volcanic eruptions	Seismometers, tiltmeters, infra-red radiation sensors, gas sample analysers	Moderately high on fully monitored volcanoes
Floods and flash floods	River stage gauges, rain gauges, meteorological radar and instrumentation, weather satellite images	Generally high if monitoring is adequate, only short lead times can be obtained for flash floods
Drought	Meteorological data, crop production data, agricultural market sales	Low to moderate
Hurricanes (typhoons, tropical cyclones)	Meteorological satellite images, coastal radar	High, although imprecise
Tornadoes	Doppler radar, weather satellite images, spotter networks	Moderate
Lightning and severe thunderstorms	Thunderstorm monitoring by radar, airborne instrumentation and weather satellite images	Low to moderate
Hailstorms	Thunderstorm monitoring by radar, airborne instrumentation and weather satellite images	Low to moderate
Avalanches	Monitoring and forecasting of snowpack stability using meteorological methods and snow physics	Variable
Glacier hazards	Creep meters, site surveys	Moderate

Hazard	Examples of monitoring methods	General predictability of impacts
Snowstorms	Snowstorm monitoring by radar, airborne instrumentation and weather satellite images	Variable, but high if weather forecasts are accurate
Frost hazards	Standard meteorological monitoring and forecasting	Variable
Soil erosion	Site survey, sediment traps, remote-sensing imagery	Moderately high, but only if monitoring is intensive
Desertification	Remote-sensing imagery, ground truthing surveys, social and economic indicators	Subject to controversy
Landslides	Piezometers, creep meters, laser survey, aerial photograph interpretation	Variable to quite high if monitoring is adequate
Subsidence	Creep meters, tiltmeters, aerial survey, ground survey	Variable, depending on knowledge of subsurface conditions
Soil heave and collapse	Geotechnical testing	High, if knowledge of site is adequate
Coastal erosion	Site survey, coastal storm monitoring	Moderate to high if monitoring is adequate
Wildfires	Infra-red sensors, visual monitoring	Variable, as phenomenon is highly volatile
Dam disasters	Engineering and geotechnical surveys	Low

This led Coates (1976b, p. 6) to advocate a field of geomorphic engineering as combining

> ... the talents of the geomorphology and engineering disciplines. It differs from environmental geology, wherein man is studied as one of the typical surface processes that change the landscape, and instead brings knowledge of physical systems to bear on systems that may require construction for their solution. The geomorphic engineer is interested in maintaining (and working towards the accomplishment of) the maximum integrity and balance of the total land–water eco-system as it relates to landforms, surface materials and processes.

A final group of approaches concerns *prediction, design and policy-related issues*. Whereas evaluation of the environment is primarily devoted to the uses of contemporary environments, prediction, design (Chapter 10, p. 262) and global change impacts (Chapter 9, p. 243) are more concerned with the future and associated with the policy issues necessary to influence developments. In considering examples of these four broad types of applied approach in the branches of physical geography, it has to be remembered that what is viewed as physical geography in one country may elsewhere be partly or completely embraced by other disciplines. This is exemplified by *environmental geology* and, in the seventh edition of one book with that title (Keller, 1996), the contents list illustrates the point that whereas certain chapters would be associated with the work of geologists in many countries, such as the UK, the remaining topics are elsewhere traditionally covered by other disciplines including physical geography. Perhaps some flexibility is inevitable in applied areas: *Managing the Human Impact on the Natural Environment: Patterns and Processes*, edited by the physical geographer Newson (1992a), covered pollution, conservation, environmental law and environmental economics and led to radiation and the environment, future energy use, natural environments of the future and planning, control or management. It is also significant that applied problems are not always easily dealt with by the techniques and approaches conventionally associated with the established branches of physical geography.

In *geomorphology* it has been argued (Vitek and Giardino, 1993, p. x) that:

> Unlike doctors, dentists, or lawyers whose work is recognised by the general public, the importance of geomorphological work is relatively obscure.

But textbooks have been produced such as *Geomorphology in Environmental Management* (Cooke and Doornkamp, 1990) authored by two physical geographers, and *A Handbook of Engineering Geomorphology* (Fookes and Vaughan, 1986) edited by a consulting engineering geologist and a civil engineer but including contributions by 15 physical geographers among the 25 authors. The latter book aimed to produce a basic yet authoritative handbook for geotechnical engineers and others, seeing geomorphology as the study of the Earth's landforms and landform change. They provided a diagram to summarize the relationship between engineering, geomorphology and geology (Fig. 8.1), and contended that: 'The application of geomorphological knowledge to civil engineering will come about by the education of both engineers and geomorphologists' (Fookes and Vaughan, 1986, p. 8). The education of both sides has now advanced.

Conferences, often with contributions from other disciplines, have provided a useful way of engendering discussion of possible applications and have subsequently generated edited book collections. Thus in *Geomorphology in Environmental Planning* (Hooke, 1988) the intention was to cover the interaction between geomorphology and public policies, by contributions organized into rural land use and soil erosion, urban land use, slope management, river management, coastal management, and policy formulation, and it was suggested, from the contributions focused largely on

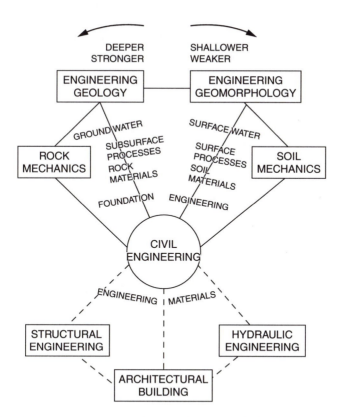

Figure 8.1 Some engineering, geology and geomorphology relationships (after Fookes and Vaughan, 1986).

Britain, that one of the reasons for the lack of geomorphology in environmental planning may be the perception that Britain is a relatively stable environment, that most geomorphological processes are not rapid, and that the land is not subject to the extreme events and hazards of other environments. Hooke (1988, p. 263) concluded that:

> There is an increasing realization on the part of both planners and geomorphologists of the need for involvement but strategies for successful collaboration still need to be worked out in many cases ... The environmental consequences of planners and policy makers remaining ignorant of geomorphological issues may be damaging and even dire. Geomorphologists owe it to society to communicate their findings and use their expertise.

Seven years later, and also deriving from a conference, *Geomorphology and Land Management in a Changing Environment* (McGregor and Thompson, 1995) emphasized the changing environment. Other volumes have reflected two trends: firstly the tendency to concentrate upon particular themes, with floods obviously being a major focus (e.g. Baker *et al.*, 1988; Beven and Carling, 1989); and secondly the need to work with other disciplines as exemplified by *River Conservation and Management* (Boon *et al.*, 1992) in which the preface to a volume of 29 chapters by international authors contends:

> From a human perspective, wild rivers are hazards; water flowing to the sea is wasteful; shallows are obstructions to navigation. These traditional views are slowly being replaced by recognition of the ecological and landscape values of natural rivers. The dilemma is to reconcile our immediate practical needs with long-term sustainability in terms of an ecologically sound and aesthetically acceptable environment ...

In addition to edited volumes arising from conferences, increasing interest in applied aspects has been reflected in volumes such as *Impounded Rivers* (Petts, 1984), *Channelized Rivers* (Brookes, 1988), *Coastal Problems: Geomorphology, Ecology and Society at the Coast* (Viles and Spencer, 1996), and *Beach Management* (Bird, 1996). One approach uses an explanation of process dynamics as the basis for an understanding of management requirements and in this, Bird (1996, p. ix) avowed:

> This book is an account of the origin, evolution and changing features of beaches, prepared as a reference work for people concerned with coastal planning, management, engineering and development.

He concluded (p. 243) that:

> Satisfactory beach management requires an understanding of the nature and dynamics of beach systems, the various physical, chemical, biological and social interactions that take place on and around them,

and the aims and perception of people who come to use them ...
beaches will continue to be seen as playgrounds at the edge of land and
sea, places of opportunity, entertainment, exercise and enjoyment, and
beach management will be necessary to optimize these values.

This epitomizes the approach of the geomorphologist, to espouse the environmental system and its processes as a basis for underpinning the way in
which management takes place (Jones, 1980, 1983). However, some volumes have been much more specifically focused. For example, in the case of
rivers, the adverse effects of channelization, some of which have produced
'ecological disasters', stimulated the search for alternative methods which do
not have such dramatic downstream effects and which minimize the degradation of environmental quality, particularly the aesthetic appearance of the
channelization scheme. Whereas early work focused on the extent of channelization and the alternatives available (Brookes *et al.*, 1983) and on the
review of these alternatives (Brookes, 1988), an edited volume on *River
Channel Restoration* (Brookes and Shields, 1996) was appropriately subtitled *Guiding Principles for Sustainable Projects*, with 14 authors of whom
at least five were trained as geographers, with chapters organized to relate
to a flow chart which can be used by the river manager in evaluating the
potential for river restoration in a particular region, and in developing strategies for their management organization. The book identifies critical issues
and proposed solutions (Brookes and Shields, 1996, Table 14.6, p. 399) and
indicates that an integrated approach to river channel restoration, involving
engineers and environmental scientists, is a recurrent theme throughout the
volume, that multifunctional teams are required to tackle the key issues
scoped for an individual project, and that a constraint to multifunctional
working in the past may have been that individual engineers or scientists
have anticipated little payoff or recognition in their area of expertise. They
therefore (Brookes and Shields, 1996, p. 400) concluded that:

> Cross training is advocated, whereby each member of the team devel-
> ops a practical understanding of one or more disciplines while main-
> taining strength in a base discipline ... Potentially this will lead to
> negotiation and the formulation of more sustainable solutions which
> integrate all requirements.

In addition to the work contained in volumes of this kind, there have
been significant achievements from individual studies. On the Rio Grande,
plutonium was released during the Second World War from the Los Alamos
laboratory and became attached to sedimentary particles accumulated on
canyon floors. In subsequent decades some of the sediments with the plutonium moved to the Rio Grande river, raising concerns about environment
and health hazards, but Graf (1994) was able to show how 90% of the plutonium moving through the system was from sources other than laboratory
discharges, although in some critical years the laboratory source amounted

to 80% of the total. This is an excellent example of the way in which research on a specific area can address a substantial problem which requires a solution. A further specific example is provided by Gurnell (1998) who reviewed the hydrogeomorphological effects of beaver dam-building activity in relation to the proposal by the UK Government to consider the reintroduction of certain species of beaver, threatened in mainland Europe, extinct in the UK and non-existent in Scotland for some 300 years. Areas of geomorphology have produced similar types of potential problem; research on aeolian dust and dust deposits stimulated a number of applications (Pye, 1987). These need to be clarified in the way that Thomas (1993) showed that desertification had been subject to considerable confusion, misinterpretation and lack of clarity regarding its characteristics and occurrence. He showed that improved satellite-based monitoring of environmental changes in Africa and a recent global assessment of human-induced land degradation indicated that earlier assessments of desertification may have overestimated the worldwide extent of the phenomenon by a factor of three. In the Mediterranean, Brandt and Thornes (1996) related desertification to land use in the first phase of the multidisciplinary and international Medalus project. Techniques have been collated for rivers and beaches (e.g. Thorne *et al.*, 1995) and also for desert reclamation (Goudie, 1990a).

One contribution to the future environment is through the conservation and explanation of particular sites, exemplified in the UK by Sites of Special Scientific Interest covered in Geological Conservation Review volumes (e.g. Gregory, 1997), although Murray Gray (1997) concluded that geomorphological conservation has been the Cinderella of nature conservation (see Chapter 10, p. 269).

There are many instances where research by physical geographers is concerned with **hydrology** and this has inspired edited volumes on *Hydrological Forecasting* (Anderson and Burt, 1985) and, more specifically, on *Floodplain Processes* (Anderson, Walling and Bates, 1996). Major interbasin transfers of water in Russia have been investigated, including the planting of shelterbelts (Rostankowski, 1982) and consideration of the environmental (L'vovich *et al.*, 1982) and climatic (Chubukov *et al.*, 1982) consequences. In hydrology it is necessary to take account of policy; in *Land, Water and Development*, Newson (1992c) subtitled the book *River Basin Systems and Their Sustainable Management,* including river basin development and sustainable river basin management. In many cases it is necessary to explore the extent to which integrated basin management is as integrated as it purports to be. Downs *et al.* (1991) compared the content of integrated basin management systems and suggested how elements of geomorphological change should feature more significantly. Geomorphological involvement, according to Jones (1995), may be open to question because of the lack of coherence in applied geomorphology, the need to interact with scientists in other disciplines and the need to market potential expertise. However, it is clear that there is an established range of applications, that interaction with other scientists is increasing, and that applications continue to develop from pure and applicable research.

A similar pattern of trends can be found in **climatology**; in *Applied Climatology: Principles and Practice* Thompson and Perry (1997) have 27 contributors of whom 16 are geographers, focusing upon the tools of research, the physical and biological environment from a climatic perspective, human and cultural environments, and humankind in the atmosphere. In relation to applied climatology, Marotz (1989) suggested that applied climatologists deal with one or more of the following types of problems:

- design and specification of material or equipment – including the effects of icings, hail storms, wind speed, and so on;
- location and use of equipment or structures – including telescopes, wind power;
- planning of particular operations – such as the scheduling of transport;
- climatic influences on biological activities – including yield predictions, energy usage.

In addition to contributions to many aspects of global change (see p. 244), there have been evaluations of the ways in which climate relates to human comfort in biometeorology; to building climatology, which involves climates within buildings and the ability of structures to withstand hazards; and to the way in which climate relates to leisure. Evaluation of the economic benefits of forecasting is a useful product of research, as reviewed by Maunder (1970) and developed as *The Uncertainty Business* (Maunder, 1986, p. 323) in which he contended:

> If we are to live within the limit of our 'climatic income' or our 'elite atmospheric resource', however, appropriate meteorological and climatological planning must be involved. For this to be accomplished, the politician and the planner must become more weather and climate oriented . . .

Appreciation of the impacts of climate change are of increasing interest, and Bryant (1997) included considerable discussion of such impacts including those on health and ecosystems. It is necessary to think of atmospheric management in a broader sense, and Thornes (1982, p. 561) explained that atmospheric management is not always understood but that the daily struggle to achieve thermal comfort is an obvious example. Thornes shows how space heating allows the atmospheric manager to have a role in an examination of the current efficiency of a heating system, and in the forecasting of effective temperatures; and how a small investment of road surface sensors could lead to greatly improved road danger warnings and so contribute to management of the effects of atmospheric conditions on winter maintenance of roads. In subsequent research, Thornes (1992) classified the impact of weather and climate upon air, rail, road and water transport by considering climatic design and required levels of maintenance; weather sensitivity and thresholds; and weather forecasting and new technology. The ways in which short-term climate variations occur are illustrated by the improved understanding of climate linkages and evaluation of investment risks in the citrus industry of the southern US (Rohli and Rogers, 1993).

In relation to *soils*, Ellis and Mellor (1995, Chapter 8) proposed applications to include physical problems of erosion, compaction, water excess and deficit; and chemical problems of acidification, salinization and sodification, agrochemical pollution, urban and industrial pollution. In *biogeography* many contributions have been developed from landscape ecology (Kupfer, 1995; Kent *et al.*, 1997) and landscape evaluation. US biogeographers consider that many applications have occurred (Palm and Brazel, 1992) in conservation, national park management, and reserve or preserve planning and management. They suggest that opportunities revolve around issues of forest productivity and regeneration, tropical deforestation and agriculture, interactions between fauna and productive vegetation regions, maintenance of plant communities in urbanized regions, soil nutrient cycling, and vegetative forcing of climatic change and responses to it. Biogeographers are therefore very aware of the opportunities, and Haines-Young (1992) suggested that they can confront ecosystem management issues in four major types of situation, namely: reclamation processes; restoration of damaged ecosystems; exploitation, control and creation of agro-ecosystems; and conservation, which is the conscious attempt to maintain the ecological system in some prevailing state.

Originating in biogeography, but evolving to other aspects of the physical environment, Battarbee *et al.* (1990) considered the causes of lake acidification with special reference to the role of acid deposition as an outcome of work on SWAP (Surface Water Acidification Programme); this work on 'acid rain' was included in a report of government-funded research involving 109 freshwater lakes in the UK, showing that the primary cause of surface water acidification is acid deposition (Battarbee, 1988). Such research originates not only within pure or applicable research but is also necessarily multidisciplinary. Also transcending the individual branches of the subject, mountain environments (Gerrard, 1990, Chapter 8) and natural hazards (Chapter 7, p. 187) provide areas in which there is considerable scope for applications of geographical research. The four categories of research outlined above (p. 207) can all be identified in the case of hazards research (Alexander, 1993; Smith, 1992). This is also an area where there is a clear intersection with human geography. *Natural Hazards: An Integrative Framework for Research and Planning* (Palm, 1990) provided a framework that helps to explain human response to environmental hazards. Such work has to some extent overcome the problem foreseen by Trudgill (1991, p. 89):

The emergence of environmental geography is noticeable in the development of school syllabuses, together with the importance of an overall appreciation of interactions. This does not seem to be paralleled at a research level, probably because it is easier to achieve at a simpler level appropriate for schools while at a research level, the systems require immense data sets and computational facilities.

In addition to the four categories suggested (Table 8.2), it is imperative that the results of research are communicated so that decision-makers are aware of the implications of environmental sensitivity.

8.3 The role of physical geographers in consultancy

The changing scope of physical geography and new techniques, especially remote sensing and geographical information systems, enabled physical geographers to contribute in two particular ways. First was the adoption of a holistic view of physical environment, a perspective not always achieved by other disciplines. This should not be interpreted to signify that environmental geomorphologists or applied physical geographers are 'super-scientists – synthesizers and integraters of everything and anything – when clearly we are not . . .' (Johnston, 1983c, p. 143), but rather that the greater opportunities for applications of physical geography are found at the interface, or lie in the interface area, with other disciplines. In summarizing four contributions that an environmental geomorphologist has to offer land managers, Coates (1982) included an eclectic approach to land–water ecosystems, a knowledge of feedback systems, the recognition of potential thresholds, and the site application of classic geomorphological principles. Therefore Coates (1982, p. 166) concluded that:

> The geomorphologist is probably the last of the science generalists, because by necessity he has to have had a proper background in not only geography and geology but also in mathematics, other sciences, and aspects of engineering in order to understand the complex relationships that operate in the dynamics of the Earth's surface. Thus, the environmental geomorphologist is in position to not only bridge the gap with peer natural scientists but also to translate various pieces of a puzzle into a composite whole.

Many examples could be quoted, but the holistic approach is epitomized by the Bahrain surface materials resources survey (Brunsden *et al.*, 1979, 1980). A survey was undertaken between 1974 and 1976 at the request of the Ministry of Development and Engineering Services, Government of Bahrain, and involved a team comprising ten geologists, seven geomorphologists, two pedologists, two surveyors and a cartographer. The survey produced a series of maps at a scale of 1:10,000 and an extensive report so that the final volume was 'probably the most intensive and comprehensive view of the surface materials of any state within the arid lands of the world' (Brunsden *et al.*, 1980). Many benefits accrued from this survey and from others undertaken by members of the same team (see Brunsden, 1999), and this has included knowledge of the consequences of environmental processes that otherwise would not have been appreciated. In drylands this is well exemplified by the salinity of groundwater and by salt weathering, which has been shown (Cooke *et al.*, 1982) to arise from a complex hazard that depends upon the relationships between local environmental conditions, the types of salt present, the nature of susceptible materials, and the design and nature of the structures built in hazardous areas.

A second type of contribution has been achieved as physical geographers have contributed specifically in relation to proposed solutions to a particular site-specific problem or question. This has been possible because the former reticence of physical geographers has been replaced by a willingness to focus upon specific developmental problems, an ability aided by enhanced modelling capability. In introducing the first issue of *Applied Geography* the editor wrote that:

> . . . the applied geographer needs to be brave. He needs to commit himself before he knows all the answers. He needs to be prepared to make public mistakes. But he must be prepared to learn from them. (Briggs, 1981, p. 6)

Increasing involvement of physical geographers in seeking solutions to specific problems and in consultancy roles is illustrated by the projects undertaken by one individual as listed by Coates (1990, Table 3). Specific consultancy projects may involve a degree of innovation which merits wider dissemination by publication of papers, for example in journals such as *Applied Geography* or *Environmental Management*. Physical geographers employed in an organization dealing with environmental management, or in institutions of higher education, have been commissioned to undertake specific applied projects, and the increasing need of institutions for overhead income has encouraged the formation of advisory units or groupings of scientists specifically founded to undertake consultancy research and to survive on 'soft money'. This advisory unit model had been rather longer established in relation to the engineering disciplines, but became more familiar in physical geography as awareness of environmental problems increased. In the UK, units such as the Flood Hazard Research Centre at the University of Middlesex, or the Geodata Institute at the University of Southampton, are examples of this trend. Such units could be entirely staffed by physical geographers but more usually are composed of scientists from a number of disciplines, including geologists, biologists, civil engineers, hydrologists and possibly planners. This model has become more common as universities have had to become more accountable for the use of staff time and more conscious of income generation potential, although it is still possible for physical geographers to participate in independent consulting firms or, as is the case in the US and some other countries, to have an academic contract which provides a salary for less than 12 months of the year because it is assumed that the balance will be made up from income from research grants or consultancy.

Involvement by physical geographers in such consultancy research has produced tensions, however. Firstly there is the problem of securing the funding necessary to ensure that the 'soft money' continues to arrive to pay the salary bill of the advisory unit, and to meet the overheads required by the host institution, or to bridge the gap between the academic salary and the remainder of the year. Secondly there is the fact that consultancy or applied research has not always been accorded the same status as 'blue skies', Research Council-funded research. Penning-Rowsell (1981a, p. 11) indicated that:

> Contract research is still seen in many quarters as inferior to Research Council sponsorship which in turn is seen as inferior to unfinanced scholarship . . . The reputation of the subject as a whole needs careful nurturing following the quantification debacle. An essential part of this nurturing involves a considered analysis of the potentials and problems of closer involvement by academic geographers in an advisory capacity with environmental groups, policy analysis and decision-making in the outside world.

This situation has been addressed since 1981, but in the UK there is still a debate as to the extent to which applied research is appropriate for the Research Assessment Exercises; if applied research requires fundamental and innovative enquiry, then it should be accorded the same status as pure or applicable research.

Many examples of specific enquiries have produced new approaches in, or adopted by, physical geography. When the Federal Power Commission studied applications for a permit to construct one or more additional dams for electric power in the Hell's Canyon area of the Snake River, Idaho, it was necessary to consider how the attributes of the landscape could be ranked so that some, possibly the most unique, could be preserved from development. Leopold (1969) therefore developed a uniqueness method which was a scoring system designed to identify which of a number of sites was the most unique, and therefore to suggest which of several alternatives could be developed without losing some particular qualities. This method was developed for the evaluation of riverscape (Leopold and Marchand, 1968) and a related matrix method can be extended to environmental impact (Leopold *et al.*, 1971). This illustrates how novel approaches can emerge from a specific applied problem, and in this case the method of evaluating scenery could complement synthetic methods (e.g. Linton, 1968) and could generate considerable subsequent discussion about the aesthetic quality of landscape and comparison of the methods available (Penning-Rowsell and Hardy, 1973; Penning-Rowsell, 1981b). Many other specific examples could be quoted, and Graf (1988) showed how geomorphologists in the dryland western US increasingly deal with legal issues surrounding the management of rivers. He cited cases (Graf, 1988, p. 300) in western Wyoming where the stability of a river-defined boundary was an issue, and whether or not locational changes would reasonably be expected in the course of normal river processes or whether recent changes were unusual. Along the Agua Fria River near Phoenix, Arizona, explanation was required from a geomorphologist of river channel change and its implications for the use of engineering models. Schumm (1994) drew attention to three types of misperception of fluvial hazards, which were:

(1) of stability – any change is not natural;
(2) of instability – change will not cease;
(3) of excessive response – change is always major.

Schumm concluded that such misperceptions can lead to litigation and unnecessary engineering works (see Chapter 10, p. 264).

8.4 Outstanding potential

In the last two decades of the twentieth century, applications of physical geography, with all categories in Table 8.1 represented, have been increasingly evident and this trend seems certain to continue. In human geography Johnston (1997, p. 323, 354) followed the questions that Harvey (1974) posed in relation to studies claiming to be 'relevant', namely 'relevant to whom' and 'for what'. However, in physical geography these questions have been implicitly answered, to some extent at least, as indicated in Sections 8.2 and 8.3, although two other areas to be explored concern global change (Chapter 9, p. 243) and the potential of design (Chapter 10, p. 262).

Research on applications of physical geography is incomplete, however, if the results are not acknowledged by other disciplines, by the public in general, and then by decision-makers so that they subsequently become reflected in policy. The ultimate objective must be to secure environmental improvements, and in a discussion of ideas about the barriers that prevent such improvements and how they can be overcome, Trudgill (1990) argued that we need to think holistically in a forward-looking way that is both remedial and preventative. With an eye to the effectiveness of what he characterized as environmental geography, not explicitly defined but surely largely coincident with physical geography, Cooke (1992) examined three interdependent research imperatives where geographers can make distinctive contributions to environmental issues before, during and after public responses to them. These three imperatives were:

(1) *The landscape imperative* – use of field-based skills of exploration and audit in real places to recognize and analyse environmental issues prior to public response, so that the environmental agenda can be influenced at an early stage from the basis of original research.

(2) *The institutional imperative* – involves research within active environmental issues and the agencies responsible for them. A weakness at the
interface between environmental advice and managerial decision-
making can account for many failures of environmental policy and arise
because of: *institutional causes* whereby scientists fail to appreciate the
institutional contexts within which the environmental issues are embedded; *translational causes* if environmental evidence is incomprehensible
to users; and *operational causes* if analyses of an environmental issue
incorrectly diagnose the often exceptionally complex two-way relationship between human activity and the physical environment.

(3) *The historical imperative* – necessary to evaluate the evolving dialogue
between people and environment in the context of environmental
change in the human domain in the recent past, so that the consequences of human actions can be judged and if necessary blame appor-
tioned, management responses improved and future needs assessed.

These three imperatives relate to the four stages of applied physical geography outlined at the beginning of Section 8.2 (p. 207) and underline the fact
that policy implications need to be considered. This theme has been underlined by Graf (1992) who contended that effective science and well-
informed public policy are the avenues to successful management of
environmental resources. He demonstrated that in the management of the
river resources of the western US, geomorphology, hydrology and public
policy have been poorly connected to each other so that, as geomorphological and hydrological understanding of river behaviour has improved
throughout the past century, the complexity of the systems has become more
apparent. Graf (1992, p. 17) concluded that:

> It is now apparent that public policy needs and can use the explana-
> tions provided by science. The remaining question is whether or not
> geomorphologists and hydrologists can address the useful and socially
> significant issues with convincing answers.

To influence policy it is necessary to be in a position to formulate it, and an
excellent example is provided by Graf as Chair of the Committee on
Watershed Management in the US, which was charged with:

- reviewing the range of scientific and institutional problems related to watersheds; especially water quality, water quantity and ecosystem integrity;
- evaluating selected examples of watershed management in a search for
  the common elements of successful management;
- recommending ways for local, state, regional, and federal water managers to integrate ecological, social, and economic dimensions of watershed management.

This led to the publication of *New Strategies for America's Watersheds*
(National Research Council Committee on Watershed Management, 1999),

which includes a number of suggestions to guide the reauthorization of the Clean Water Act and 15 conclusions 'to steer the nation toward improved strategies for watershed management'. More generally it has been argued (Owens *et al.*, 1997, p. 4) that:

> environmental policy needs to develop in ways which are consistent with the prevailing scientific knowledge, the recognized uncertainties, and the distinctiveness of the spatially distributed and temporally variable conditions within which environmental problems emerge . . .

In fact many environmental problems are highly context-dependent (Trudgill and Richards, 1997) and if policy-making is to be based upon environmental science, it needs to reflect the nature and characteristics of that science so that policies should be based on a dialogue between generalization and specific contexts rather than simply on generalization. Trudgill and Richards (1997, p. 11) concluded that:

> Implementation of broad policy guidelines should be cautious and flexible, expect (and promote) local calibrations of principles and processes, and permit parallel development of preventative and curative policies with a changing balance between them in response to the continuing information supplied by science as the policies are applied. What may emerge from a debate about these issues is a philosophical grounding for an environmental policy framework which combines the precautionary principle with subsidiarity, leading to policy practice which allows developed decision-making through a choice of instruments that are applicable at local scales in time and space.

Thus in applications of physical geography it is necessary to be aware of several scientific approaches to a problem, even if a single one is adopted; to propose solutions where these are required but with the appropriate precautionary caveats; and to have an awareness of the implementation of recommendations and the implications that these may have in relation to policy. Such background is even more pertinent to global change, as considered in the next chapter.

Further reading

An example of the range of applications in geomorphology is given by:
COOKE, R.U. and DOORNKAMP, J.C. 1990: *Geomorphology in environmental management. A new introduction.* Oxford: Clarendon Press.

The general context and some of the debates are given by:
KENZER, M.S. (ed.) 1989: *Applied geography: issues, questions and concerns.* Dordrecht: Kluwer Academic.

Potential in climatology is illustrated in:
   THOMPSON, R.D. and PERRY, A. (eds) 1997: *Applied climatology: principles and practice*. London: Routledge.

A geomorphological perspective with consultancy examples is given by:
   COATES, D.R. 1990: Perspectives on environmental geomorphology. *Zeitschrift für Geomorphologie* Supplementband **79**, 83–117.

Topics for consideration

(1) Have all possible applications of physical geography research been identified?
(2) Is applied research appropriate in higher education, or should it be undertaken elsewhere?
(3) Do physical geographers have skills enabling them to make a distinctive contribution in employment?
(4) Should environmental management seek to produce approximations of what is natural rather than what public perception would like (see also Chapter 10, p. 264)?

TRENDS FOR THE MILLENNIUM

|9|

Global physical geography

Lack of a global vision has been corrected progressively, at least to some extent, in a number of ways. This is appropriate and timely as globalization has become such a frequently discussed issue and a driving force for change which physical geographers must acknowledge. In fact there is a long tradition of a global focus in physical geography, for example in climatic, soil, and vegetation classification. More recently there have been some explicit moves towards a more comprehensive global approach (9.1), for example in mega-geomorphology, reinforced by the trends emerging from studies of environmental process (Chapter 5), from environmental change (6), from human activity (7) and from environmental management applications (8). The potential should be realized using scientific and technological advances (9.2), facilitated by developments in remote sensing, GIS, information technology and, particularly, the advent and utilization of global databases. Particular emphasis is being placed upon global change scenarios, and the involvement of physical geographers is reviewed (9.3) together with the way in which the environment has been 'rediscovered'. This leads, finally, to consideration of the potential for contribution by physical geographers in an interdisciplinary context at the global level (9.4).

9.1 A more global physical geography

It might seem odd to suggest that a global approach to physical geography is a new development, because in certain branches of physical geography there has always been a global dimension, and in climatology, including emphasis upon world climatic classification, the global scale was always evident. This became even more real, however, when developments in computing unleashed the potential so that general circulation models (GCMs) could be employed more routinely. In soil geography, awareness of the need for a global approach was always underlined by the need to relate field classification of soils to world classification levels, and so was a theme adopted by Bridges (1970) and more recently by Paton *et al.* (1995) in *Soils: A New Global View* in which the zonal approach is challenged by their view inspired by antipodean experience. In biogeography, the trend towards ecosystems and energy patterns (Simmons, 1991) stimulated a more global outlook. In geomorphology the move towards a global focus was perhaps less obvious but came about for two main reasons. Firstly, the necessary focus upon the detailed scale, seen to be required in the investigation of earth surface processes, temporal change, human activity and applications of geomorphology, led inexorably towards attempts to collate results at a more global scale. Thus erosion rates, based upon analyses of small areas, gave data that could facilitate the revision of world patterns of sediment yield and erosion (Walling, 1996a), and analyses of coastal dynamics could provide the foundation for world classifications of coastal systems. Secondly, it was appreciated that some areas had been relatively ignored by geomorphologists so that the previous balance had to be redressed. Geomorphologists had not focused significantly upon continental-scale problems because they were not in a position to contribute to tectonics and geophysics, which must surely be a prerequisite for an effective and accepted research contribution at the world level. Thus when Ollier (1981) made a plea for a more evolutionary geomorphology, he did not cite the quantitatively expressed geophysical research and, in order to contribute in such fields, geomorphology had to be prepared to adopt the methods and language of practitioners of other disciplines. Summerfield (1981) made a plea for macro-geomorphology in arguing for a more secure basis of geophysical, sedimentological and geochronometric data; and potential for mega-geomorphology still exists (Gardner and Scoging, 1983). This field was developed by Summerfield (1991) in *Global Geomorphology*, which emphasized global-scale processes and phenomena and is described as the first textbook fully to integrate global tectonics into the study of landforms, and to incorporate planetary geomorphology as a major component, including (in Chapter 19) planets and large moons composed of a solid crust on which landforms can develop and be preserved. In addition, the impact of global change research meant that the opportunities for a global geomorphological future could begin to be identified (Goudie, 1990b).

Whereas Johnston (1984) had lamented the lack of global vision in British geography, this may have been corrected to some extent by a revival of a form of regionalization (Haggett, 1990) and (largely for human geographers) by *Geographies of Global Changes: Remapping the World in the Late Twentieth Century* (Johnston *et al.*, 1995). The move was not unique to human geography because in other social sciences, globalization was increasingly evident, as shown by Giddens (1998) and reflected in his 1999 Reith lectures. In physical geography, Clayton (1991, p. 13) argued that geographers are well-based to offer a rational and sound debate on environmental issues in teaching and writing their geography, but after the British Association meeting at Swansea in 1990, he despaired of a subject that allows others to discuss natural hazards, environmental economics and acid rain while itself spending the day describing the local environment. The outstanding need to encompass the global scale was expressed by Simmons (1990a, p. 385):

> The natural science of the moment is global in scale, and in general, that is an empirical focus abandoned by geographers in recent years as we have looked at processes at a small spatial scale. As a result, only a few of us are tooled up to consider the great flows of the planet ...

Some physical geographers were concerned that the focus of research had become too detailed and, for example, Kennedy (1993, p. 134) avowed:

> ... I find many of the current preoccupations of geography and especially physical geography surprisingly parochial ... Why are we so concerned with the small scale, the short term and the present state of things as the best or most appropriate ... ?

Physical geographers responded to the need for increasing awareness of the global perspective by a series of textbooks, and several give the flavour of the increasingly global trend. *Global Environmental Issues: A Climatological Approach* (Kemp, 1990) was one of the first geographical texts and its climatological approach was particularly appropriate because climatology had been so fundamental in research on the global environment. It covered drought, famine and desertification (Chapter 3), acid rain (4), atmospheric turbidity (5), the threat to the ozone layer (6), the greenhouse effect (7) and nuclear winter (8). Kemp (1990, p. xv) noted:

> If the experts are right, the full effects of current global environmental problems will be felt within the lifetime of the students using this book [so that it is necessary to] recognize the importance of facing up to the issues before it becomes too late.

Trudgill (1991, p. 89) concluded that this book 'shows that geographers can make a real contribution to environmental issues by the synthesis of apposite knowledge'. A year later *Global Environmental Change* (Mannion, 1991) was published to provide a synopsis of the natural and cultural environmental history during the last 3 million years, embracing the impact of tourism, biotechnology and genetic engineering (Chapter 7). Subsequently an edited volume on *The Changing Global Environment* (Roberts, 1994) involved 22 contributors, of whom 20 would be recognized as geographers, and was prompted from a standpoint of deficiencies of existing student texts in physical geography which were 'stuck in the rut of a rather abstract and mechanistic "systems" approach to analysing environmental processes' (Roberts, 1994, p. ix). Even more specifically directed towards the implications of global change was *The Global Casino: An Introduction to Global Issues* (Middleton, 1996, 1999), a title adopted because Middleton saw many parallels between the issues discussed and the workings of a gambling joint! Issues covered ranged from deforestation, desertification and acid rain, to waste management, energy production and war, concluded by a chapter on conservation and sustainable development, providing a novel perspective related to global change. This timely volume (Middleton, 2nd edn, 1999) could lead to the discipline being perceived in a new way, but raises the problem that the implications of global change could be perceived to be topic-based and insufficiently inter-related as part of a holistic view. A final example of *Physical Geography and Global Environmental Change* (Chapter 1, p. 4) lays down four challenges for physical geography which could enable physical geographers to return to 'the holistic interpretation of environmental systems and to capitalize on the insights and understandings of the environmental systems with which they have become familiar' (Slaymaker and Spencer, 1998, p. 91). The four challenges identified are:

- Biogeochemical cycles should occupy a more central role in the physical geography curriculum and environmental mass fluxes cannot be understood without a deeper understanding of environmental chemistry. Whereas sedimentary cycles have been studied as exclusively physical systems, such sedimentary cycles and their interaction with the lithosphere, hydrosphere, biosphere, atmosphere and society are difficult to analyse unless they are viewed in a biogeochemical framework.
- The capacity for modelling complex environmental systems should be further developed.
- The interpretation of remote sensing and the emphasis upon GIS and related technology should be further developed.
- The role of society in changing the face of the Earth and the importance of culture and human value systems as drivers of global environmental change need to be recognized more explicitly by physical geographers. It is an obligation on physical geographers to try to inform policy-makers of the global environmental implications of their strategies.

In Part II of their book, Slaymaker and Spencer (1998) dealt with responses of the environment to global change to signal the way in which environmental systems may change in the future, and this has been explored in several other contributions such as that on *Global Warming, River Flows and Water Resources* (Arnell, 1996).

Whereas the above volumes are all authored by geographers, other books involve collaboration with practitioners of other disciplines. Thus *Global Environmental Change* (Moore *et al.*, 1996) derived from a course mounted in the University of London and was authored by an ecologist, a geologist and a geographer, and concludes with a chapter on paradigms and politics and a plea for realism and objectivity. In addition to a great surge of volumes on the subject of global environment change, there have been three other developments. First have been books dealing with major areas of the Earth's surface. These have been particularly notable in geomorphology and have included *Geomorphic Systems of North America* (Graf, 1987), written to bring selected research problems to the attention of the earth science community, to emphasize questions of modern processes, and to identify the emerging theoretical developments in each region. In the foreword the editor explained:

> This is a marvellous time to be a geomorphologist. Formally the science is not yet a century old, but it is in one of its most exciting periods. New ideas, new techniques, new data, even new extraterrestrial worlds are in greater abundance than ever before. (Graf, 1987, p. vii)

The Physical Geography of Africa (Adams *et al.*, 1996) covers the whole of physical geography, following conventional physical geography by an integrated and detailed analysis of geomorphology, biogeography, environmental change and hydrology, and concludes with four chapters on the impact of human agency on soil erosion, desertification, biodiversity and conservation. Such volumes begin to realize an opportunity which exists for physical geography and is expressed for geomorphology:

> Geomorphologists have a very real challenge to meet, particularly as we gain access to a larger slice of the research pie under the rubric of the 'global change' movement presently sweeping the funding agencies. Scientists in other disciplines are starting to see landscape evolution as a fundamental component to the reconstruction of past environments and the prediction of future change. We must meet this challenge with innovative approaches designed to create synergism between studies of geomorphic processes and landscape evolution. (Dorn *et al.*, 1991, p. 302)

Edited volumes also provided an important vehicle for the dissemination of significant developments (e.g. Michener *et al.*, 1994; Stewart *et al.*, 1996).

A second development was the creation of new journals (see Table 5.1) including *Global Environmental Change* (1990–), *Global Environmental Outlook* (1997–) and *Global Ecology and Biogeography Letters* (1991–) which were initiated to complement those longer established journals, such as *Climatic Change* (1978–), which themselves increasingly included contributions on global change. Thirdly, physical geographers have played a significant role in big interdisciplinary environmental programmes.

Physical geography was becoming more global in its focus at a time when international developments, and the thrust of other disciplines, were very much focused on global change. Thornes (1995) felt that geographers had been passed by and that big science was needed to solve the problems of global dimensions. However, special sessions of meetings which were organized, for example, at the annual meeting of the Canadian Association of Geographers at St Johns in 1997, prompted papers in a special issue of *Canadian*

Geographer, including estimations of regional environmental change in northern Canada (Jacobs and Bell, 1998, Table 1). Such meetings had to take account of the international imperatives and the way in which the ICSU (International Council of Scientific Unions) had established the International Geosphere–Biosphere Programme (IGBP) in 1986 with the objective:

> To describe and understand the interactive physical, chemical and biological processes that regulate the total earth system, the unique environment it provides for life, the changes that are occurring in this system and the manner in which they are influenced by human action. (ICSU, 1986, p. 6)

Subsequently a plan for action was published (IGBP, 1988) and the initial core projects were publicized in 1990. This established the multidisciplinary agenda and a series of subsequent meetings were recorded in volumes of proceedings, such as Walker and Steffen (1996), which included papers recording the scientific achievements of the first 4–5 years of the science conferences of Global Change and Terrestrial Ecosystems (GCTE), begun in 1994 as a core project of IGBP.

Specifically in relation to climate change, the Intergovernmental Panel on Climate Change (IPCC) produced a report in three parts in 1990 (Houghton *et al.*, 1990, quoted in Henderson-Sellers, 1991, p. 53), which was presented to the United Nations and formed the focus of the Second World Climate Conference in November 1990. Discussion at that conference asserted the reality of humanity's disturbance of the natural climate system, called for improvement in knowledge of processes vulnerable to climate change, especially integrated research into the processes associated with agriculture, forestry and water resources, and demanded policy responses to mitigate and to adapt to projected climatic changes. Therefore all nations were required to address two fundamental issues (Henderson-Sellers, 1991, p. 69):

(1) How will global climate change affect their natural resources and human population?
(2) How will the international impetus towards policy responses, particularly greenhouse gas emission reduction treaties, affect their industry, their economy and trade?

Further progress followed the second report of the IPCC (1996).

With the advent of textbooks, appreciation of the scientific climate and awareness of global climate change, together with research extended to the global scale, some of the deficiencies were being addressed. One of the pervasive themes in American geography was highlighted as the 'local–global continuum', and Meyer *et al.* (1992, p. 270) noted that:

> As physical geographers have become more confident of their concepts and techniques at the microscale (site scale), two forces are drawing

attention back to the meso- to macroscales. First, there is a growing understanding of the interaction of factors across different scales. . . . Second, explanations of landscape phenomena are held by most geomorphologists . . . and by many biogeographers . . . to be the basic goals of research.

And furthermore (p. 273):

. . . if disjunctures in scale pose problems, they also offer opportunities . . . Geographers are unlikely to abandon their interest in the micro-to-macro continuum of spatial scales . . . geographers will preserve the insights gained at the extremes of the scale spectrum and, ideally, move more nimbly up and down the continuum. The risk that they run is that of finding a comfortable resting place in the mesoscale, and remaining so grounded there that they never grasp the levels of understanding found at either end. The challenge to geography is to devise a means for embedding the local and the global in the regional.

In *Rediscovering Geography* (Rediscovering Geography Committee, 1997), the chapter on 'Strengthening geography's foundations' included a directive to focus on six research challenges which were set out as:

(1) disequilibrium and dynamics in complex systems;
(2) the expanded concepts of global change;
(3) the local–global continuum;
(4) comparative studies using longitudinal data;
(5) effects of geographical technology on decision-making;
(6) geographical learning.

Four of these challenges (2–5) are directly associated with the more global approach and it was suggested that comparative studies of global-scale processes will benefit from sustained and coordinated attention from multidisciplinary teams of geographers and experts from allied natural and social science fields. Thus there was an awareness of the global opportunity and need for a world focus in geography as a whole, which could be realized in physical geography with technological developments (Section 9.2), was imperative in view of the increasing emphasis upon global change (9.3) and required greater multidisciplinary cooperation (9.4).

9.2 Developments in technology

Progress within physical geography and its branches, in the context of much greater geographical awareness of the global environment and global change, meant that physical geography was equipped to participate in the global movement. However, to realize the opportunities available it was necessary to be conversant with, and to be heavily involved in, developments in technology. The major strands of geoinformatics, of the developments in remote sensing, GIS, global databases, global positioning systems, geostatistics and geocomputation for global-scale research made geographical involvement feasible. Physical geographers have taken a greater interest in each of these areas and so have been equipped to contribute to, and be recognized in, the investigation of global change. It is not appropriate to detail here the recent developments at great length, but it is necessary to appreciate the major areas of progress and potential (e.g. Smith, 1984). The World Climate Programme developed by the World Meteorological Office and the United Nations Environment Programme in conjunction with the World Health Programme led to the creation of the Global Environment Monitoring System (GEMS) group as part of the Earthwatch priority. The GEMS has the assessment of major environmental issues and the provision of early warning through the GRID (Global Resources Data Base) as its objective. The United Nations Environment Programme (UNEP) provided a direct link to the Global Environment Facility (GEF) controlled by the United Nations Development Programme (UNDP) and the World Bank. The GEF, established in 1990, provides environmental grants and low-interest loans to developing countries to assist them in carrying out environmental work to address global programmes. This therefore indicates one aspect of the global context that existed and warranted increasing attention from physical geographers.

Progress in remote sensing (Curran, 1989) that has been particularly significant arises from other developments associated with spectral, spatial and temporal resolution (Curran *et al.*, 1998). As the number of platforms and sensors has increased, so the regions of the spectrum that can be sensed to determine environmental characteristics have increased. Airborne imaging spectrometers are a fundamental technological advance in remote sensing, and the outstanding challenge is to turn this technological advance into a useful tool for understanding and managing the environment (Curran, 1994). Fig. 9.1 illustrates the regions of the spectrum within which environmental characteristics (rock and soil, vegetation, snow and ice, water, atmosphere) can be determined using imaging spectrometry. Curran (1994) reviewed the range of applications that are possible, and concluded that the future lies in the space-borne sensors onboard the forthcoming polar platforms. A great amount of new data will be provided with the launch of the Earth Observing System (EOS) platforms (e.g. Terra) in the US, and the Envisat satellite in Europe, and it could well be that satellite sensor technology will continue to change faster than the theoretical development of surface climate models. Obtaining land surface climatology values (including solar radiation, surface albedo, surface temperature, outgoing long-wave radiation, cloud cover, net radiation, soil moisture, latent and sensible heat flux, surface cover and leaf area index) and linking them to surface clima-

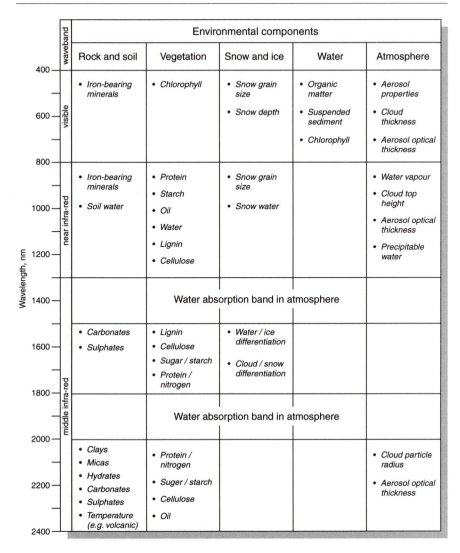

waveband	Environmental components				
	Rock and soil	Vegetation	Snow and ice	Water	Atmosphere
visible (400–800)	• Iron-bearing minerals	• Chlorophyll	• Snow grain size • Snow depth	• Organic matter • Suspended sediment • Chlorophyll	• Aerosol properties • Cloud thickness • Aerosol optical thickness
near infra-red (800–1400)	• Iron-bearing minerals • Soil water	• Protein • Starch • Oil • Water • Lignin • Cellulose	• Snow grain size • Snow water		• Water vapour • Cloud top height • Aerosol optical thickness • Precipitable water
	Water absorption band in atmosphere (1400)				
middle infra-red (1600–2000)	• Carbonates • Sulphates	• Lignin • Cellulose • Sugar / starch • Protein / nitrogen	• Water / ice differentiation • Cloud / snow differentiation		
	Water absorption band in atmosphere (1800)				
(2000–2400)	• Clays • Micas • Hydrates • Carbonates • Sulphates • Temperature (e.g. volcanic)	• Protein / nitrogen • Suger / starch • Cellulose • Oil			• Cloud particle radius • Aerosol optical thickness

Wavelength, nm (left axis: 400, 600, 800, 1000, 1200, 1400, 1600, 1800, 2000, 2200, 2400)

Figure 9.1 Regions of the spectrum within which environmental characteristics can be determined using imaging spectrometry (after Curran, 1994).

tology models is very important and fast developing (Greenland, 1994), as is driving ecosystem models with remotely sensed data. The extent of the use of remote sensing to provide information on the atmosphere, oceans, land deforestation, desertification and snow accumulation, and human-induced environmental changes including urbanization, forest fires and oil spills, has been demonstrated in an atlas (Gurney *et al.*, 1993), whose authors noted that:

The advent of remote sensing technology and space borne sensors is a critical advance, making it possible to monitor from space a large number of the Earth's vital signs, from atmospheric ozone to vegetation cover to sea level and glacial ice, and therefore to provide a background for prediction studies.

Advances in satellite instruments and their calibration and in methods of processing and archiving remotely sensed data have led to numerous time series of geophysical data relevant to global change. Application to global change (Curran, 1989) has also been stressed by Curran and Foody (1994) who noted the movement of remote sensing from field and forest to the landscape and the continent, and more significantly the use of remotely sensed data to infer the immeasurable, extrapolate point data to areas, and parameterize and drive environmental models. The resolution achievable now provides excellent opportunities for research, and an impressive example of the utility of remote sensing is shown by the Amazon Basin map in Fig. 9.2 (Foody *et al.*, forthcoming).

Advances in geographical information systems have also been possible, and these can be coupled to models and supported by input from remote sensing as appropriate (e.g. Atkinson and Tate, 1999). Since the Earth Summit in Rio in 1992, many of the actions approved by governments include references to spatial data and geographical information in areas such as soil erosion, marine pollution, biological diversity and breeding stocks, drinking water supplies, climate change, poverty, urbanization and demographic change. GIS will continue to provide answers, but successful use depends upon questions posed, data available, institutional and political support necessary to bring about skills, and insight of those who use them (Burrough and McDonnell, 1998). A particular development has been the development of three-dimensional GIS (Raper, 1989) which affords considerable potential. Thus in geomorphology it is possible to calculate the movement and storage of slope material in relation to the determination of the sediment delivery ratio (see Raper, 1991).

A further significant aspect of development concerns the greater availability of databases and their accessibility, which have been exemplified in relation to global continental palaeohydrology (Branson *et al.*, 1995, 1996). The far-reaching implications as foreseen by Clark *et al.* (1991, p. 217) are that:

> The study of the earth is undergoing a revolution . . . Until recently, much of the research effort was tied up in the definition and research of each discipline-specific problem. It is now realized that research in the earth sciences has progressed to a point where all of the many parts of the Earth system must be studied together and integrated if a reasonable attempt is to be made to understand fully the problem under study and to predict the nature of the processes involved. This concept has been called Earth System Science . . . or, simply known by the

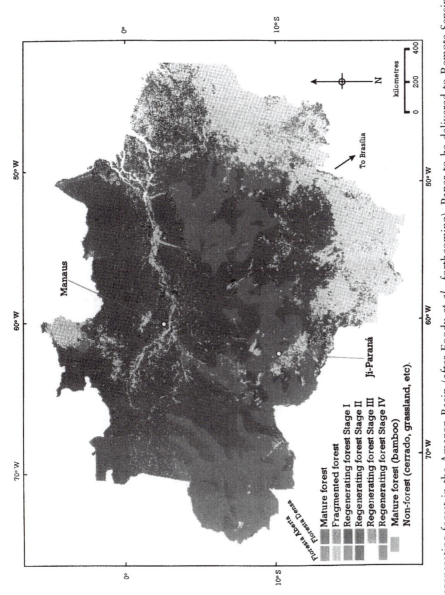

Figure 9.2 Regenerating forest in the Amazon Basin (after Foody *et al.*, forthcoming). Paper to be delivered to Remote Sensing Society Conference 2000, Leicester, September 2000.

phrase which describes its predictive goal, Global Change . . . To make this a reality, there are three necessary links to be forged: between narrowly focussed disciplines and interdisciplinary studies; and between local and global scales.

Clark *et al.* (1991) provided a table (Table 50.1) that lists variables of which measurements are required for the long-term study of global change as recommended by the Earth System Science Committee in 1988.

In the further use of technology, in addition to the links between disciplines that are stressed by Clark *et al.* (1991), there are two other issues that are paramount, namely real-time analysis and changes of scale. For effective management of environmental problems, especially at the national or global scale, it is imperative that data can be obtained so rapidly that analysis can be undertaken immediately, facilitating rapid decision-making so that appropriate solutions are reached and applied. This is particularly pertinent in relation to managing natural disasters. Microcomputers, earth resources satellites, communications satellites and GIS all offer considerable potential for natural disaster management, particularly if real-time uses are developed by integrating these technologies (Alexander, 1991). The range of information technology resources relevant in the case of natural disasters is wide-ranging, and Alexander (1991, Table 1) included world-based communications (including e-mail, fax, postal services, radio, telegraph, telephone, telex), image-based communications (motion picture film, photography, radar, satellite, television, video), computer facilities and data storage (data banks, micrographics and microform). Alexander (1991) then considered the predictability of different types of disaster and the monitoring techniques that can be utilized for each (summarized in Table 9.1).

The second problem that has to be addressed relates to scales, because both upscaling and downscaling are necessary. Upscaling is necessary, for example, where soil–vegetation–atmosphere models (SVAT) need to be linked so that surface, biophysical and biochemical models can be related to global climate models (Van Gardingen *et al.*, 1997). In the other direction downscaling can be required, for example when the output from GCMs is downscaled to explore relationships between large-scale circulation and local- and regional-scale climate conditions for present and future climates (Rogers, 1995). Downscaling techniques emerged to bridge the gap between what climate modellers are able to provide and what impact assessors require (Wilby and Wigley, 1997). Technological advance has now enabled research to take place at a spatial and temporal scale that could not have been envisaged several decades ago. Thus passive microwave-derived mapping of continental and regional-scale snow cover extent and the water equivalent has now been possible from remotely sensed data for just over two decades, revealing a significant decrease in snow cover associated statistically with higher temperatures (Barry *et al.*, 1995). The decrease is particularly evident in spring and early summer, but such anomalies occurred

Table 9.1 Types of disaster, predictability and monitoring techniques (after Alexander, 1991)

Type of disaster	Predictability and forecast potential	Monitoring techniques
Accelerated erosion	Potentially high over broad areas, but risk often not identified in time	Aerial photographs, field surveys, erosion pins, sediment traps, GIS, satellite images
Avalanche	Potentially high for forecasting appropriate weather and snow conditions, high for identifying common avalanche tracks, low for timing individual avalanches	Meteorological forecasting, snow sampling, track and runout mapping
Coastal erosion	Generally high for average rates, moderate for individual events	Aerial photogrammetry, field surveying
Crop blight	Varies with specific environmental and pathological conditions	Plant pathology, satellite remote sensing
Desertification	Varies with social, economic, political and ecological conditions; phenomenon is often not recognized until too late	Satellite image processing, agronomic survey, social survey of affected populations
Drought	Frequency of droughts is moderately or highly predictable, timing is often less so	Meteorological forecasting, water-level and water-consumption monitoring, satellite image interpretation of land surface and vegetational change
Earthquake	Low or zero predictability for particular events, as monitored earthquake precursors are unreliable; high predictability for seismic zones	Seismometers, accelerometers, tiltmeters, magnetometers, electrical resistivity meters, water quality analysis
Environmental fire	Timing and consequences are both predictable	Meteorological forecasting, vegetation sampling, aerial overflights, ground-based look-outs
Expansive soil	Risks are highly predictable if monitoring is adequate	Geotechnical testing of soil samples, hazard mapping, infiltration testing
Flood	Short-term recurrence intervals are often highly predictable, long-term return periods and timing of individual events are less so	Meteorological forecasting, remote sensing of snow-pack melting, satellite image interpretation, fluvial hydrological monitoring, water-level sensing alarms
Fog	Highly predictable	Meteorological forecasting, ground-based observation

Type of disaster	Predictability and forecast potential	Monitoring techniques
Frost or ice-storm	Moderate predictability	Meteorological forecasting, ground-based weather recording
Hail	Seasonality and timing of storms are highly predictable, exact location of damage is not	Meteorological forecasting, Doppler radar monitoring of storms
Hurricane	Seasonality and recurrence are highly predictable, tracking of hurricanes after they have been identified is more than 75% predictable	Meteorological forecasting, satellite monitoring, radar tracking of storms
Insect infestation	Progress of phenomenon after detection is predictable, but occurrence rather less so	Aerial survey, ground-based field monitoring
Intense rainstorm	Seasonality and timing are very predictable, but damage not easy to forecast	Meteorological forecasting, raingauges
Landslide	Susceptible terrain can be mapped easily, occurrence can be forecast in real time	Mapping and surveying, extensiometers, piezometers, rainfall monitoring, infra-red photography
Lightning	Disaster potential is seldom predictable	Limited scope for meteorological forecasting
Snowstorm	Variable predictability for both meteorological conditions and impact	Meteorological forecasting, ground-based observation, telecommunications
Subsidence	Easier to predict once it has started than before it manifests itself	Ground-based surveys, extensiometers, tiltmeters
Tornado	Moderate predictability, especially if monitoring is widely and carefully used	Meteorological forecasting, spotter networks, Doppler radar, radio signals
Tsunami	Moderate predictability before first wave is born, high predictability thereafter	Seismometers, tide gauges, taut-wire buoys, bottom pressure transducers, computer simulation models of run-up
Volcanic eruption	Moderate to high predictability if monitoring is comprehensive during pre-eruptive phase	Seismometers, infra-red remote sensing by satellite or aerial camera or ground-based telescope, tiltmeters, gas analysers, magnetometers, electrical resistometers, gravimeters
Windstorm	Low to moderate predictability	Meteorological forecasting, anemometry

before anthropogenic warming trends, so a better understanding of the variability of snow cover and the response to climate change is essential (Barry *et al.*, 1995). Remotely sensed sea-ice distribution can also provide important data for climate change studies, and Piwowar and LeDrew (1995) found some evidence for decreasing ice extent and concentration.

9.3 Global change scenarios

The greenhouse effect, which may be something of a misnomer, has probably been recognized since 1827 (Jones and Henderson-Sellers, 1990), but it is only towards the end of the twentieth century that there has been worldwide consensus on the direction of change (Houghton, 1997), namely an increase of temperature resulting from an equilibrium doubling in greenhouse gases over pre-industrial levels (Henderson-Sellers, 1994a, p. 124).

Despite increasing attention given to global change, it has been difficult to raise general awareness of the problems, although *Global Environmental Outlook 2000* (United Nations, 1999) was widely reported as indicating that global warming will trigger a series of disasters with serious world implications. An increasing global focus within physical geography (Section 9.1), together with the technology available (9.2) means that geographers should be actively involved in the surge of interest in global change. Liverman (1999, p. 107) reminds us that:

> The study of the human relationship to the environment is one of the major traditions of geography, bridging physical and human geography and echoing many of the debates in the discipline.

She provides a review of the development of environmental policy from the 1972 United Nations Stockholm conference to the 1992 Rio Earth Summit, which focused on issues of climate change, biodiversity protection and sustainable development, and also reviews how the Agenda 21 Report produced for the meeting (based on scientific assessments convened by Professor Tim O'Riordan) laid out an agenda for research and action over the next century. In view of the significant involvement of geographers, Liverman (1999, p. 116) concludes that:

> Geographers are engaging a wide range of issues in global change as defined mainly by the earth science community, including model downscaling, climate impacts, land use, regional emissions and integrated assessment . . . Geography has much to offer the study of international environmental and global change, and to the development of related policies.

Subsequent world meetings in Kyoto (1997) and Buenos Aires (1998) were equally significant. In recognizing that increased awareness of the possible consequences of global-scale pollution has spawned research programmes loosely termed 'global change', Henderson-Sellers (1992) noted how global change is in many ways a synonym for global geography: the study of processes and their consequences at the human–environment interface.

Such directions therefore indicate the scope for geographical engagement in research on topics of contemporary interest. Inevitably the first major focus has been the contribution to climate models; Huggett (1991, p. ix) addressed the potential in one of his recent books and commented:

> Our knowledge and understanding of the workings of the world climate system has taken a quantum leap in the last couple of decades. The reasons for this lie partly in the building of sophisticated climate models for computers, partly in the vast amounts of data sensed by satellites, and partly in the establishment of a reliable calendar of geological events.

Some physical geographers have therefore been able to contribute to the way in which results are obtained from climate models (Henderson-Sellers, 1994a; Howe and Henderson-Sellers, 1997), and changes of surface air temperatures, precipitation rates, and soil water world patterns have been produced for different seasons (Henderson-Sellers, 1994b). In addition, research has focused upon the ways in which climate change, as indicated by the outputs from GCMs, might impact on the environmental system. The implications for sea-level rise are an obvious subject for attention which has been considered generally (e.g. Warrick and Farmer, 1990; Spencer *et al.*, 1995) and also in relation to the detailed impact on specific areas such as the European coastal lowlands (Tooley and Jelgersma, 1992) and coral reefs (Spencer, 1995). As in the case of applications of physical geography in general (Section 8.4), this can identify further research needs which may involve field verification, historical data and model development, and lead to a focus upon the possible responses that may be made at the coastline, including fortification of the shore, retreat from the coast, or nourishing the beach (Leatherman, 1990). In addition there has been research undertaken on the hydrological consequences of global change; but GCMs are not well-suited to answering questions of primary interest to hydrologists concerning regional hydrological variability (Xu, 1999). GCMs were originally designed to predict average synoptic-scale general circulation patterns of the atmosphere, so their outputs cannot easily be harnessed by downscaling to provide inputs to hydrological models. Although estimation of hydrology and water resources changes continues to pose problems for hydrologists (Nemec, 1995), analyses of scenarios have been considered for river flow extremes and fluvial erosion in England and Wales (Newson and Lewin, 1991) and for water resources in the UK (Arnell, 1996). Against a background of methodologies for climate change impact assessments, techniques for developing climatic change scenarios and hydrological models, Arnell (1996) effectively used a case study of 21 catchments in the UK to explore changes in river flows (Fig. 9.3) and groundwater recharge in Britain that

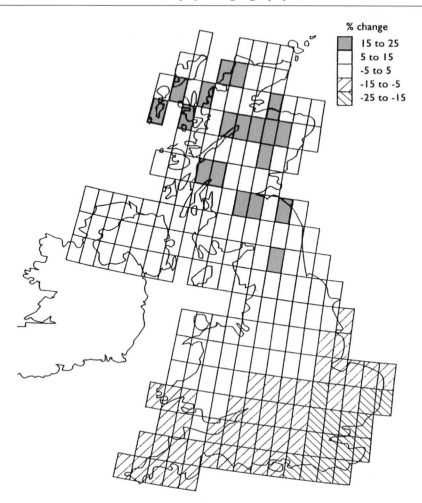

Figure 9.3 Suggested percentage change in average annual runoff for the 2050s in the UK (after Arnell, 1996), under one feasible climate change scenario, estimated by 0.5 × 0.5 grid cell.

might occur by the 2050s, and analysed the possible impacts of the changes on water use and the management of water resources. This offers an excellent illustration of the potential role that a physical geographer can fulfil and one which meets the opportunities outlined by Liverman and by Henderson-Sellers above (p. 243).

In the conclusion of his analysis, Arnell (1996, p. 200) identified five important gaps related to the potential effects of climate change on water resources in Britain and worldwide:

(1) Credible climate change scenarios at the catchment scale – requires credible simulation at the global and regional scale but also reliable techniques for downscaling climate model simulations to the catchment scale.
(2) Robust models to translate local climate changes into biophysical impacts which require more integrated models (as noted below).
(3) Studies to estimate the impacts of these changes, in economic and financial terms, in different parts of the water sector – this will extend to other areas such as navigation and power generation.
(4) Studies into adaptation to change, especially decision-making in an uncertain environment.
(5) Identification of critical sensitivities, which would require looking at the problem in reverse to identify the type and degree of change that would be necessary to have some defined impact upon a system.

Despite the opportunities available and the need for research, the amount of activity by physical geographers in this area has arguably been rather small. One reason for a reluctance to be involved may be the uncertainty which is inevitably an element in any forward-estimated scenarios and is noted by Arnell (1996, p. 198) and more generally by Goudie (1993a, p. 138):

> There are almost certainly mechanisms of climate change which we do not yet suspect or which we cannot qualify to an adequate level. Some of these might accentuate the greenhouse effect, while others might serve to counter it.

Goudie (1993a) offered ten possible explanations of this uncertainty which really devolve upon complex interactions of systems with large numbers of variables, possibly non-linear and difficult to model; paucity of long-term data and difficulty in defining some human response factors, desertification and deforestation, etc.; and the fact that there are bound to be factors that have been ignored, with nature possibly having surprises up its sleeve in the form of catastrophic or extreme events. It is generally agreed that regional climate prediction is not an insoluble problem, although it is characterized by inherent uncertainty which derives from two sources: the unpredictability of the climatic system as a result of deterministic chaos; and the global system which renders climate predictions uncertain through unpredictability of external forcings superimposed on the climate system (Mitchell and Hulme, 1999). A number of people have therefore argued (e.g. Airey and Hulme, 1995) that model simulations of future changes to the magnitude, timing and spatial pattern of global precipitation should be viewed as scenarios, not predictions. Even ahead of recent research, White (1989; quoted by Rayner and Hobgood, 1991, p. 216) had noted the general problem that exists:

> We are confronted with an inverted pyramid; a large and growing mass of proposals for policy action is balanced upon a handful of real factors. Data or likely causes are robust, though future emission projections vary widely. Projections based upon mathematical approxi-

mations of atmosphere and oceanic conditions are credible but uncertain ... Let us not confuse selected observation with representative samples. Let us not confuse scenarios with predictions. Let us not confuse short term fluctuations with long term implications. Above all, let us not confuse our friends and colleagues who must make the political decisions that will ensure the inhabitability of this planet.

In any future contributions, therefore, physical geographers have to be aware of the uncertainties, of the outstanding need to utilize more integrated models, and to undertake research on impact scenarios. In proceeding towards more integrated models, the Holdridge life zone classification has been examined by Henderson-Sellers (1994b) as an example of an impacts model: there have been at least ten versions of a modified Holdridge model, and, according to which version is used, very different conclusions can arise about the future distributions of continental vegetation. The coupling of a terrestrial vegetation model to a global climate model has the potential to generate transient vegetation changes, but there are many outstanding difficulties in converting the continental vegetation maps into policy advice (Henderson-Sellers, 1994b). The combination of remotely sensed data and soil–vegetation–atmosphere (SVAT) schemes is an important priority for future research to develop from what has already been achieved (Burke *et al.*, 1997). More sophisticated and complex methodologies are being used to evaluate impacts and to develop informed decision-making strategies (Parry, 1993), so that attention has moved from global Domesday scenarios to predictions (scenarios) of potential effects on individual nations, regions and industries. However, the spatial patterns that could result need to be more fully explored by physical geographers with some urgency because as Perry (1992, p. 99) prophesied:

> Initial results strongly suggest that global warming will worsen rather than reduce inequalities between the rich North and the poor South nations. The developed temperate countries are not only most likely to develop technologies for adaptation to climate stress, but they also stand to derive most benefit from reduced cold temperature stress.

9.4 Future contributions by physical geographers

The scope for physical geographers to contribute in a multidisciplinary context seems very considerable, and several physical geographers have emphasized the opportunities. Thus Thornes (1995, p. 366) asserted:

> As geographers we stand to gain much if we can harness this regional-scale analysis of the effects of change. The tools are already there. Are there still the demanding intellectual skills and interdisciplinary sense of purpose which such analysis requires?

In one of his books noted above (p. 244), Huggett (1991, Chapter 8) identified three pivotal issues to be resolved, namely:

(1) *Relations in the biosphere* – the biosphere system is composed of a huge number of elements and needs to be modelled.
(2) *The problems of scale* – upscaling and downscaling are referred to above (p. 240).
(3) *Origin of cyclicity* – many variables are periodic in nature with period lengths varying from days to millions or billions of years.

Although the physical geographer should be able to contribute in these areas in considering modelling the human impact on nature, Huggett (1993, p. 184) argued that although geographers are eclectic and interested in spatial matters in the terrestrial sphere, nevertheless most modelling of global and regional systems has been carried out by specialists in departments of atmospheric, ecology and earth sciences and applied system sciences. Huggett (1993, p. 185) concluded that:

> Geographers then are in a curious position. The core of their discipline is the interaction between the human species and its environment; yet some of the giant scientific strides in this topical area are being made by non-geographers. And, to add insult to injury, many of the calibration techniques used in complex spatial models are borrowed from geography.

He went on to make the even more salutory point (Huggett, 1993, p. 186) that:

> There may be a case for studying the spatial distribution of fish-and-chip shops in Weston-super-Mare, but there is an even more urgent case . . . for tackling the environmental consequences of global warming.

What particular problems should physical geographers address, as distinct from the contributions to be made by human geographers, and how should they achieve a significant impact on studies of global warming? Perhaps eight major thrusts have to be pursued or developed from present initiatives.

(1) To continue to research on data provision related to global change, especially remote sensing data, because understanding of the environment is reasonably sound at the local scale, but there have been comparatively few observations and much less research at the regional and global scales.

(2) To contribute to establishing links between GCMs and models of parts of the environmental system. This embraces many aspects of climatological study and what Henderson-Sellers (1989b) termed atmospheric physiography (see p. 287). In many disciplines it is now appreciated (e.g. Cotton and Pielke, 1995, p. 219) that:

... many scientists are grossly underestimating the complexity of interactions among the earth's atmosphere, ocean, geosphere, and biosphere

and physical geographers have a role in clarifying and modelling that complexity. This can be at the national, regional or local-site scales, as exemplified by Knapp and Soule (1996) at a site in central Oregon where they concluded that changes in vegetation composition and dominance should be viewed not as unusual or unnatural but as a normal component of ecosystem dynamics.

(3) To develop research investigations on the spatial impacts of global change. Whereas GCMs produce output at the global scale, a number of researchers have demonstrated the necessity of relating these outputs to regional or national levels, the kind of spatial analysis that a physical geographer is particularly trained to achieve. This can continue the approach effectively achieved by Arnell (1996), and geographers could have a major part to play in informing political decision-making in relation to a contemporary hydrological science which is imbued with notions of sustainability, holism and ecocentricity (Wilby, 1997). There are many cases, arising from the type of investigations referred to in Chapters 5 to 7, where future impacts of global change need to be analysed in the context of spatial and temporal models of geomorphological and ecological response to specified environmental forcing (e.g. French *et al.*, 1995).

(4) To make the results of their research available and to ensure that the contribution that physical geographers can make is proselytized. This involves collaboration with other disciplines and, after contending that 'There are a variety of views about the success of interdisciplinary climate change research caricatured, at the extremes, as either a totally dysfunctional or fully harmonious system' (p. 453), Henderson-Sellers (1996) advocates a climate change information exchange to improve the mechanisms of information exchange and to encourage greater transparency; and also a cyclic social cooperative which would involve the exchange of scenario development. This could contribute significantly to integrated assessment by improved communication, which transcends hooking models together to deliver information and end results to climate researchers for the benefit of society as a whole (Henderson-Sellers, 1996).

(5) To forge links with social sciences for the benefit of impact assessment, and if appropriate to modify the training of physical geographers. Thus it is necessary to keep under review the leading climate indices that can be used for impact assessment, which Easterling and Kates (1995) suggest should include a climate extremes index, a greenhouse climate response index, hazard warning, ecosystem health, and energy demand and renewable natural resources; and to think of the vulnerability and adaptive responses possible in the context of climate and climatic change (Burton, 1997). In this area it has been suggested (Easterling and Kates, 1995, p. 646) that:

> Climate observing systems are at a crossroads in planning the archi-
> tecture and attributes of future climate data and information retrieval
> systems. If public support is to be obtained and sustained, then the
> fruits of such systems need to be useable knowledge.

This raises a particular opportunity for the geographer climatologist
because there will be a need for a broader training than hitherto which is
founded in atmospheric and social sciences; and there will be an increasing
requirement for social scientists with an environmental interest to take climatology courses and to become climatologically aware (Perry, 1995). Perry
(1995, p. 281) develops the argument to suggest that:

> If the 'complete climatologist' of the future is an individual who not
> only has had a thorough grounding in atmospheric science studies but
> is also conversant in the social sciences, this has major implications for
> the training and perhaps the background of the new generation of
> climatologists.

(6) It is imperative that the results of impact studies are communicated
beyond the discipline to address the position stated by O'Riordan (1994,
p. 10) that:

> ... science on its own cannot solve global environmental change. This
> is not even a matter of multidisciplinarity, the welding together of
> disciplines yet maintaining their ideological and conceptual integrity
> ... interdisciplinarity is much more than a fusing of disciplines.
> Interdisciplinarity covers both the similarities in the natural and social
> worlds as well as the subtleties of how science is manipulated in its
> civic contexts.

It is therefore important to be aware of multidisciplinarity and interdisciplinarity (Fig. 9.4) and what O'Riordan describes as civic science, which is
negotiated science where future states are envisaged through open structures
of learning, consensus-seeking and bargaining:

> Above all ... civic science needs its application in the real world. Its
> achievements will not be in the universities ... its successes will come
> through the tough bargaining that will now be a feature of the post
> Rio World, as the sustainability transition is properly addressed.
> (O'Riordan, 1994, p. 10)

We have to consider the position relative to all interested disciplines.

(7) There is the need to be aware of the overall interlinkages which have
led some geographers such as Park (1997) to focus students' attention upon
the environment as a whole, with others considering the ingredients for a
green(er?) geography, as when Simmons (1990) stated that:

> What seems to be necessary here is an understanding not solely of
> synchronous patterns ... but of changes and interlinkages ... Our

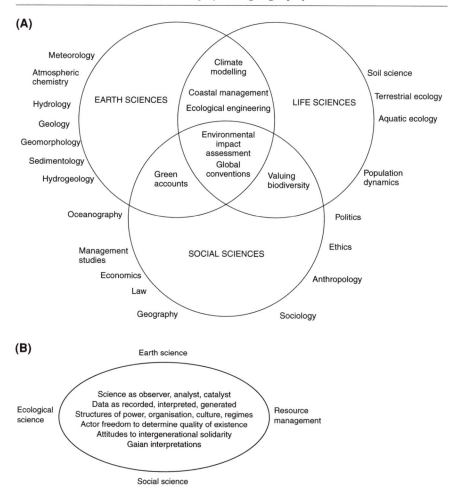

Figure 9.4 Depiction of (A) multidisciplinary and (B) interdisciplinary environmental science (after O'Riordan, 1994, with permission from *Scottish Geographical Magazine*).

physical geography needs also to be aware of the interlinking between the major components of the global biophysical systems. Geographers' tendency to study the oceans, land surface, and atmosphere in isolation has led to insufficient emphasis in our teaching on interactions.

This could extend to some areas which have hitherto been given comparatively little attention by geographers, and especially physical geographers. Ethical issues were referred to in the context of applications of physical geography (Chapter 8, p. 222) and this is pertinent in relation to global change, so that we must find a way of cutting through the complexities of institutions and communication channels to ensure that the science and ethics issues are revealed:

> The goal of environmental geography could be to sustain clear communications among the complex organizations debating policy implications that pertain to the human/physical interface. (Henderson-Sellers, 1998, p. 306)

(8) Physical geographers can take the initiative in bringing specialists in other disciplines together for edited volumes and conferences. A special issue of *Climatic Change* (1993, vol. 25) followed two workshops sponsored by the Australian Federal Department of the Environment, Sport and Territories (DEST) in May 1991. A workshop convened by two scientists, one of whom was a geographer (Henderson-Sellers, 1993), had goals to focus upon severe weather hazards and urban planning, transport and pollution; to address seven objectives including identification of the main areas of cost and benefit to Australia of reducing emissions of greenhouse gases; and to identify the principal impacts of global warming in Australia and strategies in response to these impacts.

There are therefore opportunities waiting for physical geographers to seize, and although one should be aware of the danger of maintaining funding for earth scientists (Blaikie, 1996), there is no doubt that the research field available will be cultivated by other disciplines if physical geographers do not respond to the opportunity. Thus Ke Chung Kim and Weaver (1994) addressed *Biodiversity and Landscapes* in a book derived from an international discussion meeting at Pennsylvania State University Center for Biodiversity Research, with leading contributors from fields including ecology, landscape management and sociology, but not geography or physical geography. Great opportunities exist for the branches of physical geography, and in relation to geomorphology Goudie (1990b, p. 59) predicted:

> ... humans may be on the brink of a great geomorphological experiment, in that there is a probability that human atmospheric gas modification will create geomorphologically significant climatic changes.

Considering areas for the future siting of physical geography as a whole, Slaymaker and Spencer (1998, p. 252) in their final chapter concluded that 'identification of the global impacts of society on the environment or an updating of society's role in changing the face of the Earth is in itself a worthy goal for physical geography'. They refer to this theme as identification of global impacts and global actors, and their other concluding themes are sustainability of biogeochemical systems, decision-making in an environment of uncertainty, and a renewed emphasis on biogeochemical fluxes and protection of the atmosphere, biosphere, hydrosphere and lithosphere. These are pertinent to the future of physical geography (Chapter 11), after consideration of cultural trends in Chapter 10.

Further reading

SLAYMAKER, O. and SPENCER, T. 1998: *Physical geography and global change*. London: Longman.

ARNELL, N. 1996: *Global warming, river flows and water resources*. Chichester: John Wiley.

ROBERTS, N. (ed.) 1994: *The changing global environment*. Oxford: Blackwell.

Topics for discussion

(1) Ways in which physical geographers should develop further research in relation to global change.
(2) Devices to overcome the separation of physical and human geography in relation to global change.
(3) Are realist research investigations conducted at the microscale compatible with, and related to, global-scale analyses?

|10|

Cultural physical geography

Previous chapters have outlined approaches used by physical geographers and it could be argued that those approaches are sufficient. However, a number of trends, including those towards holism as well as those that question whether all the objectives of physical geography can be fulfilled by the existing branches, mean that a more cultural physical geography (10.1) can be contemplated. This should not in any way detract from, or reduce, the core content of physical geography, but it may reflect roots similar to those that prompted cultural geography in the human part of the subject. This cultural trend is not entirely new; it was anticipated in much of the work of Yi Fu Tuan, and is supported by strands emerging from the branches of physical geography (10.2) including cultural ecology. A series of linked themes appropriate for a focus at the beginning of the twenty-first century prompt the question of the extent to which physical geographers should become involved in implications of environmental design (10.3), which is pertinent to future cultural themes (10.4).

10.1 Towards a cultural physical geography

Preceding chapters may have offered sufficient approaches to physical geography without adding one more – and one which is less coherently developed than earlier ones. However, in the final part of several chapters (e.g. 5, 6, 7, 8, 9) current trends are directed towards a more holistic approach. Probably applicable to physical geography as a whole was the view of Cooke (1992, p. 132) that the subject of geomorphology:

> ... commonly only entertains human activity where people manipulate variables in complex physical systems, and it usually ignores cultural, economic and managerial contexts, and fundamental feedback mechanisms.

These issues have been rectified to some extent by a more global focus, but physical geographers may not have fully seized the opportunities available (Chapter 9, p. 246), although the need for links between the physical and human parts of the discipline has been appreciated once more. Particular approaches in physical geography, although often envisaged as increasingly divergent as they developed, have tended to become increasingly convergent: a convergence that provides a singular opportunity. The sparsely inhabited niche between the earth system science approach to global environmental change and the prevailing approach to physical geography was perceptively identified by Slaymaker and Spencer (1998, p. 247) and this niche (Fig. 4.3) is surely the one that geographers have always sought to occupy (see also p. 100, p. 286). The opportunity is in some way analogous to the way in which human geography saw the regrowth of cultural geography partly as an antidote or complement to the more statistical and quantitative positivist approaches prevalent since the mid-1960s. McDowell (1994) traced the evolution of cultural geography through three schools and noted the resurgence of the landscape school. Perhaps in the year 2000 we are at a similar position in physical geography? A Davisian approach dominated up to the 1950s, and was succeeded by a plethora of approaches involving a range of quantitative, laboratory and field techniques destined to do justice to the environment as a whole. However, having assimilated a range of new methods and encompassed a broader perspective, it has been timely to adopt a more global vision: is the time now ripe to include a more cultural stance? 'Culture' is a term used by social scientists for a way of life, and includes the total of the inherited ideas, beliefs, values and knowledge which constitute the shared bases of social action. Physical geography has focused on environment largely oblivious of the way in which culture conditions what we see as environment and how research results are received.

If physical geography does not respond to current trends, then other disciplines will, and some have already done so. Thus a historian approached *Landscape and Memory* (Sharma, 1995) by a very stimulating and wide-ranging approach which attracted much general interest, and Nash (1982, 1989) reviewed *Wilderness and the American Mind* and *The Rights of Nature*. From an artist's standpoint, Fuller (1986) argued for a post-modern aesthetic in art and suggested developing a more imaginative response to nature, a renewed interest in natural forms, and a clearer understanding of the ecological interdependence between people and their habitats. Within science, botanists and ecologists have become involved in the cultural landscape and Birks *et al.* (1989) viewed the cultural landscape from the perspective of palaeoecology, such that one geographer reviewer (Williams, 1990, p. 270) suggested that botanists feel that they have 'discovered' the cultural landscape. Environmental ethics derives from different disciplines, and *People, Penguins and Plastic Trees: Basic Issues in Environmental Ethics* (Pierce and Van de Veer, 1995) embraced issues of which physical geographers should be cognizant (Rhoads and Thorn, 1996). Sociologists and social theorists also have an interest, and Adam (1998) in *Timescapes of Modernity: The Environment and Invisible Hazards* advocated that 'Thinking of environment as a timescape allows us to see the hazards of an industrial way of life'.

In selecting a sample of books from several disciplines, there is no suggestion that the study of all aspects of environment is particular to physical geography. On the contrary, in view of the multidisciplinary approaches that are increasingly advocated (p. 167 and p. 249), it is imperative that the physical geographer's expertise and insight is appreciated, understood, and, where appropriate, used alongside that of other disciplines. In view of the great interest in aspects of environmental matters reflected in the disciplines of books quoted, there is obviously a need for research and writing on new cultural areas, with physical geographers responding to the opportunities available. In an earlier phase of development, Hare (1977, p. 263) commented on the poverty of geographical science training because we had:

> ... swept geography departments into the Social Science division of faculties of arts and sciences where, from playing second fiddle to geologists or literary critics, we learned to play second fiddle to economists and sociologists.

It would be unfortunate if physical geography were not in tempo with present cultural developments. In relation to geomorphology, Thorn (1988) looking along the yellow brick road concluded that there must be a preoccupation with covering laws as the primary goal, with a methodology in four steps:

(1) develop, maintain and sustain a strong personal knowledge of geomorphological theory;
(2) formulate research questions embedded in geomorphological theory and pose them in a falsifiable way;
(3) build and attempt to falsify a characteristic form hypothesis–equilibrium;
(4) build and attempt to falsify a relaxation form hypothesis.

To this scientific foundation could be added the inter-relationship with the human environment and its culture, and communication of the results of research in a cultural context. Although perhaps overstated, focus upon the human environment was stressed by Kates (1987, p. 533):

> Our physical geography, an important part of our claim to knowledge of the human environment, is in a shambles. It needs strengthening, attention, and self-revision. At a time when the physical and life sciences begin to work together, when they attend to the great biogeochemical flows of nature, too much of our knowledge of the Earth begins and ends separately as geomorphology and climatology.

In a broad view relating to geography as a whole, Stoddart's (1987, p. 334) view was that 'We need to claim the high ground back . . .' (Chapter 1, p. 6).

10.2 Cultural strands

Cultural strands began to appear in several ways in the research literature written by, and available to, physical geographers, and five are proposed here. The first must be associated with Yi Fu Tuan who began his academic career as a geomorphologist, published on pediments and the geomorphological problems of the South West US including arroyos, but subsequently became interested in environment in its broadest sense and developed his career in a way which led him to produce some discipline-shaping contributions (Tuan, 1974, 1992). Tuan (1989, p. 233) considered that appreciation of nature or landscape is a principal reason for becoming a geographer and, although aesthetic impulse and experience is not confined to any particular class of individuals, it is geographers who 'like to know not only where things are, but also how it feels to be in a place'. Having identified the importance of the aesthetic, Tuan (1989, p. 240) concluded that:

> Scientists strive to stand far above their material, for a view from nowhere . . . with the hope that they will thereby be able to plunge well below the surface of reality. By contrast, cultural geographers-cum-storytellers stand only a little above their material and move only a little below the surfaces of reality in the hope of not losing sight of such surfaces, where nearly all human joys and sorrows unfold.

It is not intended to suggest that physical geography should stray into the domain of human geography, or overlap with disciplines in the social sciences such as sociology, but we do need to ask to which aspects of culture is the expertise of the physical geographer particularly relevant and where can physical geographers contribute to a more informed use of knowledge of, and insight about, the physical environment. This first strand therefore derives from origins similar to those of cultural geography.

A second strand is the way in which a more cultural approach emerged from particular branches of physical geography. For example, in *Coastal Environments: An Introduction to the Physical, Ecological and Cultural Systems of Coastlines*, Carter (1988) illustrated that just as research tended towards a more applied stance (Chapter 8), so the cultural element also appeared, particularly in books, and the two were often inter-related. A particular environmental theme was sometimes used; in relation to floods, Baker (1994) contended that human perception needs to be stimulated with real understanding. *Water, Earth and Man* (Chorley, 1969a) was an excellent example using water, later used in a more cultural context by Cosgrove and Petts (1990) in *Water, Engineering and Landscape*, bridging the traditional gulf between human and environmental approaches to produce an integrated geographical view of the relationship between technology, culture and natural resources. In the introduction Cosgrove (1990, p. 9) suggested that:

> ... there is growing practical recognition of the magnitude and interlocking nature of long-term social and environmental problems associated with large scale intervention. This is part of a broader change in consciousness. Recognition of the benefits of ecological and cultural diversity and acknowledgement of the integrity of 'otherness' are world-wide phenomena ... The favoured term for the new way of conceiving the hydro-environment and its management is 'holism'.

Climatology also provides examples including a perspective on *John Constable's Skies* (Thornes, 1999).

A third strand may be identified from the development of human ecology, or its synonym cultural ecology (Butzer, 1989), which deals with the complex interaction between the ecological system and the human social system. Whereas thermodynamics and the biogeochemical cycles govern the web of transfers of energy and mass in ecosystems, and organisms grow by symbiosis with each other and with the inorganic parts of the system by utilizing energy and by reproducing themselves, two additional factors (Alexander, 1999a, p. 326) are the use of technology which enables the social system to exist and to propagate itself, and human ecology responses to decisions to seek new sources of fuel, food, medicines and lebensraum. The origin of human ecology is ascribed to the geographer Kropotkin (1842–1921) and also to Elisee Reclus (Stoddart, 1986). The term 'human ecology' came into use about 1910 as a label for man–environment relationships which had become a significant focus for study by geographers (Alexander, 1999a, p. 327), and in 1923 (see also p. 173) Harlan Barrows used his presidential address to the Association of American Geographers to propose that human ecology should be the central theme of the discipline (Barrows, 1923). Although not pursued at that time, the theme was referred to half a century later by Chorley (1973) and Porter (1978). Geographers emphasized ecological aspects of resource usage and natural hazards from the ecological perspective by investigating the human adaptation to environmental extremes (e.g. White, 1973) but other disciplines also became interested in this interface. Anthropologists studied the use of natural resources, cultural behaviour (cultural anthropology) and the impact of environmental constraints on human biophysical functions (physical anthropology), whereas sociologists examined human behaviour under complex social conditions (Alexander, 1999a). In a definitive summary, Butzer (1990, p. 685) concluded:

> Societies can be regarded as interlocking, human ecosystems. They operate on the basis of individual initiatives and actions, embodied

in aggregate community behaviour and institutional structures. Decisions are made with respect to alternative possibilities, within a social system.

Although aspects of relationships between environment and human activity need to be considered by physical geographers, the dangers of a simplistic determinism need to be borne in mind. As the systems approach enables human ecological relationships to be studied at a wide variety of scales, from the individual level to the level of large social groups, Butzer (1990) argued that these can be examined in diachronous mode by historical synthesis, or in synchronous mode by analysis of contemporary case studies. Cultural considerations are thus appropriate because the physical environment is perceived through a cultural filter made up of attitudes and limits set by past experience, observational ability, and ideology. Social norms can determine the choice of what is significant, although Butzer (1990) sees ethics and ecological behaviour as separate issues, and outlines cultural ecology (Butzer, 1989, p. 193) as concerned with the manipulation of resources within ecosystems.

Alexander (1999a) quotes Brislin (1980, p. 47) who argued that because of the multiplicity and variety of human cultures, cross-cultural studies are necessary for the development of theories in environmental research since no one culture embraces all environmental conditions that can affect human behaviour. Brislin (1980, p. 57) therefore used the term *etic* to designate that which is common to many cultures, and can be verified by scientific analysis or transferred from one culture to another; and *emic*, meaning that which has full meaning only within the bounds of a particular culture and cannot easily be transferred. Thus science and technology tend to be etic, but legends and myths, associated with agrarian cultures for example, are emic in character. It is this dichotomy that prompts the issues raised in this chapter. Hitherto we have tended to think of physical geography in etic terms, but increasingly have to be aware of more emic considerations, for example in relation to landscape management.

The human ecology approach features in other disciplines such as anthropology and sociology, and overlaps with archaeology, which provides a fourth strand. In the introduction to *Landscape and Culture – Geographical and Archaeological Perspectives,* Wagstaff (1987) argued that geography and archaeology are both concerned with two dimensions of a single field and that, whereas geography takes the spatial dimension, archaeology and history take the time dimension. Butzer (1982) viewed archaeology as human ecology, more ecologically based than geoarchaeology which can be viewed as the study of environmental changes in so far as they influence the interpretation of archaeological remains (Vita Finzi, 1999, p. 274). Fruitful research links built with archaeology (e.g. Thornes, 1987; Brown, 1997) advanced the study of environmental change (Section 6.5, p. 169).

A fifth strand arose because the writing by biogeographers on the man–environment interaction led not only towards human impact but also towards the way in which nature is described and interpreted. Thus Simmons (1993a) imaginatively reviewed cultural constructions of the environment, which was the subtitle for *Interpreting Nature* in which he argued:

> Since this book is aimed primarily at geographers, it also serves notice on them that I think they have to break out of some of their traditional notions of what Geography can deal with if they are to have anything to say about environment. (Simmons, 1993a, p. xii)

To review ways in which nature has been interpreted, Simmons considered four views: those of the natural sciences and the realism which they espouse; those of the human sciences; the world as seen from the viewpoint of the individual person; and what we ought to do in philosophical and legal terms. His book cited a large multidisciplinary literature, and from sociology he quoted Cotgrove (1982) who identified two social groups with particular environmental attitudes, namely *Cornucopians,* who have faith in technology and economic development, including resources which can be available if investment in technology is high; and *catastrophists* who see the physical limits to resources so that the planet's systems can be greatly degraded by environmental contamination, and require reform in controlling wastes and reducing consumption in industrialized nations. Simmons (1993a) concluded with two alternative models which are:

(1) Based on realism, acknowledging that there is a real world outside our bodies. It can be affected by human actions but can go its own way, our part is small and possibly accidental, and we are therefore not a necessary component. This leads through to the values which have been deduced from a Darwinian view of organic evolution.

(2) Based on idealism, so that everything we claim to know is an extension of our minds. This has resulted in giving our species a dominance of natural processes so that all is eventually possible. This coincides with a basic humanistic view.

Subsequently Simmons (1997) utilized the approach in a textbook which offered an integrative approach to understanding human–environment interaction. A further part of this strand, fostered in biogeography, has been that of landscape ecology which focuses explicitly on the inter-relation between landscape structure (i.e. patterns) and landscape function (i.e. processes) and can be the basis for application of principles to issues such as nature reserve design and functioning. Kupfer (1995, p. 30) proposed that:

> The theoretical shift towards a landscape ecological perspective should be viewed as a fresh opportunity for biogeographers to be on the cutting edge of new theory development. Many of the central landscape ecological principles focus explicitly on spatial relations and the links between humans and the environment . . .

Because landscape ecological studies show a greater dependence upon geographical tools such as geographical information systems (e.g. Haines-Young *et al.*, 1993), there should be contributions for biogeographers to make, for example, in improving conservation and management plans (see Chapter 8, p. 218).

These five strands indicate that physical geographers should be aware of the ways in which the physical environment that they study and research is portrayed and perceived, and it is therefore germane to know how environmental meanings are conveyed in the media. Burgess (1990) argued that the media are an integral part of a complex cultural process through which environmental meanings are produced and consumed, so that physical and human geographers could usefully collaborate in research with both producers and consumers of media text so as to better understand contemporary discourses about human–environment relations. In the research questions that are part of an agenda for media research in geography, Burgess (1990, p. 157) proposed:

> ... I would ask how landscapes, places and nature are encoded in the press, television, radio, the cinema and advertising, and what do they signify for different groups of consumers? ... How is environmental science transformed by different media for non-specialist audiences and what sense do people make of different areas of scientific research which may affect their daily lives or the world which their children and grandchildren might inherit? How is an increased awareness of environmental degradation and landscape change accommodated within people's codes of values, attitudes, experiences and social practices and how are these meanings constrained by the structural contexts in which people live?

Physical geography research can be extended to embrace the way in which the environment is perceived as an integral part of different cultures, and to appreciate how these affect the way in which the environment is managed. Hitherto, research across the various branches of physical geography has not explicitly addressed the fact that the physical environment is perceived and portrayed in different ways which can have relevance for management and decision-making. Environmental consciousness is distinctive and Cooke (1992, p. 132) speculated that:

> At a time when geography enjoys a public recognition far greater than ever before precisely because of its perceived role in studying environmental issues ... and the distinction of some its recent research contributions, there is a danger that once again we shall allow our fissiparist tendencies to deflect our attention away from our difficult, but fundamentally important integrative role on the common ground where human and physical geography overlap.

As developments have embraced some of the subjects studied by physical geographers, *Ecumene* (1993), a new journal for the humanities, cultural studies and social sciences, may be of interest.

10.3 Towards environmental design?

In view of the strands outlined in Section 10.2, how can we outline an agenda for a more cultural physical geography appropriate for the twenty-first century? All that is attempted here is to place emergent approaches within a general framework.

Present cultural themes can be thought of as a series of successive levels, summarized with examples in Table 10.1. The first *information level* includes the responsibility of the physical geographer to inform and educate about the characteristics of the physical environment. This encapsulates the need to inform those concerned with development and management of the environment in a specific cultural context about the current understanding of the dynamics of environment and the implications of long-term change, always remembering how our knowledge and understanding are time-dependent. Decisions have been made without fully appreciating the holistic nature of the problem under consideration. Thus integrated drainage basin management (IBM) has been employed in various guises in many countries, but Downs *et al.* (1991) considered how integrated the various schemes really were, demonstrating that appreciation of geomorphological implications of river channel change, a fully holistic view and an understanding of change over time were insufficiently incorporated in many existing schemes. In reviewing landscape indices available to quantify landscape structure, Haines-Young and Chopping (1996) commented that an implicit assumption of many environmental decisions is that some pattern, or combination of land cover, is optimal or more preferable to others, and it is important to be able to characterize and to convey the character of existing landscape structures. In the course of suburban development at Fountain Hills, Arizona, the initial road pattern was not drained by a stormwater system, although such a system was installed later (Chin and Gregory, forthcoming), perhaps reflecting an initial lack of awareness of environmental process responses. The best example of information and education relates to natural hazards and their impact. It is imperative that decision-makers are aware not only of the physical incidence of hazards (e.g. Table 8.3, p. 210), of the meaning and significance of their occurrence and return periods, but also of their impact – a holistic view which is not confined to impact in selected areas. Indeed, in reviewing the emphases present in investigations of natural hazards, Alexander (1999b, p. 424) suggests that traditionally there has been an emphasis on individual perception and choice of actions, particularly regarding location and use of geographical space.

At a second *maintenance level* it is desirable to maintain the physical environment and processes as closely as possible to some 'natural' or

Table 10.1 Themes for cultural physical geography (as suggested in Sections 10.3 and 10.4)

	Definition	Example
Level of application		
Information level	To inform and educate the general population about characteristics and processes of the physical environment	Integrated drainage basin management, urban drainage
Maintenance level	During the course of development to maintain physical environment and processes as closely as possible to natural condition or state	River channel management
	To maintain conserved areas as representative of natural conditions	SSSIs and conservation sites
Restoration level	To restore landscape or part of it to a condition perceived to be 'natural'	River restoration
Element in environment – created entirely by human action		
Individual landscape components	Features of landscape created by human activity	Golf courses, airports
Sections of landscape	A significant area of landscape created by human activity	Climate comfort indices, gardens
Museums	A building or a tract of land exhibiting an aspect or a tract of environment to demonstrate its characteristics	Environmental displays in museums, countryside museums
Processes engineered	Instances where environmental processes, or processes that have a bearing upon the environment, are engineered by human action. This introduces new elements into the ecosystem	Biotechnology, mirrors in space

'normal' condition. Nature has to be maintained despite changes that occur due to environmental impacts, and the heart of environmental impact assessment is to evaluate exactly how a particular impact has changed the 'original' environmental condition. Many opportunities exist for physical geographers to estimate the precise consequences of global change and thence to suggest how impacts can be minimized. This is not necessarily as easy as one may first think because of the several players often involved in management decisions. Research on coarse woody debris accumulated along streams in the New Forest (see Chapter 7, p. 196) enabled a series of general management strategies for debris in streams to be devised (Gregory and Davis, 1992), but in actual management the interests included forestry, conservation and field sports representatives, the latter being reluctant to accept that debris accumulations were 'natural' and should be maintained. A further way in which the maintenance level is expressed concerns selection of areas to conserve. In the UK, when fluvial sites of special scientific interest were decided (Gregory, 1997), it was necessary to determine which sites should be preserved. The decision to some extent reflected the availability of sites that have been researched and so a selection was made from a compilation of potential sites in the UK, although the national criteria of international importance, presence of exceptional features, and representativeness were all kept in mind.

A third *restoration level* embraces all those cases where it is necessary to restore the landscape, or part of an environment, to a condition perceived to be 'natural' or 'normal'. After open-cast mineral mining is complete and the landscape is restored, exactly how are the contours configured and the landscape recreated? Is advice sought from a geomorphologist or a physical geographer with an understanding of the landscape appropriate for that area, or is it undertaken by someone without such training? This issue is increasingly important in relation to river channel management because after many years with the view prevailing that 'technology can fix it' (Leopold, 1977), whereby many rivers were 'engineered' in concrete channels, there has now been a return to what has been described as 'working with the river rather than against it' (Winkley, 1972). Alternatives to stream channelization (Chapter 8, p. 215) require restoration of rivers to a more natural state. Sear (1994) summarized the terms used for different strategies (Table 10.2), and Rhoads and Herricks (1996) and Rhoads *et al.* (1999) defined stream naturalization as the creation of a restored stream resulting from discussion between scientists, the river managers, and the representatives of the local community. Such developments allow for variations from one area to another by obtaining views from the local population. Whereas at the information level it is important to communicate information about natural environment processes to the local population, at the restoration level it can be significant to obtain views to inform the management strategy adopted.

Terms used (Table 10.2) raise the question of what is 'natural', and as environment can be seen as a social construct, in any one area there may not be a single natural condition which should be recreated. There is also the question of environmental change. A small river in southeast London was placed in a culvert in 1937, and in the 1990s it was decided (Tapsell, 1995) to excavate the river and recreate it through Sutcliffe Park, Eltham. Is it appropriate to recreate the river as it was in the 1930s or as it would have developed to the 1990s? Even if it was agreed to recreate the natural envi-

Table 10.2 Some terms used in river management in 'working with nature', developed from a table of dictionary definitions (in italics) produced by Sear (1994, p. 170)

Term	Definitions
Recovery	*The act of restoration (of a river) to an improved/former condition*
Re-establishment	*To make (a river) secure in a former condition*
Enhancement	Any improvement of a structural or functional attribute (NRC, 1992) Any improvement in environmental quality (Brookes and Shields, 1996)
Rehabilitation	*To help (a river) adapt to a new environment* 'A partial structural or functional return to the predisturbance state' (Cairns, 1991; NRC, 1992) Partial return to a pre-disturbance structure or function (Brookes and Shields, 1996)
Reinstatement	*To restore (a river) to a former condition*
Restoration	*The act of restoring (a river) to a former or original condition* 'The complete structural and functional return of a biophysical system to a predisturbance state' (NRC, 1992)
Full restoration	The complete structural and functional return to a pre-disturbance state (Brookes and Shields, 1996)
Creation	Development of a resource that did not previously exist at the site. Includes the term 'naturalization' which determines morphological and ecological configuration with contemporary magnitudes and rates of fluvial processes (Brookes and Shields, 1996)
Naturalization	Recognizes that the concept of 'natural' is defined by the community relative to the modified state of the system and that the goal of naturalization is to drive the system as a whole toward a state of increasing morphological, hydraulic, and ecological diversity, but to do so in a manner that is acceptable to the local community and sustainable by natural processes, including human intervention (Rhoads *et al.*, 1999)

ronment, should one impose the views of scientists or should one heed public preference? In a study of several streams in Hampshire, Gregory and Davis (1993) showed that the majority of people did not prefer the most natural riverscape, but opted for what they perceived to be most visually attractive – despite the fact that it was heavily managed. The meanings that English beach experiences have for people have also been explored (Tunstall and Penning-Rowsell, 1998). In mapping multiple perceptions of wilderness in New Zealand, Kliskey and Kearsley (1993) demonstrated that, whereas several wilderness areas are defined in strictly objective, physical terms, wilderness conditions are perceived differently by different people. Also in New Zealand, Mosley (1989) showed how wild riverscapes are less valued than managed rivers. Professionals in waste management are reported to view soil as having a limited ability to degrade wastes (Colten, 1998), and a conclusion from a study of attitudes of environmental management personnel involved in surface coal mine reclamation in Alberta and British Columbia, Canada (Smyth and Dearden, 1998, p. 293) was:

> . . . results indicate that reclamation practitioners in these jurisdictions do not understand the concepts that underlie their management practices, particularly with respect to ecosystem processes. The relationship between biophysical processes and management practices and ecosystem monitoring should be stressed in a formal training programme.

This issue has been considered by Graf (1996, p. 469) who concluded:

> The primary consideration in establishing policy for river restoration is to address the problem of 'what is natural'. It is unlikely that long reaches of American rivers will ever return to their original, truly natural states, and they are likely to remain in their fragmented condition . . . Policy-making for rivers is often a perceptual issue . . . but the decision about how far to go in making rivers more natural is partly political (what does society want?), partly scientific (what is possible?), and partly economic (what can society afford?).

This question of what is natural is one issue considered in relation to the physical integrity of America's rivers (Graf, forthcoming). Bren (1993) contended that in the last two decades, the effects of forest management on streams, riparian zones, and floodplains have received much interest, that there is agreement that such areas should be maintained in a state approximating naturalness, but it is recognized that definition of this state is usually difficult or impossible. Harper *et al.* (1995) suggested that we need 'wild/pristine examples' to tell us what many rivers once were. This could apply to other aspects of the physical environment because a common theme in the legislative development of almost all countries is concern for the natural environment. But is the meaning of 'natural environment' always sufficiently clear? This leads to two outstanding issues: ethics and environmental design. There are obvious ethical considerations in relation to environmental restoration, and Eden *et al.* (1999) consider the ethical and political issues raised by two environmental restoration schemes in England and question how far restoration is an acceptable option for environmental management. More generally, the need to embrace ethical considerations in geomorphology (see Chapter 11, p. 280) was emphasized by Rhoads and Thorn (1996), and shallow ecology, which considers the values of nature to be instrumental to humans and is strongly anthropocentric, and deep ecology, which maintains that all species have an intrinsic right to exist in the natural environment (Lemons, 1999), should find parallels in physical geography.

Greater involvement of physical geographers in environmental design (Gregory, 1985, p. 207), perhaps with training typically involving a design project (e.g. Gregory, 1992), was taken up by Newson (1995, p. 420) who suggested that geomorphological research in the last decade had made substantial progress in four areas worthy of application to practical river management, namely:

(1) long profile aspects including water and sediment storage and routing;
(2) planform elements including meander development and the role of the floodplain;
(3) sources of sediment, including land use and the channel bed itself;
(4) the role of the 'rare' event and of climate change.

In relation to discussion of management of the catchment as a critical component, because few geomorphological designs will be 'stand alone', Newson (1995, p. 427) concluded:

> It is the author's belief that the basis of Gregory's term 'environmental design' is just this: a set of clear prescriptions which, however, must be developed, presented and approved in context. Like natural historians we are not located at the core of physical science, but our advantage is an acute knowledge based on the boundary conditions within which physical laws operate.

The idea of a design science was introduced in relation to geomorphology by Rhoads and Thorn (1996, p. 132) who argued that:

> the growth of geomorphology as a practical profession requires that geomorphologists continue to devote effort to developing and refining a design science to support this profession. Such a design science can provide the basis for professionalization of the discipline by codifying a body of information, tools and skills for licensing or certification programs.

Should physical geographers become involved in a design science and in suggesting environmental design for whole landscapes? This could be the natural extension and logical outcome of many of the investigations covered in Chapters 5–9. *Design with Nature* (McHarg, 1969, 1992) written from the standpoint of the landscape architect, prompted the question of exactly where the boundary with physical geography occurs. Writing from an ecological viewpoint and perceived as an ecological planner, McHarg (1992) saw that energy is the currency for the ecological value system (compare with Chapter 4, p. 94) and urged us to:

> Consider a time when ecological inventories have been completed. A sample survey of the population is made. Those interviewed are asked

to enumerate the ideal characteristics of climate, scenery, recreation, employment, residence that constitute their Utopian preferences. The results are searched for concurrences from which emerges a social program for a new urban America. The ecological inventory is searched for sites that conform to the aspirations of those interviewed ... In the quest for survival, success and fulfillment, the ecological view offers an invaluable insight. It shows the way for the man who would be the enzyme of the biosphere – its steward, enhancing the creative fit of man–environment, realizing man's design with nature.

The ecological view certainly offers an important approach, but is it sufficiently holistic, and would physical geographers have contributed more if they had been more numerous in universities in the US? Specific examples illustrate how environmental design could be developed. Thus Thornes (1992) referred to climatic design and to required levels of maintenance, and cited airport layout in relation to atmospheric conditions and hazards. In the design of the urban environment of Tumbler Ridge, a new coal-mining town in British Columbia, Canada, it was suggested that a planning approach which embodies environment behaviour research was an exciting innovation and that: '... planners need to consider more carefully how to integrate or adapt design elements to reflect unique "senses of place" more specifically' (Gill, 1989, p. 177). This may be pertinent to the need perceived by other disciplines. Thus according to Ke Chung Kim and Weaver (1994), global environmental problems are much worse and more difficult to solve than the public believes; their book is concerned with the paradox that humanity depends upon biodiversity and landscape systems for its survival, but the current burden of humanity's use of living resources places the existence of these natural systems at risk. They concluded (Ke Chung Kim and Weaver, 1994, p. 21) that: 'The challenge is to find and adapt cultural means through which the present paradox is resolved'. That book arose from an international discussion meeting at the Pennsylvania State University Center for Biodiversity Research attended by leading contributors from different fields – ecology, landscape management and sociology were mentioned, but not geography or physical geography. Physical geographers are qualified to contribute, should become more involved, and greater involvement is a challenge for the next generation (see also Section 9.4, p. 251).

10.4 Further cultural themes – how far do we go?

In addition to the three levels referred to in Section 10.3, there is also the question of environments, or environmental components, entirely created by human action. Should the interest of the physical geographer extend this far? A series of levels is suggested in Table 10.1. Firstly, *individual landscape items* can include golf courses, screening by bunding, and low-level pit restoration schemes which in design often show insufficient awareness of geomorphological concepts or topographic distinctiveness, according to Murray Gray (1997), who concluded:

> ... there is still no widespread recognition of the importance of land-form conservation and design. Geomorphology in the wider country-side ought to be part of the 'sustainability' and 'local distinctiveness' movements, yet it is too often undervalued, subsumed or ignored.

Secondly, there are *sections of landscape* such as gardens, urban areas or restored land which create completely new environments. In her review of gardens, earthworks and environmental art, Ross (1992) argued that gardens force us to think deeply about nature itself, our relationship to nature, and nature's relationship to art. Does physical geography have an interest here? Also in this category we should include comfort indices in the case of climate, and these comprise Part 3 of the book by Thompson and Perry (1997).

Thirdly, *museums*, usually thought of as buildings used for storing and exhibition of objects, can embrace the way in which aspects of the physical environment now feature in the traditional museum. Such developments include the Eden project in Cornwall, UK, which is constructing in a disused china clay pit a partially enclosed Eden which will house all existing world plant communities, including tropical rainforest; already opened is the Biosphere 2 near Tucson, Arizona; and Centre Parcs, which began in the Netherlands and extended to three localities in the UK, provide types of development where the expertise of a physical geographer could be relevant.

Finally, *processes engineered* (Table 10.1) could involve gene developments which extend to the possibility of what has been described as Frankenstein food. Biotechnology is the manipulation of living organisms and their components, for example genes or gene components, to undertake specific tasks which, in principle at least, are beneficial to society. Mannion (1995) suggested that the environmental implications must be taken seriously in view of the huge investment in biotechnological research, and the rapidity with which it is proceeding. She included these developments in chapter 8 of her book *Global Environmental Change* (Mannion, 1991). The possibility of mirrors in space that could significantly impact on our climate is being entertained, illustrating that the Homosphere or the Noosphere (see p. 174) has arrived and we need to consider the extent to which the physical geographer could be a part of it.

10.5 Extending the physical geography equation

When the physical geography equation was proposed in 1978 (Gregory, 1978a, 1985) it was focused on the 'natural environment', although allow-ing for human impact. When presenting a paper in the symposium in honour of Yi Fu Tuan in Madison, Wisconsin, in April 1998, I suggested that although the equation was still valid, it was now appropriate to extend the equation to a fifth level. That could be described as the cultural level and necessitates focus upon the interpretation of the physical environment with all the implications that it has and that have been suggested in this chapter. To the equation (Chapter 1, pp. 18–19) we can add:

- *Level 5: appreciating the equation* – involves a cultural physical geography approach embracing ways in which human reaction to the physical environment and physical landscape influence how environment is managed and designed.

Further reading

It is always stimulating to read writings by Yi Fu Tuan, and an appropriate chapter is:

TUAN, YI FU. 1992: Desert and ice: ambivalent aesthetics. In S. Kemal and I. Gaskell (eds) *Landscape, natural beauty and the arts*. Cambridge: Cambridge University Press, 136–57.

A wide-ranging book that links to human geography and to other disciplines is:

SIMMONS, I.G. 1993: *Interpreting nature: cultural constructions of the environment*. London: Routledge.

To see how another discipline views the prospect of environmental design and includes much that will be familiar to a physical geographer:

McHARG, I.L. 1992: *Design with nature*. Chichester: John Wiley.

A thought-provoking read is provided by:

GRAF, W.L. 1996: Geomorphology and policy for restoration of impounded American rivers: what is 'natural?' In B.L. Rhoads and C.E. Thorn (eds) *The scientific nature of geomorphology*. Chichester: John Wiley, 443–73.

Topics for consideration

(1) Strands of recent development are taken (in Section 10.2) to indicate a move towards a greater involvement of physical geographers in a more cultural approach. Are there other examples of these approaches that vindicate this cultural trend or other trends?

(2) Ways in which cultural approaches have been advocated and pursued are suggested in Sections 10.3 and 10.4 and summarized in Table 10.1,

but are there other approaches that should feature on the physical geography agenda?

(3) Environmental design (Section 10.3) could be a priority as defined on p. 267, but do you agree that physical geography should go this far?

PART

V

CONCLUSION

|11|

Future physical geography

Various suggestions about where physical geography should go now (11.1) derived from preceding chapters, from published views of physical geographers, and from developments external to physical geography can be used as the basis for three major options for the future (11.2). It is inevitable that, completely independently of the direction that physical geography takes, certain tenets must apply and these are suggested (11.3), and final words (11.4) relate to the prologue of Chapter 1.

The thrust of this volume has been to direct attention to where physical geography has come from and where it is now, so that we should also think about the way in which it may develop. Jensen and Dahlberg (1983) saw the growth of any discipline (see Curran, 1985, pp. 6–7) as sigmoidal and composed of:

- a preliminary growth period with small absolute increments of literature and little or no social organization (Stage 1, Fig. 11.1);
- a period of exponential growth when the number of publications double at regular intervals and specialist research units are established (Stage 2);
- a subsequent period when the growth rate begins to decline and although annual increments remain constant, specialization and controversy increase (Stage 3);
- a final period when the rate of growth approaches zero, specialist research units and social organization break down, and the subject reaches maturity (Stage 4).

Applying this model to remote sensing, Curran (1985, pp. 6–7) suggested that in 1985 Europe was in Stage 2 and the US was beginning to enter Stage 3. Physical geography is already in Stage 3 because of the way in which it now interacts with other disciplines, and has probably reached Stage 4.

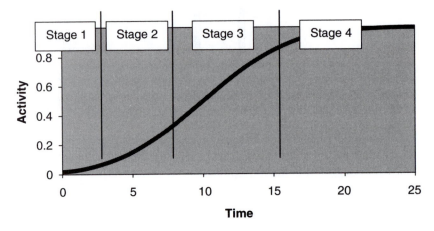

Figure 11.1 Stages of development of a discipline (developed from Jensen and Dahlberg, 1983; also used by Curran, 1985)

11.1 Issues for the future

Views about future directions of physical geography can be gleaned from three sources: firstly from the current trends in each of the preceding ten chapters (11.1.1); secondly from statements made by physical geographers (11.1.2); and thirdly from views in science and other disciplines (11.1.3).

11.1.1 Current trends and topics

Critical topics which have emerged in the preceding ten chapters can be summarized as follows:

- *Chapter 1* – the inevitability of different, pluralist, approaches emerging according to the character of the environment in particular countries, the academic traditions of the country, and the way in which the physical environment has been studied. *Pluralist approaches are inevitable.*
- *Chapter 2* – the need to maintain awareness of the evolution of the discipline. *No single approach or paradigm will dominate forever.*
- *Chapter 3* – physical geography must maintain an awareness of philosophical developments in science. Investigations that are very detailed and reductionist, and those that relate to environment as a whole, should both be embraced within a spatial approach to analysis of the physical environment. *Research should not get too embroiled in detailed investigations – they should be realist in all senses.*
- *Chapter 4* – physical geographers must have a basic understanding of physical principles, energy flows, energetics and global biogeochemical

cycles to appreciate the types of models available, to apply them in research, and be innovative in further model development. *Models are fundamental for physical geography.*

- *Chapter 5* – further analysis of environmental processes, using increasingly sophisticated techniques, will relate processes to environmental change, and focus upon controls of spatial patterns of environmental processes – an approach unique to physical geography. *Spatial patterns should be interpreted from a secure foundation built on physical principles and applied to environmental change.*
- *Chapter 6* – continue collaboration with the wide range of disciplines necessarily involved in investigations of environmental change and participate in multidisciplinary research programmes. *All disciplines now have to be multidisciplinary to some extent.*
- *Chapter 7* – benefit from the position of physical geography and its relationship with human geography to identify, document and analyse human impact. *Study of human impacts is not just about effects on the environment and processes, but can be developed to forecast future environmental change.*
- *Chapter 8* – extend applications of results of environmental research, be involved in consulting as appropriate, and engage with the policy- and decision-makers. *Take opportunities to educate opinion-formers and others about the character and potential of physical geography.*
- *Chapter 9* – maintain physical geography research and scholarship at the global level and contribute, especially in team research, on global change with particular reference to spatial impacts and the ways in which they can be managed. *Physical geography has a distinct role in analysing spatial impacts and should not pretend to be too eclectic.*
- *Chapter 10* – include, as appropriate, the cultural dimensions of investigation of the environment which extend not only to natural and human-modified environments, but also human-made environments. *Physical geography does not have to be restricted to the 'natural' environment and is well equipped to propose desirable physical environment in relation to environmental design at specific locations.*

Critical issues are highlighted in italics.

11.1.2 Recent statements

A second approach is to consider five issues (in italics) raised by physical geographers in the recent literature.

The first relates to the question of *whether a core exists*, and Gardner (1996, p. 99) argued that:

... physical geography has increasingly come to resemble a polo mint. The debate as to whether this is perceived as good or bad for the 'subject' is largely academic and arguably not productive. It exists and the profession needs to go forward from here.

Although there may not be a core in the sense of a discrete heart of physical geography, there are core areas for each of the sub-branches, and there should be a core in terms of the focus and approach that physical geography adopts. The issue is not unique to physical geography, of course, and Johnston (1989) in a review of the Study Groups of the Institute of British Geographers characterized geography as a whole as a discipline without a core. Considering the focus of geography and its relationship with physical geography, Guelke (1989) saw a historical-related intellectual approach to the human–environment relationship as central to geography. In many disciplines it is a fact that some of the most exciting developments occur near the extremes of the subject, close to interfaces with other disciplines.

A second question, *the relation of physical geography to human geography*, is an issue to which there have been several notable contributions. In 1986 Johnston directed attention to what he saw as four fixations of human/physical, integration, spatial and regional, all pertinent to the quest for unity in geography. This was supplemented by comments from Goudie (1986b, p. 458) that:

> ... the fragmentation of geography into two discrete sub-disciplines would be prejudicial to its continued existence and vitality. Such a fragmentation would go against many of the traditions of geography and create a narrowness that would be antithetical to the view of geography expressed since the beginning of the last century ...

Human geography has an influence on physical geography (Douglas, 1987), and Douglas (1986, p. 462) asserted that:

> Some of us must escape from the narrow research followed to gain PhDs and broaden our horizons to explore the greatest challenges to our discipline, those big, fundamental, important issues in which the unity of geography is obvious.

However, Stoddart (1987) robustly saw the opportunity of claiming the high ground for the end of the century (see p. 6) with a prescription (Stoddart, 1987, p. 334):

> ... Land and life is what geography has always been about. It is time we got out again into the great wide world, met its challenges ...

Part of the global focus (Chapter 9) may have met some of the aspirations in these statements, but it may be easier to write rhetoric than to take effective action, and Johnston (1997, p. 344) assessed the situation as follows:

> ... while connections have been made across the human–physical geography interface, there has been no integration of the study of physical and societal processes; for human geographers their links

with other social scientists are very much stronger than those with environmental scientists . . .

We should resist the temptation to engage in somewhat introspective debate, because a positive opportunity exists (Tickell, 1992, p. 323):

In the past the weakness of academic geography has been its spread. A subject ranging between supposedly hard physical sciences and supposedly soft human studies looked fuzzy to those who liked locking learning into impermeable boxes. But this spread has now become a strength . . . Understanding of natural things requires an intellectual continuum not a series of subdivisions. Geography is exactly such a continuum.

Thirdly, there is the question of *the internal organization of physical geography*. Traditionally subdivided into the three major subdivisions of geomorphology, climatology and biogeography, the subject structure adjusted to embrace areas of hydrology and developments in Quaternary science. The view of Stoddart (1996) in correspondence (Chapter 1, p. 6) was stimulated by a subheading *The end of systematic physical geography* used by Richards and Wrigley (1996, p. 50), who concluded:

In Quaternary Science, both in its 'pure' and its 'applied' forms, the key developments have occurred through the interdisciplinary integration of several techniques, some recently developed and others more established palaeontological methods such as palynology and diatom analysis, to produce multi proxy, high resolution records of environmental change.

Agnew and Spencer (1999, p. 8), considering why more physical geographers do not publish in the *Transactions of the Institute of British Geographers*, attempted to stimulate debate across the 'unfortunate human/physical divide' to ensure that geography makes real and lasting contributions to many of the pressing issues of the late twentieth and early twenty-first centuries and contended that:

One of the many difficulties is that physical geography has fragmented from a previously more unified state into a series of subdisciplines, including geomorphology, biogeography, climatology and hydrology. It is, of course, important that physical geography continues to reach audiences in other sciences, so publication elsewhere does have real value to the discipline; we must not create our own intellectual ghetto.

Recent trends in research on environmental processes and environmental change have shown that the traditional subdisciplines are no longer as clear as they were in the 1980s, and in a review of geomorphology Spedding (1997, p. 265) suggested:

Arguably, human geography became the richer because of its willingness to reinvent itself, to explore beyond its traditional boundaries and
ask new questions within a contextual framework of dynamic socio-
spatial organization ... Is now not the time for geomorphology to try
the same?

If physical geography reinvented itself there could be difficult issues to
address and difficult choices to make, which Harman *et al.* (1998, p. 277)
articulated:

> ... many environmental problems present us with very difficult
> choices today. Whether the issues be regional (e.g. aquifer depletion,
> land use zoning, river management) or global (climate change), much
> of this difficulty stems from the combination of scientific, economic,
> and ethical uncertainty entangled in the details.

Ethical considerations were mentioned in Chapter 10 (p. 266), and there is
also the question as to how far the subject matter of physical geography
should extend. Summerfield (1991, p. 507), at the end of his chapter on
planetary geomorphology, proposed:

> Perhaps the most significant stimulus provided by the exploration
> of planetary landscapes, is one which has yet to be fully realized; it
> is an enlargement of the conceptions of the spatial and temporal
> scales with which geomorphologists should be concerned. Coupled
> with major advances in other fields within the earth sciences, such
> as those represented by plate tectonics and contemporary theories
> of long-term climate change, the investigation of planetary land-
> scapes provides us with an exciting global perspective for geomor-
> phology.

Baker and Twidale (1991, p. 90) commented that planetary geomorphology
was not mentioned in *The Nature of Physical Geography* (Gregory, 1985)
but at that time there were relatively few contributions by physical geographers or geomorphologists dealing with other planets. Is this something that
should be embraced to a greater degree?

Fourthly, physical geography should continue to *investigate holistic and
global problems*. Newson (1992b, p. 219) argued that:

> ... society is likely to require geographers who are drafted to front-
> line environmental management to be both practical and professional:
> they will need to carry a clear and distinct disciplinary label.

He therefore suggested (Newson, 1992b, p. 219) that we should:

> ... use both a wealth of experience with natural systems analysis and
> a contemporary research familiarity with some of nature's surprises to
> develop a more enthusiastic, socially responsible and committed phys-
> ical geography ...

To this end it is necessary to clarify and improve the way in which physical geographers are perceived. Thus Thornes (1990) lamented the role that geographers play in relation to the atmospheric sciences in the UK, not redressed by recent reports by NERC and the UGC, and he commented that Canadian experience had been more effective.

Fifthly, we need to maintain and *foster good relations with other disciplines*. Collaborative research is increasingly undertaken, certainly at a global level where multidisciplinary collaboration is a prerequisite for the solution of some of the current major world problems. In looking at the changing geomorphology of the humid tropics, Gupta (1993) concluded that 'it is no longer fashionable to wander alone in the rainforest'. Research in rainforest areas increasingly requires teams of researchers who style themselves as geomorphologists, hydrologists, forestry scientists, ecologists or physical geographers, and in other environments appropriately constituted multidisciplinary teams are needed.

It would be possible to continue this list of selective quotations, and perhaps to demonstrate more than five themes from the recent literature, but even if we do not agree with Kates (1987) that physical geography is in a shambles (see p. 256), it would surely be generally agreed that physical geography should keep its nature, purpose and methods under review with an awareness of what is happening external to the discipline.

11.1.3 Views from other disciplines

External developments, including those in other subjects, can be invigorating, and four are mentioned here.

Firstly, basic trends in the methods of science necessarily have a bearing upon approaches in physical geography. In his book on *Climate, Earth Processes and Earth History*, Huggett (1991, p. 218) noted:

A radical aspect of the new and revolutionary branch of cause and effect relationships by sets of embedded systems (a systems hierarchy) which resound with resonances ... The implications of this idea are momentous ... There is no need, arguing from the basis of non-linear dynamics, for events in the biosphere and galactic or geological events to be precisely correlated ... A challenge for the future is to develop dynamical models for the world climate system as a whole, to consider the conjoint action of all parts of the biosphere.

The advent of major new approaches will continue to impact, and non-linear dynamics (see Phillips, 1999a, b) is to be added to chaos theory and to fractals as well as to others. Technical developments in remote sensing and in the increasing use of micro-level technology will further impact on physical geography.

Secondly, there are philosophical developments in science of which many could be quoted. An example is John Horgan's (1996) provocative and thought-stimulating *The End of Science: Facing the Limits of Knowledge in the Twilight of the Scientific Age*. Informed by interviews with major players, Horgan argued that disciplines, including geography, are bounded, and he foresaw the end of sciences and social science. However, is such a finite view of knowledge realistic?

Thirdly, how is physical geography placed not only in relation to science but to other disciplines? Simmons (1993) drew attention to the ways in which different groups of scholars within particular disciplines adopt fundamentally distinct positions (see Chapter 10, p. 260). His discussion underlined the fact that not only do views differ between disciplines but they also change over time (Simmons, 1993, p. 150):

> I suspect this may translate for present purposes to something like the notion that 'environment' and 'nature' can never be used as terms to postulate some primitive state of the biosphere: what we can relate about the biosphere of say two million years ago is still as much a reflection of the present use of words and especially of metaphor and the shifting entities we call 'self' as of any truth of what the world might have been like two million years ago.

Fourthly, there are secular developments external to the discipline of which many could be quoted, but perhaps increasing public awareness of the environment must continue to be the major one. Potentially influential could be decisions such as that by the Kansas Board of Education reported in *The Times* on 12 August 1999, under the headline 'Kansas Schools delete Darwin from Curriculum'. The Board had decided to eliminate references to evolution as a central principle of biology from the state school syllabus, and it is salutory that such instances, together with the way in which emphases are placed by the media, could become increasingly pervasive.

11.2 Options available

In the light of all the issues set out under the headings above, there are at least three major alternatives available. The first alternative is evolution of the discipline, whereby present trends continue, involving some increasingly reductionist investigations, accompanied by research at a variety of scales including global, and interaction with other disciplines in multidisciplinary research. In this *status quo* scenario the research links with human geography are not particularly strong, and could weaken even further, although the structural bond would remain strong in certain countries.

It is not easy to obtain data on the present structure of physical geography, and throughout this book reference has often been made to the traditional branches of geomorphology, meteorology/climatology, biogeography, soil geography and hydrology. Some broad idea of the structure in the UK in

1996 was obtained from the submissions to the Geography panel of the Research Assessment Exercise. Although the detailed submissions were confidential, a report was produced for the community after that exercise and it is possible to see, from the four publications cited by each research-active academic submitted, how they perceived their important research in the period 1992–1996. Although it was not exclusively geography departments that were considered, and it is not always easy to decide who would regard themselves as physical geographers, on the basis of the publications cited it was possible to recognize in the 1083.5 full-time equivalent (headcount 1119) academic staff submitted those who are researching on the physical environment. That gave a total of 454 or 41.9% (40.6% of the headcount) of the research-active community in 1996. From the subjects of more than 1700 publications submitted, this UK snapshot showed how physical geographers represented themselves in the period 1992–1996. A broad classification of the subject categories included is shown in Table 11.1 and leads to provisional conclusions:

(1) Geomorphology and biogeography together with hydrology dominate the field (40.2%),which is even more dominant if Quaternary studies (18.8%) are added to make a total of 59%.
(2) Many academics are not confined in their research activity to a single branch of physical geography (19%) and tend to range across branches – despite reductionism there is greater interaction between specialized branches.
(3) The traditional divisions of physical geography are no longer as clear as they were (cf. E.H. Brown, 1975). Although broad categories are used in Table 11.1, the academic publications of individual staff not only range across categories, but remote sensing (11%) represents those staff who have a primary interest in remote sensing but also in physical geography or the physical environment.
(4) Many publications are directed towards applications of physical geography. Although (Chapter 8, p. 198) there was once a reticence to become involved in applications, this is certainly not borne out by the UK situation from 1992–1996 where 22% of the publications are clearly related to applied topics.

Several tensions that can be identified include:

• A tendency towards reductionism and very detailed investigations (see quotations on p. 52, p. 124).
• A perceived need to return to evolutionary geomorphology as suggested by Baker and Twidale (1991). They envisaged either the path to disenchantment aping the 'hard' sciences with their dogmatic ideologies, or a reenchanted future. Although they were describing geomorphology, their sentiments could apply to physical geography as a whole when they contended (Baker and Twidale, 1991, pp. 94–5):

Table 11.1 Publications submitted to 1996 Research Assessment Exercise for geography (454 academic staff, more than 1700 publications submitted 1992–1996)

Branch of physical geography	Topics included	Percentage
Coastal	Sea-level rise Sediments Management and risks	6
Arid and semi-arid landscapes	Landscapes Processes	3.6
Hillslope processes, weathering	Modelling	3.4
Tectonic, volcanicity, karst		1.7
Water-related research	Models Hydrological processes Sediment and water quality Channel form and process Channel classification and changes Management	19.7
Glacial processes and glacial hydrology	Glacier fluctuations Glacier budgets and processes Glacial morphology Proglacial and glacial hydrology	5.8
Quaternary	Models and approaches Quaternary sediments River sediments Sea-level change and uplift Palaeolimnology Reconstruction, techniques and correlation	18.8

Biogeography	Palynology
	Vegetation change, climate
	   change and acidification
	Fauna and plant geography
	Rainforest and wetland
	   ecosystems
	General ecology and soil
		14.3

Soils	Soil processes and profiles Soil erosion Land reclamation	5.6
Atmosphere	Modelling Climatology Applied aspects	5.6
Remote sensing	Related to physical environment	11
Other	Resource management Environmental management Hazard management	4.2

... in the lawless disorder of nature lies the potential reenchantment of Geomorphology with its subject matter, leading to a period when studies of landform and landscape evolution at all scales will lead to a reinvigoration of its disciplines ... A new Golden Age awaits in which the geomorphologist will discover an orderly evolution in a real world of apparent confusion and contradiction ... We see geomorphology as an holistic, chronological integrative field-based science, that is integral to the study of a dynamically vibrant planet.

We need to take the opportunity of greater applications of physical geography as suggested in Chapters 8, 9, 10 (see quotations on p. 220, p. 250).

• Particularly in some countries, the need to bolster the discipline to prevent a decline in status taking place (e.g. Turner, 1989) (see Chapter 10, p. 268).

Some physical geographers have subscribed to the notion that physical geography is what physical geographers do, so that concern with approaches, philosophy and paradigms is unnecessary. That notion has been largely abandoned (Bird, 1989, p. 224) because of concern with philosophical and systems understanding (Chapters 3 and 4), and because without some thought being given to the direction in which the subject goes, there is a danger of missing potential opportunities. This first option is probably insufficient.

A second alternative is for physical geography and its sub-branches to disappear and to become part of independent subjects of geomorphology, hydrology, soil science, biogeography and climatology. This has already happened to some extent as the branches have expanded and interacted with other disciplines. Although the disappearance of physical geography might seem a retrograde step, it is inevitable that modifications evolve in the way in which knowledge is organized, but however appropriate reorganization and multiplication of disciplines might be, many imperatives and existing structures militate against change. These include the existing HE degree structure in which geography is increasing in its attraction. In the UK in 1999 geography comprised 5.38% of the papers taken and was the fifth most popular subject in the A-level examination for 17–18-year-olds, being exceeded only by general studies, English, mathematics and biology. The comparable 1999 UK figures for the GCSE examination, usually taken at 16, show that geography had 4.78% of the papers taken and was the seventh most popular subject at that level. Disappearance of the subject in the UK is therefore unlikely in view of its strength at the public examination levels, despite national curriculum challenges. However, it should be remembered that new areas of research have developed very successfully: environmental science incorporates many physical geographers, Quaternary science has become truly interdisciplinary (Table 11.1), and the links with biology and ecology have now become at least as strong if not stronger than those with geology, which were significant in the early days of the development of physical geography. One example of this trend is the focus upon hydrosystems in fluvial research (e.g. Petts and Amoros, 1996). For geography as a whole, Haggett (1996, p. 972) colourfully contends: 'Today the same academic space is heavily occupied and the campfires of near neighbours ... burn brightly around us'. But Haggett also argues against sweeping away old disciplinary boundaries and merging geography with its disciplinary neighbours.

The third alternative is, by an element of revolution, to reaffirm the need for physical geography, and to reinvent the discipline to some extent. This would maintain sub-branches, possibly redefined in the light of recent trends (e.g. Table 11.1), allow them to evolve as appropriate and interact with related disciplines in multidisciplinary endeavour, but it would also reinforce the links with human geography, and provide the basis to respond to the opportunities identified in Chapters 5–10. In a situation where non-linear models are becoming more frequent than the linear ones that have served physical geography well for the last three or four decades, where the links between the branches of physical geography are stronger rather than weaker, and where there is also a pragmatic awareness of current world problems especially associated with global change, it is perhaps timely to reinvent physical geography, or at least refashion it, to occupy the niche (Fig. 4.4) that Slaymaker and Spencer (1998) identified. This would be as a discipline equipped to analyse and model multiple interactive environmental and human systems sensitive to adaptation to changing global conditions. Stott (1998) analysed the situation in biogeography to be at a crisis level. He suggested that our metalanguage is deeply flawed, with key signifiers such as forest ecology and equilibrium notions, and that our thinking is being driven largely from Europe and North America. He argued that we continue to speak and think in terms including climaxes, optima, balance, harmony, equilibria, stability, ecosystems and synecology, and therefore averred (Stott, 1998, p. 2):

> We have to replace our northern-derived, historic metalanguage of equilibrium, sustainability and balance with a different metalanguage more accepting of change and comprising a new range of 'key signifiers' including adaptation, migration, movement, opportunism, flexibility and resilience. We will also require a new and more radical approach to the political ecology and economics of risk assessment. The language of non-equilibrium will then come to the fore ...

When appointed to my first lecturing post at the University of Exeter in 1962, one of my tasks was to prepare a course on World Physiography, imaginatively conceived by my predecessor Ron Waters who had departed to a Chair in New Zealand before the course had been taught. Hence I had to create a course on World Physiography for which there were no texts at that time other than Huxley's approach at the end of the nineteenth century, or the *Cours de Geomorphologie* of Jean Tricart and the French physical geographers. It was a challenge to create such a course, to challenge generations of students, with what they came to know as World Fizz, and perhaps at the beginning of the twenty-first century the time is ripe for a return to world fizz! Physiography was regarded as broader than physical geography (Holt-Jensen, 1999), and after giving twelve lectures on physiography at the London Institution in 1869 and repeating these in South Kensington in 1870, T.H. Huxley subsequently wrote his book and commented (Huxley, 1877, p. vii):

> ... the conviction dawns upon the learner that, to attain even an elementary conception of what goes on in his parish, he must know something about the universe; that the pebble he kicks aside would not be what it is and where it is unless a particular character of the earth's history, finished untold ages ago, had been exactly what it was.

A resurrection of physiography, as described in Chapter 2 (p. 32) and in the sense used by Huxley (1877), in which the discipline returns to its roots, could accommodate the most recent developments, including needs in climatology, aptly styled atmospheric physiography (Henderson-Sellers, 1989b). If physical geography was reconfigured as physiography, this would be equivalent to going from Stage 4 to Stage 1 of Fig. 11.1 to accommodate the radical and exciting developments on the horizon.

11.3 Tenets for physical geography

However physical geography develops, there are certain tenets, in the sense of principles or doctrine rather than dogma, which might be considered, and these are offered to stimulate further thought and debate, organized under two specific headings.

Tenets for the pursuance of physical geography

(1) Emphasize the spatial perspective.
(2) Attempt to model past and present systems behaviour and interactions, and focus on future likely trends.
(3) Enhance an awareness of the way in which the subject has developed so that investigations at any one time are perceived and understood according to the knowledge available at that time.
(4) Do not lose sight of the pressing environmental problems especially at the global scale, and participate in a distinctive way in the analysis and solution of those problems, developing approaches to environmental design as appropriate.
(5) Avoid an eclectic approach that imputes that physical geography is

somehow an over-arching discipline which integrates others, but at the same time appreciate that it must be a compound discipline in the sense of Chapter 3 (p. 49).

(6)   Realise the benefits of a discipline that focuses upon the totality of physical environment and straddles the interface between the physical and human sciences.

(7)   Maintain awareness of, and interaction with, other disciplines, collaborating to ensure that research investigations are conducted as multidisciplinary or interdisciplinary as appropriate.

(8)   Clarify the specific objectives of the discipline of physical geography and ensure that these are communicated to other disciplines, to decision-makers, and through the media to the general public.

Tenets for the way in which individual research investigations are pursued – pleas after reading a range of physical geography literature

(9)   Know the literature as fully as possible and acknowledge previous work, giving appropriate citation to earlier contributions. This is not easy in view of the volume of literature that now exists and because similar ideas emerge in different countries, possibly in different languages, but is facilitated by databases etc.

(10) Resist the temptation to write up the same research results in several different articles, despite apparent encouragement by national and international pressures.

(11) Avoid changing the way in which your name is referred to because computer databases or library catalogues do not easily discriminate!

Although it may seem presumptuous to make points of this kind, they are offered to stimulate thought, debate and reaction.

11.4 Final conclusion

Two points refer back to Chapter 1. In the light of the alternatives suggested in Section 11.2, the tentative definition of physical geography (p. 9) might be amended, with the new parts emphasized in bold (and physiography tentatively inserted):

*Physical geography/**Physiography** focuses upon the character of, and processes shaping, the land-surface of the Earth and its envelope, emphasizes the spatial variations that occur and the temporal changes necessary to understand the contemporary environments of the Earth. Its purpose is to* *understand how the Earth's physical environment is the basis for, and is affected by, human activity. Physical geography* **was** *conventionally subdivided into geomorphology, climatology,* **hydrology** *and biogeography, but is* **now more holistic in systems analysis of recent environmental and Quaternary change.** *It uses expertise in mathematical and statistical modelling and in remote sensing,* **develops research to inform environmental management and environmental design,** *and benefits from collaborative links with many other disciplines such as* **biology (especially ecology),** *geology and engineering. In many countries, physical geography is studied and researched in association with human geography.*

In Chapter 1 reference was made to a Catechism of Geography published in 1823 (p. 21). In 1811 an equally slim volume of 50 pages introduced *A Complete Course of Geography by Means of Instructive Games*, invented by the Abbé Gaultier. Despite the vast increase of knowledge in almost two centuries since these volumes were published, we still need to deal with the world's environment holistically. If this book has encouraged the reader to focus upon the questions that need to be asked as well as the ways in which answers have been attempted, it will have been successful. Hypothesis creation and emphasis upon retroduction from the present to the past, however recent (Baker, 1996a), will surely continue at the very heart of physical geography.

Further reading

It could be instructive to look at:
> HUXLEY, T.H. 1877: *Physiography: An introduction to the study of nature*. London: Macmillan.

And as thought-provoking as ever:
> BAKER, V.R. 1996: Hypotheses and geomorphological reasoning. In B.L. Rhoads and C.E. Thorn (eds) *The scientific nature of geomorphology*. Chichester: John Wiley, 57–85.

Topics for consideration

(1) Does the model of stages of development in a discipline suggested in Fig. 11.1 apply to physical geography, and if so when would the stages apply to development of the discipline in specific countries?
(2) In Section 11.1.2 (p. 277) a series of issues raised by physical geographers is given. Should other issues be included – perhaps in preference to those that are included?

(3) How do you perceive the advantages and disadvantages of the three alternative futures sketched very briefly in Section 11.2, and are other scenarios possible?

(4) With the caveat that it is dangerous to propose tenets for physical geography (Section 11.3), would you add others to the list or perhaps delete some of those included?

References

ABLER, R.F., MARCUS, M.G. and OLSON, J.M. (eds) 1992: *Geography's inner worlds. Pervasive themes in contemporary American geography.* New Brunswick, NJ: Rutgers University Press.

ACKERMANN, W.C. 1966: *Guidelines for research on hydrology of small watersheds.* US Department of Interior OWRR 26.

ADAM, B. 1998: *Timescapes of modernity. The environment and invisible hazards.* London: Routledge.

ADAMS, G.F. (ed.) 1975: *Planation surfaces. Peneplains, pediplains and etchplains.* Stroudsburg, PA: Dowden, Hutchinson & Ross.

ADAMS, J., MASLIN, M. and THOMAS, E. 1999: Sudden climate transitions during the Quaternary. *Progress in Physical Geography* 23, 1–36.

ADAMS, W.M., GOUDIE, A.S. and ORME, A.R. (eds) 1996: *The physical geography of Africa.* Oxford: Oxford University Press.

AGNEW, C. and SPENCER, T. 1999: Editorial: Where have all the physical geographers gone? *Transactions of the Institute of British Geographers* NS 24, 5–9.

AHLMANN, H.W. 1948: Glaciological research on the north Atlantic coasts. *Royal Geographical Society Research Series* 1.

AIREY, M. and HULME, M. 1995: Evaluating climate simulations of precipitation: methods, problems and performance. *Progress in Physical Geography* 19, 427–48.

ALDEN, W.C. 1927: Discussion; channeled scabland and the Spokane flood. *Washington Academy of Science Journal* 17, 203.

ALEXANDER, D. 1979: Catastrophic misconception? *Area* 11, 228–30.

_____ 1991: Information technology in real-time for monitoring and managing natural disasters. *Progress in Physical Geography* 15, 238–60.

_____ 1993: *Natural disasters.* London: UCL Press.

_____ 1999a: Human ecology (Cultural ecology). In D.E. Alexander and R.W. Fairbridge (eds) *Encyclopedia of environmental science.* Dordrecht: Kluwer Academic Publishers, 326–8.

_____ 1999b: Natural hazards. In D.E. Alexander and R.W. Fairbridge (eds) *Encyclopedia of environmental science*. Dordrecht: Kluwer Academic Publishers, 421–5.

ALLAN, A. and FROSTICK, L.E. 1999: Framework dilation, winnowing and matrix particle size: the behaviour of some sand–gravel mixtures in a laboratory flume. *Journal of Sedimentary Research* **69**, 21–6.

ALLAN, J.A. 1978: Remote sensing in physical geography. *Progress in Physical Geography* **2**, 55–79.

ALLEN, J.R.L. 1970: *Physical processes of sedimentation*. London: Allen & Unwin.

AMIN, M.H.G., HALL, L.D., CHORLEY, R.J. and RICHARDS, K.S. 1998: Infiltration into soils, with particular reference to its visualization and measurement by magnetic resonance imaging (MRI). *Progress in Physical Geography* **22**, 135–65.

AMOROCHO, J. and HART, W.E. 1964: A critique of current methods in hydrologic systems investigation. *Transactions of the American Geophysical Union* **45**, 307–21.

ANDERSON, M.G. 1975: Some statistical approaches towards physical hydrology in large catchments. In R.F. Peel, M.D.I. Chisholm and P. Haggett (eds) *Processes in physical and human geography, Bristol essays*. London: Heinemann Educational, 91–109.

_____ 1988: *Modelling geomorphological systems*. Chichester: John Wiley.

ANDERSON, M.G. and BURT, T.P. (eds) 1985: *Hydrological forecasting*. Chichester: John Wiley.

ANDERSON, M.G. and SAMBLES, K.M. 1988: A review of the bases of geomorphological modelling. In M.G. Anderson (ed.) *Modelling geomorphological systems*. Chichester: John Wiley, 1–32.

ANDERSON, M.G. and KEMP, M.J. 1991: Towards an improved specification of slope hydrology in the analysis of slope instability problems in the tropics. *Progress in Physical Geography* **15**, 29–52.

ANDERSON, M.G. and BROOKS, S.M. (eds) 1996: *Advances in hillslope processes*. Chichester: John Wiley.

ANDERSON, M.G., WALLING, D.E. and BATES, P.W. (eds) 1996: *Floodplain processes*. Chichester: John Wiley.

ANDREWS, J.T. 1970a: *A geomorphological study of post-glacial uplift with particular reference to Arctic Canada*. Institute of British Geographers Special Publication No. 2.

_____ 1970b: Techniques of till fabric analysis. *British Geomorphological Research Group Technical Bulletin* **6**.

_____ 1972: Glacier power, mass balances, velocities and erosion potential. *Zeitschrift für Geomorphologie* **13**, 1–17.

_____ 1975: *Glacier systems: an approach to glaciers and their environments*. North Scituate, MA: Duxbury Press.

ANUCHIN, V.A. 1973: Theory of geography. In R.J. Chorley (ed.) *Directions in geography*. London: Methuen, 43–64.

APPLEBY, P.G. and OLDFIELD, F. 1978: The calculation of lead-210 dates assuming a constant rate of supply of unsupported ^{210}Pb to the sediment. *Catena* 5, 1–8.

ARNELL, N. 1996: *Global warming, river flows and water resources.* Chichester: John Wiley.

ARONOFF, S. 1989: *Geographic information systems: a management perspective.* Ottawa: WDL Publishers.

ATKINSON, B.W. 1978: The atmosphere: recent observational and conceptual advances. *Geography* 63, 283–300.

_____ 1979: Precipitation. In K.J. Gregory and D.E. Walling (eds) *Man and environmental processes.* London: Dawson, 23–37.

_____ 1980: Climate. In E.H. Brown (ed.) *Geography yesterday and tomorrow.* Oxford: Oxford University Press, 114–29.

_____ 1983: Numerical modelling of thermally-driven mesoscale airflows involving the planetary boundary layer. *Progress in Physical Geography* 7, 177–209.

ATKINSON, P.M. and TATE, N.J. (eds) 1999: *Advances in remote sensing and GIS analysis.* Chichester: John Wiley.

BAGNOLD, R.A. 1940: Beach formation by waves: some model experiments in a wave tank. *Journal of the Institution of Civil Engineers* 15, 27–52.

_____ 1941: *The physics of blown sand and desert dunes.* London: Methuen.

_____ 1960: Sediment discharge and stream power: a preliminary announcement. *US Geological Survey Circular* 421.

_____ 1979: Sediment transport by wind and water. *Nordic Hydrology* 10, 309–22.

BAKER, V.R. 1978a: Palaeohydraulics and hydrodynamics of scabland floods. In V.R. Baker and D. Nummedal (eds) *The channeled scabland.* Washington: NASA, 59–79.

_____ 1978b: Large-scale erosional and depositional features of the channeled scabland. In V.R. Baker and D. Nummedal (eds) *The channeled scabland.* Washington: NASA, 81–115.

_____ 1981: *Catastrophic flooding: the origin of the channeled scabland.* Stroudsburg, PA: Dowden, Hutchinson & Ross.

_____ 1988: Overview. In V.R. Baker, R.C. Kochel and P.C. Patton (eds) *Flood geomorphology.* Chichester: John Wiley, 1–8.

_____ 1994: Geomorphological understanding of floods. *Geomorphology* 10, 139–56.

_____ 1996: Hypotheses and geomorphological reasoning. In B.L. Rhoads and C.E. Thorn (eds) *The scientific nature of geomorphology.* Chichester: John Wiley, 57–85.

_____ 1998: Palaeohydrology and the hydrological sciences. In G. Benito, V.R. Baker and K.J. Gregory (eds) *Palaeohydrology and environmental change.* Chichester: John Wiley, 1–10.

BAKER, V.R. and BUNKER, P.C. 1985: Cataclysmic late Pleistocene flooding from Glacial Lake Missoula: a review. *Quaternary Science Reviews* **4**, 1–41.

BAKER, V.R., KOCHEL, R.C. and PATTON, P.C. (eds) 1988: *Flood geomorphology.* Chichester: JohnWiley.

BAKER, V.R. and TWIDALE, C.R. 1991: The reenchantment of geomorphology. *Geomorphology* **4**, 73–100.

BAKKER, J.P. 1963: Different types of geomorphological maps. In *Problems of geomorphological mapping, Geographical Studies* **46**, 13–31.

BAKKER, J.P. and LE HEUX, W.N. 1946: Projective-geometric treatment of O. Lehmann's theory of the transformation of steep mountain slopes. *Proceedings Koninklijke Nederlandsche Akademie Van Wetenschappen* **49**, 533–47.

——— 1952: A remarkable new geomorphological law. *Proceedings Koninklijke Nederlandsche Akademie Van Wetenschappen* Series B, **55**, 399–410, 554–71.

BALCHIN, W.G.V. 1952: The erosion surfaces of Exmoor and adjacent areas. *Geographical Journal* **118**, 453–76.

BARBER, K.E. 1976: History of vegetation. In S. Chapman (ed.) *Methods in plant ecology.* Oxford: Blackwell, 5–83.

BARNES, C.P. 1954: The geographic study of soils. In P.E. James and C.F. Jones (eds) *American geography: inventory and prospect.* Syracuse University Press: Association of American Geographers, 382–95.

BARRETT, E.C. 1970: The estimation of monthly rainfall from satellite data. *Monthly Weather Review* **98**, 198–205.

——— 1974: *Climatology from satellites.* London: Methuen.

BARRETT, E.C. and MARTIN, D.W. 1981: *The use of satellite data in rainfall monitoring.* London: Academic Press.

BARROWS, H.H. 1923: Geography as human ecology. *Annals of the Association of American Geographers* **13**, 1–4.

BARRY, R.G. 1963: Appendix. An introduction to numerical and mechanical techniques. In F.J. Monkhouse and H.R. Wilkinson (eds) *Maps and diagrams.* London: Methuen, 385–423.

——— 1967: Models in meteorology and climatology. In R.J. Chorley and P. Haggett (eds) *Models in geography.* London: Methuen, 97–144.

——— 1979: Recent advances in climate theory based on simple climate models. *Progress in Physical Geography* **3**, 119–31.

——— 1997: Palaeoclimatology, climate system processes and the geomorphic record. In D.R. Stoddart (ed.) *Process and form in geomorphology.* London: Routledge, 187–214.

BARRY, R.G. and PERRY, A.H. 1973: *Synoptic climatology. Methods and applications.* London: Methuen.

BARRY, R.G. and CHORLEY, R.J. 1976: *Atmosphere, weather and climate.* London: Methuen.

BARRY, R.G., FALLOT, J.M. and ARMSTRONG, R.L. 1995: Twentieth century variability in snow-cover conditions and approaches to detecting and monitoring changes: status and prospects. *Progress in Physical Geography* **19**, 520–32.

BARTON, T.F. and KARAN, P.P. 1992: *Leaders in American geography, Vol. 1: Geographic education.* Mesilla, New Mexico: New Mexico Geographical Society.

BASSETT, K. 1994: Comments on Richards: The problem of 'real' geomorphology. *Earth Surface Processes and Landforms* **19**, 273–6.

BATTARBEE, R.W. (ed.) 1988: *Lake acidification in the UK 1800–1986.* London: ENSIS Publishing.

BATTARBEE, R.W., MASON, J., RENBERG, I. and TALLING, J.F. (eds) 1990: *Palaeolimnology and lake acidification.* London: Royal Society.

BAULIG, H. 1935: The changing sea level. *Transactions of the Institute of British Geographers* **3**, 1–46.

BECKINSALE, R.P. 1997: Richard J. Chorley: a reformer with a cause. In D.R. Stoddart (ed.) *Process and form in geomorphology.* London: Routledge, 3–12.

BECKINSALE, R.P. and CHORLEY, R.J. 1991: *The history of the study of landforms or the development of geomorphology, Vol. 3: Historical and regional geomorphology 1890–1950.* London: Routledge.

BELASCO, H.E. 1952: *Characteristics of air masses over the British Isles.* London: HMSO.

BELL, M.G. and WALKER, M. 1992: *Late Quaternary environmental change: physical and human perspectives.* Harlow: Longman.

BENITO, G., BAKER, V.R. and GREGORY, K.J. (eds) 1998: *Palaeohydrology and environmental change.* Chichester: John Wiley.

BENNETT, H.H. 1938: *Soils and men.* Washington, DC: US Department of Agriculture Yearbook.

BENNETT, R.J. and CHORLEY, R.J. 1978: *Environmental systems: philosophy analysis and control.* London: Methuen.

BENNETT, R.J. and ESTALL, R.C. (eds) 1991: *Global change and challenge: geography for the 1990s.* London: Routledge.

BERG, L.S. 1950: *Natural regions of the USSR.* Translated by O. Adler, edited by J. Morrison and C.C. Nikiforoff. New York: American Council of Learned Societies.

BERGLUND, B.E. 1983: Palaeohydrological studies in lakes and mires – a palaeoecological research strategy. In K.J. Gregory (ed.) *Background to palaeohydrology.* Chichester: John Wiley, 237–56.

BERGLUND, B.J., BIRKS, H.J.B., WRIGHT, H.E. and RALSKA-JASIEWICZOWA, M. (eds) 1996: *Palaeoecological events during the last 15 000 years – regional synthesis of palaeoecological studies of lakes and rivers in Europe.* Chichester: John Wiley.

BEVEN, K. and CARLING, P. (eds) 1989: *Floods. Hydrological, sedimentological and geomorphological implications.* Chichester: John Wiley.

BILLI, P., HEY, R.D., THORNE, C.R. and TACCONI, P. (eds) 1992: *Dynamics of gravel bed rivers*. Chichester: John Wiley.

BIRD, E.C.F. 1979: Coastal processes. In K.J. Gregory and D.E. Walling (eds) *Man and environmental processes*. London: Dawson, 82–101.

_____ 1987: Coastal processes. In K.J. Gregory and D.E. Walling (eds) *Human activity and environmental processes*. Chichester: John Wiley, 87–116.

_____ 1996: *Beach management*. Chichester: John Wiley.

BIRD, J.H. 1963: The noosphere: a concept possibly useful to geographers. *Scottish Geographical Magazine* 78, 54–6.

_____ 1989: *The changing worlds of geography. A critical guide to concepts and methods*. Oxford: Clarendon Press.

BIRKELAND, P.W. 1974: *Pedology, weathering and geomorphological research*. New York: Oxford University Press.

BIRKS, H.J.B. and BIRKS, H.H. 1980: *Quaternary palaeoecology*. London: Arnold.

BIRKS, H.H., BIRKS, H.J.B., KALAND, P.E. and MOE, D. (eds) 1989: *The cultural landscape – past, present and future*. Cambridge: Cambridge University Press.

BISHOP, P. 1980: Popper's principle of falsifiability and the irrefutability of the Davisian cycle. *Professional Geographer* 32, 310–15.

BISWAS, A. 1970: *History of hydrology*. Amsterdam: North Holland Publishing Company.

BLAIKIE, P. 1985: *The political economy of soil erosion in developing countries*. London: Longman.

_____ 1996: Post-modernism and global environmental change. *Global Environmental Change* 6, 81–5.

BLAIKIE, P., CANNON, T., DAVIS, I. and WISNER, B. 1994: *At risk – natural hazards, people's vulnerability and disasters*. London: Routledge.

BLONG, R.J. 1982: *The time of darkness*. Seattle: University of Washington Press.

_____ 1984: *Volcanic hazards*. Sydney: Academic Press.

_____ 1997: A geography of natural perils. *Australian Geographer* 28, 7–27.

BOARDMAN, J. (ed.) 1985: *Soils and Quaternary landscape evolution*. Chichester: John Wiley.

BOON, P.J., CALOW, P. and PETTS, G.E. (eds) 1992: *River conservation and management*. Chichester: John Wiley.

BOULTON, G.S. 1986: A paradigm shift in glaciology. *Nature* 322, 18.

BOULTON, G.S., JONES, A.S., CLAYTON, K.M. and KENNING, M.J. 1977: A British ice sheet model and patterns of glacial erosion and deposition in Britain. In F.W. Shotton (ed.) *British Quaternary studies: recent advances*. Oxford: Clarendon, 231–6.

BOURNE, R. 1931: Regional survey and its relation to stocktaking of the agricultural and forest resources of the British Empire. *Oxford Forestry Memoir* 13.

BOWEN, D.Q. 1978: *Quaternary geology*. Oxford: Pergamon.

BOWMAN, I. 1922: *Forest physiography*. New York: John Wiley.

BOYGLE, J. 1993: The Swedish varve chronology – a review. *Progress in Physical Geography* 17, 1–19.

BRADLEY, R.S. and Jones, P.D. (eds) 1992: *Climate since AD 1500*. London: Routledge.

BRAITHWAITE, R.B. 1953: *Scientific explanation*. Cambridge: Cambridge University Press.

———— 1960: *Scientific explanation*. 2nd edn. New York: Harper Torch.

BRANDT, C.J. and THORNES, J.B. (eds) 1996: *Mediterranean desertification and land use*. Chichester: John Wiley.

BRANSON, J., CLARK, M.J. and GREGORY, K.J. 1995: A database for global continental palaeohydrology: technology or scientific creativity? In K.J. Gregory, L. Starkel and V.R. Baker (eds) *Global continental palaeohydrology*. Chichester: John Wiley, 303–25.

BRANSON, J., BROWN, A.G. and GREGORY, K.J. (eds) 1996: *Global continental changes: the context of palaeohydrology*. Geological Society Special Publication 115.

BRANSON, J., GREGORY, K.J. and CLARK, M.J. 1996: Issues in scientific cooperation or information sharing: the case of palaeohydrology. In J. Branson, A.G. Brown and K.J. Gregory (eds) *Global continental changes: the context of palaeohydrology*. Geological Society Special Publication 115, 235–49.

BRAZEL, A.J., ARNFIELD, A.J., GREENLAND, D.J. and WILLMOTT, C.J. 1991: Physical and boundary layer climatology. *Physical Geography* 12, 189–206.

BREN, L.J. 1993: Riparian zone stream and floodplain issues: a review. *Journal of Hydrology* 150, 277–99.

BRETZ, J.H. 1923: The channelled scabland of the Columbia plateau. *Journal of Geology* 3, 617–49.

BRIDGES, E.M. 1970: *World soils*. Cambridge: Cambridge University Press.

———— 1978a: Soil, the vital skin of the earth. *Geography* 63, 354–61.

———— 1978b: Interaction of soil and mankind in Britain. *Journal of Soil Science* 29, 125–39.

———— 1981: Soil geography: a subject transformed. *Progress in Physical Geography* 5, 398–407.

BRIDGES, E.M. and DAVIDSON, D.A. 1982: Agricultural uses of soil survey data. In E.M. Bridges and D.A. Davidson (eds) *Principles and applications of soil geography*. London: Longman, 171–215.

BRIGGS, D.J. 1981: Editorial. The principles and practice of applied geography. *Applied Geography* 1, 1–8.

BRIGGS, D.J., COLLINS, S., ELLIOTT, P., FISCHER, P., KINGHAM, S., LEBRET, E., PRYL, K., VAN REEUWIJK, H., SMALLBONE, K. and VAN DER VEEN, A. 1997: Mapping air pollution using GIS: a regression based approach. *International Journal of Geographic Information Science* **11**, 699–718.

BRIGGS, S.A. and JACKSON, M.J. 1984: Remotely observed terrains. *The Times Higher Educational Supplement,* **607,** iv.

BRISLIN, R.W. 1980: Cross-cultural research methods: strategies, problems, applications. In I. Altman, A. Rapoport and J.F. Wohwill (eds) *Human behaviour and environment.* Environment and culture 4. New York: Plenum, 47–82.

BRODA, E. 1975: *The evolution of the bioenergetic processes.* Oxford: Pergamon Press.

BROOKES, A. 1988: *Channelized rivers. Perspectives for environmental management.* Chichester: John Wiley.

BROOKES, A., GREGORY, K.J. and DAWSON, F.H. 1983: An assessment of river channelization in England and Wales. *The Science of the Total Environment* **27**, 97–111 .

BROOKES, A. and SHIELDS, F.D. (eds) 1996: *River channel restoration: guiding principles for sustainable projects.* Chichester: John Wiley.

BROWN, A.G. 1991: Hydrogeomorphological changes in the Severn Basin during the last 15,000 years: orders of change in a maritime catchment. In L. Starkel, K.J. Gregory and J.B. Thornes (eds) *Temperate palaeohydrology.* Chichester: John Wiley, 147–70.

_____ (ed.) 1995: *Geomorphology and ground water.* Chichester: John Wiley.

_____ 1996: Palaeohydrology: prospects and future advances. In J. Branson, A.G. Brown and K.J. Gregory (eds) *Global continental changes: the context of palaeohydrology.* Geological Society Special Publication **115**, 257–65.

_____ 1997: *Alluvial geoarchaeology: floodplain archaeology and environmental change.* Cambridge: CUP Manuals in Archaeology.

_____ 1999: Biodiversity and pollen analysis: modern pollen studies and the recent history of a floodplain woodland in S.W. Ireland. *Journal of Biogeography* **26**, 19–32.

BROWN, A.G. and QUINE, T. (eds) 1999a: *Fluvial processes and environmental change.* Chichester: John Wiley.

_____ 1999b: Fluvial processes and environmental change: an overview. In A.G. Brown and T. Quine (eds) *Fluvial processes and environmental change.* Chichester: John Wiley, 1–27.

BROWN, E.H. 1952: The river Ystwyth, Cardiganshire: a geomorphological analysis. *Proceedings of the Geologists Association* **63**, 244–69.

_____ 1960: *The relief and drainage of Wales.* Cardiff: University of Wales Press.

_____ 1961: Britain and Appalachia: a study of the correlation and dating of planation surfaces. *Transactions of the Institute of British Geographers* **29**, 91–100.

_____ 1970: Man shapes the earth. *Geographical Journal* **136**, 74–85.

_____ 1975: The content and relationships of physical geography. *Geographical Journal* **141**, 35–48.

_____ 1980: Historical geomorphology – principles and practice. *Zeitschrift für Geomorphologie* Supplementband **36**, 9–15.

BROWN, H.I. 1996: The methodological roles of theory in science. In B.L. Rhoads and C.E. Thorn (eds) *The scientific nature of geomorphology*. Chichester: John Wiley, 3–20.

BROWN, R.J.E. 1970: *Permafrost in Canada: its influence on northern development*. Toronto: University of Toronto Press.

BRUNSDEN, D. 1990: Tablets of stone: towards the ten commandments of geomorphology. *Zeitschrift für Geomorphologie* Supplementband **79**, 1–37.

_____ 1999: Geomorphology in environmental management: an appreciation. *East Midland Geographer* **22**, 63–77.

BRUNSDEN, D. and THORNES, J.B. 1979: Landscape sensitivity and change. *Transactions of the Institute of British Geographers* NS **4**, 463–84.

BRUNSDEN, D., DOORNKAMP, J.C. and JONES, D.K.C. 1979: The Bahrain surface materials resources survey and its application to regional planning. *Geographical Journal* **145**, 1–35.

_____ (eds) 1980: *Geology, geomorphology and pedology of Bahrain*. Norwich: Geo Books.

BRYAN, K. 1946: Cryopedology: the study of frozen ground and intensive frost action with suggestions on nomenclature. *American Journal of Science* **244**, 622–42.

BRYAN, R.B. 1979: Soil erosion and conservation. In K.J. Gregory and D.E. Walling (eds) *Man and environmental processes*. London: Butterworths, 207–24.

BRYANT, E. 1997: *Climate process and change*. Cambridge: Cambridge University Press.

BUDEL, J. 1957: Die doppelten Einebnungsflächen in den feuchten Tropen. *Zeitschrift für Geomorphologie* **1**, 201–28.

_____ 1963: Klima-genetische Geomorphologie. *Geographische Rundschau* **15**, 269–85.

_____ 1969: Das System der klima-genetischen Geomorphologie. *Erdkunde* **23**, 165–82 .

_____ 1977: *Klima-Geomorphologie*. Berlin/Stuttgart: Borntraeger.

BUDYKO, M.I. 1958: *The heat balance of the Earth's surface*. Translated by N. Steepanova from original dated 1956. Washington: US Weather Bureau.

BULL, L.J. and KIRKBY, M.J. 1997: Gully process and modelling. *Progress in Physical Geography* **21**, 354–74.

BULL, P.A. 1981: Environmental reconstruction by electron microscopy. *Progress in Physical Geography* **5**, 368–97.

BULL, W.B. 1991: *Geomorphic responses to climatic change*. Oxford: Oxford University Press.

BUNGE, W. 1973: The geography. *Professional Geographer* **25**, 331–7.

BURGESS, J. 1990: The production and consumption of environmental meanings in the mass media: a research agenda for the 1990s. *Transactions of the Institute of British Geographers* NS **15**, 139–61.

BURKE, C.J. and ELIOT, F.E. 1954: The geographic study of the oceans. In P.E. James and C.F. Jones (eds) *American geography inventory and prospect.* Syracuse University Press: Association of American Geographers, 410–27.

BURKE, E.J., BANKS, A.C. and GURNEY, R.J. 1997: Remote sensing of soil–vegetation–atmosphere transfer processes. *Progress in Physical Geography* **21**, 549–72.

BURROUGH, P.A. 1983: Multiscale sources of variation in soil. I. The applications of fractal concepts to nested levels of soil variation. *Journal of Soil Science* **34**, 599–620.

_____ 1986: *Principles of geographical information systems for land resources management.* Oxford: Oxford University Press.

_____ 1991: Soil information systems. In D.J. Maguire, M.F. Goodchild and D.W. Rhind (eds) *Geographical information systems. Principles and applications* Vol. 2. Harlow: Longman, 153–69.

BURROUGH, P.A. and McDONNELL, R.A. 1998: *Principles of geographical information systems.* Oxford: Oxford University Press.

BURT, T.P. 1994: Long term study of the natural environment – perceptive science or mindless monitoring. *Progress in Physical Geography* **18**, 475–96.

BURTON, I. 1997: Vulnerability and adaptive responses in the context of climate and climate change. *Climatic Change* **36**, 185–96.

BURTON, I., KATES, R.W. and WHITE, G.F. 1978: *The environment as hazard.* New York: Oxford University Press.

_____ 1993: *The environment as hazard* (2nd edn). New York: Guilford Press.

BUTLER, B.E. 1959: *Periodic phenomena in landscapes as a basis for soil Studies.* Soil Publication 14. CSIRO, Australia.

BUTLER, D.R. 1993: The grizzly bear as an erosional agent in mountainous terrain. *Zeitschrift für Geomorphologie* NF **36**, 179–89.

BUTTLE, J.M. 1994: Isotope hydrograph separations and rapid delivery of pre-event water from drainage basins. *Progress in Physical Geography* **18**, 16–41.

BUTZER, K.W. 1964: *Environment and archaeology.* London: Methuen.

_____ 1974: Accelerated soil erosion: a problem of man–land relationships. In I.R. Manners and M.W. Mikesell (eds) *Perspectives on environment.* Washington, DC: Association of American Geographers, 57–77.

_____ 1975: Geological and ecological perspectives on the Middle Pleistocene. In K.W. Butzer and G.L.I. Isaac (eds) *After the Australopithecines.* The Hague: Mouton, 857–74.

_____ 1976: *Geomorphology from the Earth*. New York: Harper & Row.

_____ 1982: *Archaeology as human ecology*. Cambridge: Cambridge University Press.

_____ 1989: Cultural ecology. In G.L. Gaile and C.J. Willmott (eds) *Geography in America*. Columbus: Merrill Publishing Company, 192–208.

_____ 1990: The realm of cultural–human ecology: adaptation and change in historical perspective. In B.L. Turner, W.C. Clark, R.W. Kates, J.F. Richards, J.L. Mathews and W.B. Mayer (eds) *The Earth as transformed by human action: global and regional changes in the biosphere over the past 300 years*. New York: Cambridge University Press, 685–701.

CAILLEUX, A. 1947: L'indice d'emousse: définition et première application. *Comptes Rendus Sommaires de la Société Geologique de France* 13, 165–7.

CAINE, N. 1976: A uniform measure of subaerial erosion. *Bulletin of the Geological Society of America* 87, 137–40.

CAIRNS, J. 1991: The status of the theoretical and applied science of restoration ecology. *Environmental Professional* 13, 186–94.

CARSON, M.A. 1971: *The mechanics of erosion*. London: Pion.

CARSON, M.A. and KIRKBY, M.J. 1972: *Hillslope form and process*. Cambridge: Cambridge University Press.

CARTER, R.W.G. 1988: *Coastal environments: an introduction to the physical, ecological and cultural systems of coastlines*. London: Academic Press.

CAWS, P. 1965: *The philosophy of science*. Princeton: Van Nostrand.

CHAGNON, S.A., HUFF, F.A., SCHICKEDANZ, P.T. and VOGEL, J.L. 1977: *Summary of Metromex, Vol. I: Weather anomalies and impacts*. Bulletin 62, State of Illinois Department of Registration and Education, Illinois State Water Survey, Urbana.

CHANDLER, T.J. 1965: *The climate of London*. London: Hutchinson.

_____ 1970: *The management of climatic resources*. Inaugural Lecture, University College London.

_____ 1976: The Royal Commission on Environmental Pollution and the control of air pollution in Great Britain. *Area* 8, 87–92.

CHANDLER, T.J., COOKE, R.U. and DOUGLAS, I. 1976: Physical problems of the urban environment. *Geographical Journal* 142, 57–80.

CHAPPELL, J.M.A. 1974: Geology of coral terraces, Huon peninsula, New Guinea: a study of Quaternary tectonic movements and sea level changes. *Bulletin of the Geological Society of America* 85, 553–70.

CHARLESWORTH, J.K. 1957: *The Quaternary Era*. London: Arnold.

CHERRILL, A. and LANE, M. 1995: The survey and prediction of land cover using an environmental land classification. *Applied Geography* 15, 69–85.

CHESTER, D. 1993: *Volcanoes and society*. London: Arnold.

CHIN, A. and GREGORY, K.J. forthcoming: Urbanization and adjustment of ephemeral stream channels.

CHISHOLM, M. 1967: General systems theory and geography. *Transactions of the Institute of British Geographers* 42, 42–52.

CHORLEY, R.J. 1962: Geomorphology and general systems theory. *US Geological Survey Professional Paper* 500–B, 1–10.

_____ 1965: A re-evaluation of the geomorphic system of W.M. Davis. In R.J. Chorley and P. Haggett (eds) *Frontiers in geographical teaching.* London: Methuen, 21–38.

_____ 1966: The application of statistical methods to geomorphology. In G.H. Dury (ed.) *Essays in geomorphology.* London: Heinemann, 275–387.

_____ 1967: Models in geomorphology. In R.J. Chorley and P. Haggett (eds) *Models in geography.* London: Methuen, 59–96.

_____ (ed.) 1969a: *Water, earth and man.* London: Methuen.

_____ 1969b: The drainage basin as the fundamental geomorphic unit. In R.J. Chorley (ed.) *Water, earth and man.* London: Methuen, 77–100.

_____ 1971: The role and relations of physical geography. *Progress in Geography* 3, 87–109.

_____ (ed.) 1972: *Spatial analysis in geomorphology.* London: Methuen.

_____ 1973: Geography as human ecology. In R.J. Chorley (ed.) *Directions in geography.* London: Methuen, 155–69.

_____ 1978: Bases for theory in geomorphology. In C. Embleton, D. Brunsden and D.K.C. Jones (eds) *Geomorphology. Present problems and future prospects.* Oxford: Oxford University Press.

_____ 1987: Perspectives on the hydrosphere. In M.J. Clark, K.J. Gregory and A.M. Gurnell (eds) *Horizons in physical geography.* Basingstoke: Macmillan, 378–81.

_____ 1995: Classics in physical geography revisited: Horton 1945. *Progress in Physical Geography* 19, 533–54.

CHORLEY, R.J. DUNN, A.J. and BECKINSALE, R.P. 1964: *The history of the study of landforms, Vol. I, Geomorphology before Davis.* London: Methuen.

CHORLEY, R.J. and HAGGETT, P. (eds) 1965: *Frontiers in geographical teaching.* London: Methuen.

_____ (eds) 1967: *Models in geography.* London: Methuen.

CHORLEY, R.J. and KATES, R.W. 1969: Introduction. In R.J. Chorley (ed.) *Water, earth and man.* London: Methuen, 1–7.

CHORLEY, R.J. and KENNEDY, B.A. 1971: *Physical geography: a systems approach.* London: Prentice Hall.

CHORLEY, R.J., BECKINSALE, R.P and DUNN, A.J. 1973: *The history of the study of landforms, Vol. II, The life and work of William Morris Davis.* London: Methuen.

CHORLEY, R.J., SCHUMM, S.A. and SUGDEN, D.A. 1984: *Geomorphology.* London: Methuen.

CHRISTIAN, C.S. and STEWART, G.A. 1953: Survey of the Katherine-Darwin region 1946. *CSIRO Land Research Series* 1, Melbourne.

CHUBUKOV, L.A., RAUNER, Yu L., KUVSHINOVA, K.V., POTAPOVA, L.S. and SHVAREVA, Yu N. 1982: Predicting the climatic consequences of the interbasin transfer of water in the Midland region of the USSR. *Soviet Geography* 22, 426–4.

CHURCH, M. 1980: Records of recent geomorphological events. In R.A. Cullingford, D.A. Davidson and J. Lewin (eds) *Timescales in geomorphology*. Chichester: John Wiley, 13–29.

_____ 1996: Space, time and the mountain – how do we order what we see? In B.L. Rhoads and C.E. Thorn (eds) *The scientific nature of geomorphology*. Chichester: John Wiley, 147–70.

CHURCH, M., GOMEZ, B., HICKIN, E.A. and SLAYMAKER, H.O. 1985: Geomorphological sociology. *Earth Surface Processes and Landforms* 10, 539–40.

CLAPPERTON, C.M. 1972: Patterns of physical and human activity on Mount Etna. *Scottish Geographical Magazine* 88, 160–67.

CLARK, D.M., HASTINGS, D.A. and KINEMAN, J.J. 1991: Global data bases and their implications for GIS. In D. Maguire, M.F. Goodchild and D.W. Rhind (eds) *Geographical information systems. Principles and applications* Vol. 2. Harlow: Longman, 217–31.

CLARK, M.J. and GREGORY, K.J. 1982: Physical geography techniques: a self-paced course. *Journal of Geography in Higher Education* 6, 123–31.

CLARK, M.J., GREGORY, K.J. and GURNELL, A.M. (eds) 1987: *Horizons in physical geography*. Basingstoke: Macmillan.

CLAYTON, K.M. 1980a: Geomorphology. In E.H. Brown (ed.) *Geography yesterday and tomorrow*. Oxford: Oxford University Press, 167–80.

_____ 1980b: Beach sediment budgets and coastal modification. *Progress in Physical Geography* 4, 471–86.

CLEMENTS, F.E. 1916: *Plant succession, an analysis of the development of vegetation*. Washington: Carnegie Institution.

COATES, D.R. (ed.) 1971: *Environmental geomorphology*. Binghamton: State University of New York Publications in Geomorphology.

_____ (ed.) 1972: *Environmental geomorphology and landscape conservation, Vol. 1. Prior to 1900*. Stroudsburg: Dowden, Hutchinson & Ross.

_____ (ed.) 1973: *Environmental geomorphology and landscape conservation, Vol. III. Non-urban regions*. Stroudsburg: Dowden, Hutchinson & Ross.

_____ (ed.) 1976a: *Geomorphology and engineering*. Stroudsburg: Dowden, Hutchinson & Ross.

_____ 1976b: Geomorphic engineering. In D.R. Coates (ed.) *Geomorphology and engineering*. Stroudsburg: Dowden, Hutchinson & Ross, 3–21.

_____ 1980: Subsurface influences. In K.J. Gregory and D.E. Walling (eds) *Man and environmental processes*. London: Butterworths, 163–88.

_____ 1982: Environmental geomorphology perspectives. In J.W. Frazier (ed.) *Applied geography: selected perspectives*. Englewood Cliffs, NJ: Prentice Hall, 139–69.

_____ 1990: Perspectives on environmental geomorphology. *Zeitschrift für Geomorphologie* Supplementband **79**, 83–117.

COATES, D.R. and VITEK, J.D. (eds) 1980: *Thresholds in geomorphology*. London: Allen & Unwin.

COCKS, K.D. and WALKER, P.A. 1987: Using the Australian Resources Information System to describe extensive regions. *Applied Geography* **7**, 17–27.

COFFEY, W.J. 1981: *Geography. Towards a general spatial systems approach*. London: Methuen.

COLE, J.M. and KING, C.A.M. 1968: *Quantitative geography: techniques and theories in geography*. London: John Wiley.

COLE, M.M. 1963: Vegetation and geomorphology in Northern Rhodesia: an aspect of the distribution of savanna of central Africa. *Geographical Journal* **129**, 290–310.

COLTEN, C.E. 1998: Groundwater contamination: reconstructing historical knowledge for the courts. *Applied Geography* **18**, 259–73.

COMMONER, B. 1972: *The closing circle: confronting the environmental crisis*. London: Cape.

CONACHER, A.J. 1988: The geomorphic significance of process measurements in an ancient landscape. In A.M. Harvey and M. Sala (eds) *Geomorphic processes in environments with strong seasonal contrasts, Vol. 2: Geomorphic systems, Catena* Supplement **13**, 147–64.

CONACHER, A.J. and DALRYMPLE, J.B. 1977: The nine-unit land-surface model: an approach to pedogeomorphic research. *Geoderma* **18**, 1–154.

COOKE, R.U. 1976: *An empty quarter*. Inaugural lecture. Bedford College, London.

_____ 1977: Applied geomorphological studies in deserts: a review of examples. In J.R. Hails (ed.) *Applied geomorphology*. Amsterdam: Elsevier, 183–225.

_____ 1992: Common ground, shared inheritance: research imperatives for environmental geography. *Transactions of the Institute of British Geographers* NS **17**, 131–51.

COOKE, R.U. and WARREN, A. 1973: *Geomorphology in deserts*. London: Batsford.

COOKE, R.U. and DOORNKAMP, J.C. 1974: *Geomorphology in environmental management*. Oxford: Oxford University Press.

COOKE, R.U. and REEVES, R.W. 1976: *Arroyos and environmental change in the American south-west*. Oxford: Clarendon Press.

COOKE, R.U., BRUNSDEN, D., DOORNKAMP, J.C. and JONES, D.K.C. 1982: *Urban geomorphology in drylands*. Oxford: Oxford University Press.

COOKE, R.U. and DOORNKAMP, J.C. 1990: *Geomorphology in environmental management. A new introduction* (2nd edn). Oxford: Clarendon Press.

COOKE, R.U., WARREN, A. and GOUDIE, A.S. 1993: *Desert geomorphology*. London: UCL Press.

COOPER, A. and MURRAY, R. 1992: A structured method of landscape assessment and countryside management. *Applied Geography* 12, 319–38.

COPPOCK, T. and ANDERSON, E. 1987: Editorial Comment. *International Journal of Geographical Information Systems* 1, 3.

COSGROVE, D. 1990: An elemental division: water control and engineered landscape. In D. Cosgrove and G. Petts (eds) *Water engineering and landscape*. London: Belhaven, 1–11.

COSGROVE, D. and PETTS, G. (eds) 1990: *Water, engineering and landscape*. London: Belhaven.

COSTA, J.E. and GRAF, W.L. 1984: The geography of geomorphologists in the United States. *Professional Geographer* 36, 82–9.

COSTA, J.E. and O'CONNOR, G.E. 1995: Geomorphically effective floods. In J.E. Costa, A.J. Miller, K.W. Potter and P.R. Wilcock (eds) *Natural and anthropogenic influences in fluvial geomorphology. The Wolman volume*. American Geophysical Union Geophysical Monograph 89, 45–56.

COSTA, J.E., MILLER, A.J., POTTER, K.W. and WILCOCK, P.R. (eds) 1995: *Natural and anthropogenic influences in fluvial geomorphology. The Wolman volume*. American Geophysical Union Geophysical Monograph 89.

COTGROVE, S.F. 1982: *Catastrophe or cornucopia: the environment, politics and the future*. Chichester: John Wiley.

COTTON, C.A. 1922: *Geomorphology of New Zealand*. Wellington: Dominion Museum.

—————— 1942: *Landscape as developed by the processes of normal erosion*. Cambridge: Cambridge University Press.

—————— 1948: *Climatic accidents*. Wellington: Whitcomb and Tombs.

COTTON, W.R. and PIELKE, R.A. 1995: *Human impacts on weather and climate*. Cambridge: Cambridge University Press.

COURT, A. 1957: Climatology: complex, dynamic and synoptic. *Annals of the Association of American Geographers* 47, 125–36.

COURTNEY, F.M. and NORTCLIFF, S. 1977: Analysis techniques in the study of soil distribution. *Progress in Physical Geography* 1, 40–64.

COWEN, D.J., YANG, A.H. and WADDELL, J.M. 1983: Beyond hardware and software: implementing a state-level geographical information system. In E. Teicholz and B.J.L. Berry (eds) *Computer graphics and environmental planning*. Englewood Cliffs, NJ: Prentice Hall, 30–51.

COWLES, H.C. 1911: The causes of vegetative cycles. *Annals of the Association of American Geographers* 1, 3–20.

CRAIG, R.C. and CRAFT, J.L. 1982: *Applied geomorphology.* Binghamton: State University of New York.

CROIZAT, L. 1978: Deduction, induction and biogeography. *Systematic Zoology* 27, 209–13.

CURRAN, P.J. 1985: *Principles of remote sensing.* Harlow: Longman.

_____ 1989: Editorial: global change and remote sensing. *International Journal of Remote Sensing* 10, 1459–60.

_____ 1994: Imaging spectrometry. *Progress in Physical Geography* 18, 247–66.

CURRAN, P.J. and FOODY, G.M. 1994: Environmental issues at regional to global scales. In G.M. Foody and P.J. Curran (eds) *Environmental remote sensing from regional to global scales.* Chichester: John Wiley, 1–7.

CURRAN, P.J., MILTON, E.J., ATKINSON, P.M. and FOODY, G.M. 1998: Remote sensing from data to understanding. In P. Longley, S. Brooks, W. Macmillan and R. McDonnell (eds) *Geocomputation: a primer.* Chichester: John Wiley, 33–59.

CURTIS, C.D. 1976: Chemistry of rock weathering: fundamental reactions and controls. In E. Derbyshire (ed.) *Geomorphology and climate.* London: John Wiley, 25–58.

CURTIS, L.F., DOORNKAMP, J.C. and GREGORY, K.J. 1965: The description of relief in field studies of soils. *Journal of Soil Science* 16, 16–30.

CURTIS, L.F. and SIMMONS, I.G. 1976: Man's impact on past environments. *Transactions of the Institute of British Geographers* NS 1, 1–384.

CZUDEK, T. and DEMEK, K.J. 1970: Thermokarst in Siberia and its influence on the development of lowland relief. *Quaternary Research* 1, 103–20.

DACKCOMBE, R.V. and GARDINER, V. 1983: *Geomorphological field manual.* London: George Allen and Unwin.

DALRYMPLE, J.B., CONACHER, A.J. and BLONG, R.J. 1969: A nine-unit hypothetical landsurface model. *Zeitschrift für Geomorphologie* 12, 60–76.

DANA, J.D. 1863: *Manual of geology: treating on the principle of science.* Philadelphia: Bliss.

_____ 1872: *Manual of geology: treating on the principle of science* (2nd edn). Philadelphia: Bliss.

DANIELS, R.B., GAMBLE, E.E. and CADY, J.G. 1971: The relationship between geomorphology and soil morphology and genesis. *Advances in Agronomy* 23, 51–88.

DANSEREAU, P. 1957: *Biogeography: an ecological perspective.* New York: Ronald Press.

DARBY, H.C. 1956: The clearing of the woodland in Europe. In W.L. Thomas (ed.) *Man's role in changing the face of the Earth.* Chicago: University of Chicago Press, 183–216.

DARWIN, C. 1842: *The structure and distribution of coral reefs.* London: Smith, Elder and Co.

_____ 1859: *Origin of Species.* London: John Murray.

DAVIDSON, D.A. 1980a: Erosion in Greece during the first and second millennia BC. In R.A. Cullingford, D.A. Davidson and J. Lewin (eds) *Timescales in geomorphology.*Chichester: John Wiley, 143–58.

_____ 1980b: *Soils and land use planning.* London: Longman.

DAVIDSON, D.A. and SHACKLEY, M.L. 1976: *Geoarchaeology: earth science and the past.* London: Duckworth.

DAVIDSON, D.A. and JONES, G.E. 1986: A land resources information system (LRIS) for land use planning. *Applied Geography* 6, 255–65.

DAVIES, G.L. 1968: *The Earth in decay: a history of British geomorphology 1578–1878.* London: MacDonald Technical & Scientific.

DAVIES, J.L. 1969: *Landforms of cold climates.* Cambridge, MA: MIT Press.

_____ 1973: *Geographical variation in coastal development.* New York: Hafner.

DAVIS, J.C. 1973: *Statistics and data analysis in geology.* New York: John Wiley.

DE MARTONNE, E. 1909: *Traite de geographie physique.* Paris: Armand Colin.

DEMEK, J. (ed.) 1972: *Manual of detailed geomorphological mapping.* Prague: IGU Commission on Geomorphological Survey and Mapping.

_____ 1973: Quaternary relief development and man. *Geoforum* 15, 68–71.

DEMEK, J. and EMBLETON, C. (eds) 1978: *Guide to medium-scale geomorphological survey and mapping.* Stuttgart: IGU Commission on Geomorphological Survey and Mapping.

DENT, D. and YOUNG, A. 1981: *Soils and land use planning.* London: Allen & Unwin.

DE PLOEY, J. (ed.) 1983: *Rainfall simulation, runoff and soil erosion. Catena* Supplement 4.

DETWYLER, T.R. 1971: *Man's impact on environment.* New York: McGraw Hill.

DETWYLER, T.R. and MARCUS, M.G. (eds) 1972: *Urbanization and environment: the physical geography of the city.* Belmont, CA: Duxbury Press.

DOORNKAMP, J.C. and KING, C.A.M. 1971: *Numerical analysis in geomorphology.* London: Arnold.

DOORNKAMP, J.C., GREGORY, K.J. and BURN, A.S. 1980: *Atlas of drought in Britain 1975–76.* London: Institute of British Geographers.

DORN, R.I., DIXON, J.C. and ORME, A.R. 1991: Integrating geomorphic process and landscape evolution: an editorial. *Physical Geography* 12, 301–2.

DOUGHILL, A.J., THOMAS, D.S.G. and HEATHWAITE, A.L. 1999:

Environmental change in the Kalahari: integrated land degradation studies for nonequilibrium dryland environments. *Annals of the Association of American Geographers* **89**, 420–42.

DOUGLAS, I. 1969: Man, vegetation and the sediment yields of rivers. *Nature* **215**, 925–8.

_____ 1972: *The environment game*. Inaugural lecture, University of New England, Armidale, New South Wales.

_____ 1980: Climatic geomorphology. Present-day processes and landform evolution. Problems of interpretation. *Zeitschrift für Geomorphologie* Supplementband **36**, 27–47.

_____ 1981: The city as an ecosystem. *Progress in Physical Geography* **5**, 315–67.

_____ 1983: *The urban environment*. London: Arnold.

_____ 1986: The unity of geography is obvious. *Transactions of the Institute of British Geographers* NS **11**, 459–63.

_____ 1987: The influence of human geography on physical geography. *Progress in Human Geography* **11**, 517–40.

DOWNS, P.W. and GREGORY, K.J. 1993: The sensitivity of river channels in the landscape system. In D.S.G. Thomas and R.J. Alison (eds) *Landscape sensitivity*. Chichester: John Wiley, 15–30.

_____ 1995: Approaches to river channel sensitivity. *Professional Geographer* **47**, 168–75.

DOWNS, P.W., GREGORY, K.J. and BROOKES, A. 1991: How integrated is river basin management? *Environmental Management* **15**, 299–309.

DREWRY, D.J. 1986: *Glacial geologic processes*. London: Arnold.

DRIVER, T.S. and CHAPMAN, G.P. (eds) 1996: *Time-scales and environmental change*. London: Routledge.

DULLER, G.A.T. 1996: Recent developments in luminescence dating of Quaternary sediments. *Progress in Physical Geography* **20**, 127–45.

DUNNE, T. and LEOPOLD, L.B. 1978: *Water in environmental planning*. San Francisco: W.H. Freeman.

DURY, G.H. 1964a: Principles of underfit streams. *US Geological Survey Professional Paper* 452A.

_____ 1964b: Subsurface exploration and chronology of underfit streams. *US Geological Survey Professional Paper* 452B.

_____ 1965: Theoretical implications of underfit streams. *US Geological Survey Professional Paper* 452C.

_____ 1980a: Neocatastrophism: a further look. *Progress in Physical Geography* **4**, 391–413.

_____ 1980b: Step-functional change in precipitation at Sydney. *Australian Geographical Studies* **18**, 62–78.

_____ 1981: *An introduction to environmental systems*. London: Heinemann.

_____ 1982: Step-functional analysis of long records of streamflow. *Catena* **9**, 379–96.

DUSEK, V. 1979: Geodesy and the earth sciences in the philosophy of C.S. Peirce. In C.J. Schneer (ed.) *Two hundred years of geology in America.* Hanover, NH: University Press of New England, 265–76.

DYLIK, J. 1952: The concept of the periglacial cycle in Middle Poland. *Bulletin de la Société des Sciences à des lettres de Lodz* 3.

EASTERLING, W.E. and KATES, R.W. 1995: Indexes of leading climate indicators for impact assessment. *Climatic Change* 31, 623–48.

EDEN, S., TUNSTALL, S.M. and TAPSELL, S.M. 1999: Environmental restoration: environmental management of environmental threat? *Area* 31, 151–59.

EDWARDS, K.C. 1964: The importance of biogeography. *Geography* 49, 85–97.

EDWARDS, K.J. and MACDONALD, G.M. 1991: Holocene palynology II: human influence and vegetation change. *Progress in Physical Geography* 15, 364–91.

EL-KADI, A.K.A. and SMITHSON, P.A. 1992: Atmospheric classifications and synoptic climatology. *Progress in Physical Geography* 16, 432–55.

ELLIS, J.B. 1979: The nature and sources of urban sediments and their relation to water quality: a case study from north-west London. In G.E. Hollis (ed.) *Man's impact on the hydrological cycle in the United Kingdom.* Norwich: Geobooks, 199–216.

ELLIS, S. and MELLOR, A. 1995: *Soils and environment.* London: Routledge.

ELSOM, D.M. 1987: *Atmospheric pollution – a global problem.* Oxford: Blackwell.

EMBLETON, C.E. 1982: *Glaciology in the service of man.* Inaugural lecture, Kings College, London.

_____ (ed.) 1988: Applied geomorphological mapping: methodology by example. *Zeitschrift für Geomorphologie* Supplementband 68.

EMBLETON, C. and KING, C.A.M. 1968: *Glacial and periglacial geomorphology.* London: Arnold.

_____ 1975: *Periglacial geomorphology.* London: Arnold.

EMBLETON, C. and THORNES, J. (eds) 1979: *Process in geomorphology.* London: Arnold.

EVANS, I.S. 1981: General geomorphometry. In A.S. Goudie (ed.) *Geomorphological techniques.* London: Allen & Unwin, 31–7.

EVANS, R. 1990: Water erosion in British farmers' fields – some causes, impacts, predictions. *Progress in Physical Geography* 14, 199–219.

EYRE, S.R. 1978: *The real wealth of nations.* London: Arnold.

FAIRBRIDGE, R.W. 1961: Eustatic changes in sea level. *Physics and Chemistry of the Earth* 5, 99–185.

_____ 1980: Thresholds and energy transfer in geomorphology. In D.R. Coates and J.D. Vitek (eds) *Thresholds in geomorphology.* London: Allen & Unwin, 43–50.

FELS, E. 1965: Nochmals: Anthropogen Geomorphologie. *Petermanns Geographische Mitteilungen* **109**, 9–15.

FENNEMAN, N.M. 1931: *Physiography of the western United States.* New York: McGraw Hill.

———— 1938: *Physiography of the eastern United States.* New York: McGraw Hill.

FEYERABEND, P. 1970: Consolations for the specialist. In A. Lakatos and A. Musgrave (eds) *Criticism and the growth of knowledge.* Cambridge: Cambridge University Press, 197–230.

———— 1975: *Against method.* London: Verso.

FISHER, P.F., MACKANESS, W.A., PEACEGOOD, G. and WILKINSON, G.G. 1988: Artificial intelligence and expert systems in geodata processing. *Progress in Physical Geography* **12**, 371–88.

FLEMING, N.C. 1964: Form and function of sedimentary particles. *Journal of Sedimentary Petrology* **35**, 381–90.

FLENLEY, J.R. 1979: *The equatorial rain forest: a geological history.* London: Butterworth.

FLINT, R.F. 1947: *Glacial and Pleistocene geology.* New York: John Wiley.

———— 1957: *Glacial and Pleistocene geology.* 2nd edn. New York: John Wiley.

———— 1971: *Glacial and Quaternary geology.* New York: John Wiley.

FOODY, G.M., CURRAN, P.J., BOYD, D.S., LUCAS, R.M., HONZAK, M., MILNE, R., BROWN, T., GRACE, J. and MALHI, Y. forthcoming: Amazonian tropical forests and the carbon cycle: adding value to remotely sensed data. Nottingham: Remote Sensing Society, in press.

FOOKES, P.G. and VAUGHAN, P.R. (eds) 1986: *A handbook of engineering geology.* London: Surrey University Press.

FORD, D.C. and WILLIAMS, P.W. 1989: *Karst geomorphology and hydrology.* London: Unwin Hyman.

FOSBERG, F.R. (ed.) 1963: *Man's role in the island ecosystem.* Honolulu: Bishop Museum Press.

FOSTER, I.M., GURNELL, A.M. and WEBB, B.W. (eds) 1995: *Sediment and water quality in river catchments.* Chichester: John Wiley.

FREEMAN, T.W. 1980: The Royal Geographical Society and the development of geography. In E.H. Brown (ed.) *Geography yesterday and tomorrow.* Oxford: Oxford University Press, 1–99.

FRENCH, H.M. 1976: *The periglacial environment.* London: Longman.

———— 1987: Periglacial geomorphology in North America: current research and future trends. *Progress in Physical Geography* **11**, 533–51.

———— 1996: *The periglacial environment* (2nd edn). London: Addison Wesley Longman.

FRENCH, J.R., SPENCER, T. and REED, D.J. 1995: Editorial – Geomorphic response to sea level rise – existing evidence and future impacts. *Earth Surface Processes and Landforms* **20**, 1–6.

FULLER, P. 1986: *The Australian scapegoat: towards an antipodean aesthetic*. Perth: University of Western Australia Press.

GAILE, G.L. and WILLMOTT, C.J. (eds) 1989: *Geography in America*. Columbus, Ohio: Merrill Publishing Co.

GARDINER, V. and GREGORY, K.J. 1977: Progress in portraying the physical landscape. *Progress in Physical Geography* 1, 1–22.

GARDNER, R. 1996: Developments in physical geography. In E.M. Rawling and R.A. Daugherty (eds) *Geography into the twenty-first century*. Chichester: John Wiley, 95–112.

GARDNER, R. and SCOGING, H. 1983: *Mega-geomorphology*. Oxford: Clarendon Press.

GEIGER, R. 1965: *The climate near the ground* (2nd edn). Cambridge, MA: Harvard University Press.

GERARD, R.W. 1964: Entitation, animorgs, and other systems. In M.D. Mesaroivic (ed.) *Views on general systems theory*. New York: John Wiley, 119–24.

GERASIMOV, I.P. 1961: The moisture and heat factors of soil formation. *Soviet Geography* 2, 3–12.

_____ 1968: Constructive geography: aims, methods and results. *Soviet Geography* 9, 739–55.

_____ 1984: The contribution of constructive geography to the problem of optimization of society's impact on the environment. *Geoforum* 15, 95–100.

GERRARD, A.J. 1981: *Soils and landforms*. London: George Allen & Unwin.

GERRARD, J. 1990: *Mountain environments: an examination of the physical geography of mountains*. Cambridge, MA: MIT Press.

_____ 1993: Soil geomorphology – present dilemmas and future challenges. *Geomorphology* 7, 61–84.

GERSMEHL, P.J. 1976: An alternative biogeography. *Annals of the Association of American Geographers* 66, 223–41.

GEYH, M.A. and SCHLEICHER, H. 1990: *Absolute age determination: physical and chemical dating methods and their applications*. New York: Springer-Verlag.

GIDDENS, A. 1998: *The third way. The renewal of social democracy*. Cambridge: Polity Press.

GILBERT, G.K. 1877: *Report on the geology of the Henry Mountains*. US Geological Survey, Rocky Mountain Region Report.

_____ 1909: The convexity of hill-tops. *Journal of Geology* 17, 344–50.

_____ 1914: The transportation of debris by running water. *US Geological Survey Professional Paper* 86.

GILL, A.M. 1989: Experimenting with environmental design research in Canada's newest mining town. *Applied Geography* 9, 177–95.

GJESSING, J. 1978: Glacial geomorphology: present problems and future prospects. In C. Embleton, D. Brunsden and D.K.C. Jones (eds)

Geomorphology. Present problems and future prospects. Oxford: Oxford University Press, 107–31.

GLAZOVSKAYA, N.A. 1977: Current problems in the theory and practice of landscape geochemistry. *Soviet Geography* 18, 363–73.

GOUDIE, A.S. (ed.) 1981a: *Geomorphological techniques.* London: George Allen & Unwin.

_____ 1981b: *The human impact, man's role in environmental change.* Oxford: Blackwell.

_____ 1983: Dust storms in space and time. *Progress in Physical Geography* 7, 502–30.

_____ 1986a: *The human impact. Man's role in environmental change* (2nd edn). Oxford: Blackwell.

_____ 1986b: The integration of human and physical geography. *Transactions of the Institute of British Geographers* NS 11, 454–8.

_____ 1989: The changing human impact. In L. Friday and R. Laskey (eds) *The fragile environment.* Cambridge: Cambridge University Press, 1–21.

_____ (ed.) 1990a: *Techniques for desert reclamation.* Chichester: John Wiley.

_____ 1990b: The global geomorphological future. *Zeitschrift für Geomorphologie* Supplementband 79, 51–62.

_____ (ed.) 1990c: *Geomorphological techniques* (2nd edn). London: Unwin Hyman.

_____ 1993a: Environmental uncertainty. *Geography* 78, 137–41.

_____ 1993b. *The human impact on the natural environment* (4th edn). Oxford: Blackwell.

_____ 1994: R.J. Chorley Commemorative Issue. *Progress in Physical Geography* 18, 317–18.

_____ 1995: *The changing Earth. Rates of geomorphological processes.* Oxford: Blackwell.

GOUDIE, A.S., ATKINSON, B.W., GREGORY, K.J., SIMMONS, I.G., STODDART, D.R. and SUGDEN, D.A. (eds) 1985: *The encyclopedic dictionary of physical geography.* Oxford: Blackwell Reference.

GOUDIE, A.S. and VILES, H. 1997a: *The Earth transformed.* Oxford: Blackwell.

_____ 1997b: *Salt weathering hazards.* Chichester: John Wiley.

GRAF, W.L. 1977: The rate law in fluvial geomorphology. *American Journal of Science* 277, 178–91.

_____ 1979a: Catastrophe theory as a model for change in fluvial systems. In D.D. Rhodes and G.P. Williams (eds) *Adjustments of the fluvial system.* Dubuque, Iowa: Kendall/Hunt, 13–32.

_____ 1979b: Mining and channel response. *Annals of the Association of American Geographers* 69, 262–75.

_____ 1982: Spatial variation of fluvial processes in semi-arid lands. In C.E. Thorn (ed.) *Space and time in geomorphology.* London: Allen & Unwin, 193–217.

_____ 1983: Downstream changes in stream power in the Henry Mountains, Utah. *Annals of the Association of American Geographers* **73**, 373–87.

_____ 1984: The geography of American field geomorphology. *Professional Geographer* **36**, 78–82.

_____ (ed.) 1987: *Geomorphic systems of North America*. Boulder, Colorado: Geological Society of America, Centennial Special Volume 2.

_____ 1988: *Fluvial processes in dryland rivers*. Berlin: Springer-Verlag.

_____ 1992: Science, public policy and western American rivers. *Transactions of the Institute of British Geographers* NS **17**, 5–19.

_____ 1994: *Plutonium and the Rio Grande. Environmental change and contamination in the nuclear age.* Oxford: Oxford University Press.

_____ 1996: Geomorphology and policy for restoration of impounded American rivers: what is 'natural'? In B.L. Rhoads and C.E. Thorn (eds) *The scientific nature of geomorphology.* Chichester: John Wiley, 443–73.

_____ forthcoming: Damage control: dams and the physical integrity of America's rivers. *Annals of the Association of American Geographers* **91**, in press.

GRAF, W.L., TRIMBLE, S.W., TOY, T.J. and COSTA, J.E. 1980: Geographic geomorphology in the eighties. *Professional Geographer* **32**, 279–84.

GRAF, W.L. and GOBER, P. 1992: Movements, cycles and systems. In R.F. Abler, M.G. Marcus and J.M. Olson (eds) *Geography's inner worlds. Pervasive themes in contemporary American geography.* New Brunswick, NJ: Rutgers University Press, 234–54.

GRANT, K., FINLAYSON, A.A., SPATE, A.P. and FERGUSON, T.G. 1979: Ternin analysis and classification for engineering and conservation purposes of the Port Clinton area, Queensland, including the Shoalwater Bay military training area. *Division of Applied Geomechanics Technical Paper* **29**, Melbourne, CSIRO.

GREENLAND, D. 1994: Use of satellite-based sensing in land surface climatology. *Progress in Physical Geography* **18**, 1–15.

GREGORY, D. and WALFORD, R. (eds) 1989: *Horizons in human geography.* Basingstoke: Macmillan.

GREGORY, K.J. 1976: Changing drainage basins. *Geographical Journal* **142**, 237–47.

_____ (ed.) 1977: *River channel changes.* Chichester: John Wiley.

_____ 1978a: A physical geography equation. *National Geographer* **12**, 137–41.

_____ 1978b: Fluvial processes in British basins. In C. Embleton, D. Brunsden and D.K.C. Jones (eds) *Geomorphology: present problems and future prospects.* Oxford: Oxford University Press, 40–72.

_____ 1981: River channels. In K.J. Gregory and D.E. Walling *Man and environmental processes.* London: Butterworth, 123–43.

_____ (ed.) 1983: *Background to palaeohydrology.* Chichester: John Wiley.

_____ 1985: *The nature of physical geography.* London: Arnold.

_____ (ed.) 1987a: *Energetics of physical environment: energetic approaches to physical geography.* Chichester: John Wiley.

_____ 1987b: The power of nature. In K.J. Gregory (ed.) *Energetics of physical environment: energetic approaches to physical geography.* Chichester: John Wiley, 1–31.

_____ 1987c: River channels. In K.J. Gregory and D.E. Walling (eds) *Human activity and environmental processes.* Chichester: John Wiley, 207–35.

_____ 1992: Changing physical environment and changing physical geography. *Geography* 77, 323–35.

_____ 1995: Human activity in palaeohydrology. In K.J. Gregory, L. Starkel and V.R. Baker (eds) *Global continental palaeohydrology.* Chichester: John Wiley, 151–72.

_____ 1996: Introduction. In J. Branson, A.G. Brown and K.J. Gregory (eds) *Global continental changes: the context of palaeohydrology.* Geological Society Special Publication 115, 1–8.

_____ (ed.) 1997: *Fluvial geomorphology of Great Britain.* London: Geological Conservation Review Series, Joint Nature Conservation Committee, Chapman and Hall.

_____ 1998: Applications of palaeohydrology. In G. Benito, V.R. Baker and K.J. Gregory (eds) *Palaeohydrology and environmental change.* Chichester: John Wiley, 13–25.

_____ 1999: Progress, perception and potential of physical geography. Paper presented at Symposium in honour of Yi Fu Tuan, Madison, Wisconsin, April 1998.

GREGORY, K.J. and BROWN, E.H. 1966: Data processing and the study of land form. *Zeitschrift für Geomorphologie* 10, 237–63.

GREGORY, K.J. and WALLING, D.E. 1973: *Drainage basin form and process.* London: Arnold.

GREGORY, K.J. and CULLINGFORD, R.A. 1974: Lateral variations in pebble shape in north west Yorkshire. *Sedimentary Geology* 12, 237–48.

GREGORY, K.J. and WALLING, D.E. (eds) 1979: *Man and environmental processes.* London: Dawson.

GREGORY, K.J. and WILLIAMS, R.F. 1981: Physical geography from the newspaper. *Geography* 66, 42–52.

GREGORY, K.J. and THORNES, J.B. 1987: Conclusion. In K.J. Gregory (ed.) *Energetics of physical environment: energetic approaches to physical geography.* Chichester: John Wiley, 161–4.

GREGORY, K.J. and WALLING, D.E. (eds) 1987: *Human activity and environmental processes.* Chichester: John Wiley.

GREGORY, K.J. and WHITLOW, J.A. 1989: Changes in urban stream channels in Zimbabwe. *Regulated Rivers* 4, 27–42.

GREGORY, K.J. and ROWLANDS, H. 1990: Have global hazards increased? *Geography Review* 4, 35–9.

GREGORY, K.J. and DAVIS, R.J. 1992: Coarse woody debris in stream channels in relation to river channel management in woodland areas. *Regulated Rivers: Research and Management* 7, 117–36.

―――― 1993: The perception of riverscape aesthetics: an example from two Hampshire rivers. *Journal of Environment Management* 39, 171–85.

GREGORY, K.J., STARKEL, L. and BAKER, V.R. (eds) 1995: *Global continental palaeohydrology*. Chichester: John Wiley.

GREGORY, S. 1963: *Statistical methods and the geographer*. London: Longman.

―――― 1978: The role of physical geography in the curriculum. *Geography* 63, 251–64.

GRIGORYEV, A.Z. 1961: The heat and moisture regions and geographic zonality. *Soviet Geography* 2, 3–16.

GROVE, A.T. 1977: Desertification. *Progress in Physical Geography* 1, 296–310.

―――― 1997: Classics in physical geography revisited: C.Vita-Finzi 1969, The Mediterranean Valleys. *Progress in Physical Geography* 21, 251–6.

GUELKE, L. 1989: Intellectual coherence and the foundation of geography. *Professional Geographer* 41, 123–30.

GUPTA, A. 1983: High magnitude floods and stream channel response. In J.D. Collinson and J. Lewin (eds) *Modern and ancient fluvial systems*. Oxford: Blackwell, 219–27.

―――― 1993: The changing geomorphology of the humid tropics. *Geomorphology* 7, 165–86.

GURNELL, A.M. 1998: The hydrogeomorphological effects of beaver dam-building activity. *Progress in Physical Geography* 22, 167–89.

GURNELL, A.M. and GREGORY, K.J. 1995: Interactions between semi-natural vegetation and hydrogeomorphological processes. *Geomorphology* 13, 49–69.

GURNELL, A.M. and PETTS, G.E. (eds) 1995: *Changing river channels*. Chichester: John Wiley.

GURNELL, A.M., GREGORY, K.J. and PETTS, G.E. 1995: The role of coarse woody debris in forest aquatic habits: implications for management. *Aquatic Conservation: Marine and Fresh Water Ecosystems* 5, 143–66.

GURNEY, R.J., FOSTER, J.L. and PARKINSON, C.L. (eds) 1993: *Atlas of satellite observations related to global change*. London: Cambridge University Press.

GVODETSKIY, N.A., GERENCHUK, K.I., ISACHENKO, A.G. and PREOBRAZHENSKIY, V.S. 1971: The present state and future tasks of physical geography. *Soviet Geography* 2, 3–16.

HACK, J.T. 1960: Interpretation of erosional topography in humid temperate regions. *American Journal of Science* 258, 80–97.

HAGGETT, P. 1983: *Geography: a modern synthesis* (3rd edn). London: Harper and Row.

_____ 1990: *The geographer's art*. Oxford: Blackwell.

_____ 1996: Geographical futures. Some personal speculations. In I. Douglas, R.J. Huggett and M.E. Robinson (eds) *Companion encyclopedia of geography*. London: Routledge, 965–73.

HAGGETT, P. and CHORLEY, R.J. 1969: *Network analysis in geography*. London: Arnold.

HAIGH, M.J. 1987: The holon: hierarchy theory and landscape research. In F. Ahnert (ed.) *Geomorphological models: theoretical and empirical aspects. Catena* Supplement 10, 181–92.

HAILS, J.R. (ed.) 1977: *Applied geomorphology*. Amsterdam: Elsevier.

HAINES-YOUNG, R. 1992: Biogeography. *Progress in Physical Geography* 16, 346–60.

HAINES-YOUNG, R. and PETCH, J.R. 1980: The challenge of critical rationalism for methodology in physical geography. *Progress in Physical Geography* 4, 63–77.

_____ 1986: *Physical geography: its nature and methods*. London: Harper and Row.

HAINES-YOUNG, R., GREEN, D. and COUSINS, S. (eds) 1993: *Landscape ecology and geographic information systems*. London: Taylor and Francis.

HAINES-YOUNG, R. and CHOPPING, M. 1996: Quantifying landscape structure: a review of landscape indices and their application to forested landscapes. *Progress in Physical Geography* 20, 418–45.

HANSOM, J.D. 1992: Editorial. *Applied Geography* 12, 5–6.

HARBOR, J.M. 1993: Glacial geomorphology: modelling processes and landforms. *Geomorphology* 7, 129–40.

HARDISTY, J. 1990: *The British seas: an introduction to the oceanography and resources of the north-west European continental shelf*. London: Routledge.

_____ (ed.) 1998: Technical and Software Bulletin Issue No. 7. *Earth Surface Processes and Landforms* Winter 23, No. 13.

HARDY, J.R. 1981: Data collection by remote sensing for land resources survey. In J.R.G. Townshend (ed.) *Terrain analysis and remote sensing*. London: George Allen & Unwin.

HARE, F.K. 1951a: Geographical aspects of meteorology. In G. Taylor (ed.) *Geography in the twentieth century*. New York: Philosophical Library, 178–95.

_____ 1957: The dynamic aspects of climatology. *Geografiska Annaler* 39, 87–104.

_____ 1965: Energy exchanges and the general circulation. *Geography* 50, 229–41.

_____ 1966: The concept of climate. *Geography* 5, 99–110.

_____ 1968: New wings for climatology: a science reborn. In A. Court (ed.) *Eclectic climatology*. Corvallis: Oregon State University Press, 145–62.

———— 1969: Environment: resuscitation of an idea. *Area* **4**, 52–5.

———— 1973: Energy-based climatology and its frontier with ecology. In R.J. Chorley (ed.) *Directions in geography*. London: Methuen, 171–92.

———— 1977: Man's world and geographers: a secular sermon. In D.R. Deskins *et al.* (eds) Geographic humanism, analysis and social action: a half century of geography at Michigan. *Michigan Geographical Publications* **17**, 259–73.

———— 1980: The planetary environment: fragile or sturdy? *Geographical Journal* **146**, 379–95.

———— 1996: Climatic variation and global change. In I. Douglas, R.J. Huggett and M.E. Robinson (eds) *Companion encyclopedia of geography*. London: Routledge, 482–507.

HARMAN, J.R., HARRINGTON, J.A. and CERVENY, R.S. 1998: Balancing scientific and ethical values in environmental science. *Annals of the Association of American Geographers* **88**, 277–86.

HARPER, D., SMITH, C., BARHAM, P. and HOWELL, R. 1995: The ecological basis for the management of the natural river environment. In D.M. Harper and A.J.D. Ferguson (eds) *The ecological basis for river management*. Chichester: John Wiley, 219–38.

HARRIS, D.R. 1968: Recent plant invasions in the arid and semi-arid southwest of the United States. *Annals of the Association of American Geographers* **56**, 408–22.

HARRISON CHURCH, R.J. 1951: The French school of geography. In G. Taylor (ed.) *Geography in the twentieth century*. New York: Philosophical Library, 70–90.

HARRISON, C.M. 1980: Ecosystem and communities: patterns and processes. In K.J. Gregory and D.E. Walling (eds) *Man and environmental processes*. Folkstone: Dawson, 225–40.

HARRISON, S. and DUNHAM, P. 1998: Decoherence, quantum theory and their implications for the philosophy of geomorphology. *Transactions of the Institute of British Geographers* NS **23**, 501–14.

HART, J.K. 1995: Subglacial erosion, deposition and deformation associated with deformable beds. *Progress in Physical Geography* **19**, 173–91.

HART, J.K. and BOULTON, G.S. 1991: The interrelationship between glaciotectonic deformation and deposition. *Quaternary Science Reviews* **10**, 335–50.

HART, J.K. and MARTINEZ, K. 1997: *Glacial analysis: an interactive introduction*. London: Routledge, CD-ROM.

HARVEY, A.M. and SALA, M. (eds) 1988: *Geomorphic processes in environments with strong seasonal contrasts*. Vol. II *Geomorphic systems*. *Catena* Supplement 13.

HARVEY, D.W. 1969: *Explanation in geography*. London: Arnold.

———— 1974: What kind of geography for what kind of public policy? *Transactions of the Institute of British Geographers* **63**, 18–24.

HENDERSON-SELLERS, A. 1980: Albedo changes – surface surveillance from satellites. *Climatic Change* 2, 275–81.

_____ 1989a: Climate, models and geography. In B. Macmillan (ed.) *Remodelling geography.* Oxford: Blackwell, 117–46.

_____ 1989b: Atmospheric physiography and meteorological modelling: the future role of geographers in understanding climate. *Australian Geographer* 20, 1–25.

_____ 1990: The 'coming of age' of land surface climatology. *Palaeogeography, Palaeoclimatology and Palaeogeology* 82, 291–319.

_____ 1991: Policy advice on greenhouse-induced climatic change: the scientist's dilemma. *Progress in Physical Geography* 15, 53–70.

_____ 1992: Greenhouse, Gaia and global change: a personal view of the pitfalls of interdisciplinary research. *Australian Geographer* 23, 24–38.

_____ 1993: An antipodean climate of uncertainty? *Climatic Change* 25, 203–24.

_____ 1994a: Numerical modelling of global climates. In N. Roberts (ed.) *The changing global environment.* Oxford: Blackwell, 99–124.

_____ 1994b: Global terrestrial vegetation prediction: the use and abuse of climate and application models. *Progress in Physical Geography* 18, 209–46.

_____ 1996: Enhancing climatic-change information sharing. *Climatic Change* 33, 453–7.

_____ 1998: Communicating science ethically: is the 'balance' achievable? *Annals of the Association of American Geographers* 88, 301–7.

HENGEVELD, R. 1993: Ecological biogeography. *Progress in Physical Geography* 17, 448–60.

HEWITT, K. (ed.) 1983: *Interpretations of calamity from the viewpoint of human ecology.* Boston: Allen & Unwin.

HEWITT, K. and BURTON, I. 1971: *The hazardousness of a place: a regional ecology of damaging events.* Toronto: University of Toronto Press.

HEWITT, K. and HARE, F.K. 1973: *Man and environment: conceptual frameworks.* Association of American Geographers, Commission on College Geography, Resource Paper No. 20.

HEY, R.D., BATHURST, J.C. and THORNE, C.R. (eds) 1982: *Gravel bed rivers: fluvial processes, engineering and management.* Chichester: John Wiley.

HEYWOOD, I., CORNELIUS, S. and CARVER, S. 1998: *An introduction to geographical information systems.* Harlow: Longman.

HIGGINS, C.G. 1965: Causes of relative sea-level changes. *American Scientist* 53, 464–76.

_____ 1975: Theories of landscape development: a perspective. In W.N. Melhorn and R.C. Flemal (eds) *Theories of landform development.* Binghamton: State University of New York, Publications in Geomorphology, 1–28.

HILL, A.R. 1987: Ecosystem stability: some recent perspectives. *Progress in Physical Geography* **11**, 315–32.

HJULSTRØM, F. 1935: Studies of the morphological activity of rivers as illustrated by the River Fyris. *Bulletin of the Geological Institute of Uppsala* **25**, 221–527.

HOBBS, J.E. 1980: *Applied climatology: a study of atmospheric resources.* London: Butterworth.

HOLLIS, G.E. 1975: The effect of urbanization on floods of different recurrence intervals. *Water Resources Research* **11**, 431–4.

_____ (ed.) 1979: *Man's impact on the hydrological cycle in the United Kingdom.* Norwich: Geobooks.

HOLMES, C.D. 1941: Till fabric. *Bulletin of the Geological Society of America* **52**, 1299–354.

HOLT-JENSEN, A. 1981: *Geography: its history and concepts.* London: Harper & Row.

_____ 1999: *Geography: its history and concepts* (3rd edn). London: Sage Publications.

HOLZNER, L. and WEAVER, G.D. 1965: Geographic evaluation of climatic and climatogenetic geomorphology. *Annals of the Association of American Geographers* **55**, 592–602.

HOOKE, J.M. (ed.) 1988: *Geomorphology in environmental planning.* Chichester: John Wiley.

_____ (ed.) 1998: *Coastal defence and earth science conservation.* Bath: Geological Society.

HOOKE, J.M. and KAIN, R.J.P. 1982: *Historical change in the physical environment: a guide to sources and techniques.* London: Butterworth.

HOOKE, R.L. 1999: Spatial distribution of human geomorphic activity in the United States: comparison with rivers. *Earth Surface Processes and Landforms* **24**, 687–92.

HORGAN, J. 1996: *The end of science. Facing the limits of knowledge in the twilight of the scientific age.* New York: Addison Wesley.

HORTON, R.E. 1945: Erosional development of streams and their drainage basins: hydrophysical approach to quantitative morphology. *Bulletin of the Geological Society of America* **56**, 275–370.

HOUGHTON, J.T. 1997: *Global warming: the complete briefing.* Cambridge: Cambridge University Press.

HOWE, W. and HENDERSON-SELLERS, A. (eds) 1997: *Assessing climate change: results from the model evaluation consortium for climate assessment.* Roseville, NSW, Australia: Gordon and Breach.

HUGGETT, R.J. 1975: Soil landscape systems: a model of soil genesis. *Geoderma* **13**, 1–22.

_____ 1976a: Lateral translocation of soil plasma through a small valley basin in the Northaw Great Wood, Hertfordshire. *Earth Surface Processes* **1**, 99–109.

_____ 1976b: A scheme for the science of geography, its systems, laws and models. *Area* **8**, 25–30.

_____ 1980: *Systems analysis in geography*. Oxford: Clarendon Press.

_____ 1982: Models and spatial patterns of soils. In E.M. Bridges and D.A. Davidson (eds) *Principles and applications of soil geography*. London: Longman, 132–70.

_____ 1985: Earth surface systems. Berlin: Springer-Verlag.

_____ 1988: Terrestrial catastrophism: causes and effects. *Progress in Physical Geography* **12**, 509–32.

_____ 1989a: *Cataclysms and earth history: the development of diluvialism*. Oxford: Clarendon Press.

_____ 1989b: Superwaves and superfloods: the bombardment hypothesis and geomorphology. *Earth Surface Processes and Landforms* **14**, 433–42.

_____ 1990: *Catastrophism: systems of earth history*. London: Arnold.

_____ 1991: *Climate, earth processes and earth history*. Berlin: Springer-Verlag.

_____ 1993: *Modelling the human impact on nature*. Oxford: Oxford University Press.

_____ 1994: Fluvialism or diluvialism? Changing views on superfloods and landscape change. *Progress in Physical Geography* **18**, 335–42.

_____ 1997: *Environmental change. The evolving ecosphere*. London: Routledge.

HULME, M. 1998: Global warming. *Progress in Physical Geography* **22**, 398–406.

HUTTON, J. 1795: *Theory of the Earth*. Edinburgh: William Creech.

HUXLEY, T.H. 1877: *Physiography. An introduction to the study of nature*. London: Macmillan.

IGBP 1988: *The International Geosphere–Biosphere Programme: a study of global change; IGBP: a plan for action*. Stockholm: IGBP.

International Committee of Scientific Unions (ICSU) 1986: *The International Geosphere–Biosphere Progamme: a study of global change*. Final report of the Ad-Hoc Planning Group, ICSU 21st General Assembly, Berne, Switzerland, 14–18 September.

IPCC 1996: *Climate change 1995. The science of climate change*. Cambridge: Cambridge University Press.

ISACHENKO, A.G. 1973a: *Principles of landscape science and physico–geographic regionalization*. Trans. from Russian by J.S. Massey. Melbourne: University of Melbourne Press.

_____ 1973b: On the method of applied landscape research. *Soviet Geography* **14**, 229–43.

_____ 1977: L.S. Berg's landscape – geographic ideas, their origins and their present significance. *Soviet Geography* **18**, 13–18.

JACKS, G.V. and WHYTE, R.O. 1939: *The rape of the Earth*. London: Faber & Faber.

JACOBS, J.D. and BELL, T.J. 1998: Regional perspectives on 20th century environmental change: introduction and examples from northern Canada. *Canadian Geographer* **42**, 314–18.

JENNINGS, J.N. 1966: Man as a geological agent. *Australian Journal of Science* **28**, 150–6.

_____ 1973: 'Any millenniums today, Lady?' The geomorphic bandwaggon parade. *Australian Geographical Studies* **11**, 115–33.

JENNY, H. 1941: *Factors of soil formation, a system of quantitative pedology*. New York: McGraw Hill.

_____ 1961: Derivation of state factor equations of soils and ecosystems. *Proceedings of the Soil Science Society of America* **25**, 385–8.

JENSEN, J.R. and DAHLBERG, K.A. 1983: Status and content of remote sensing education in the United States. *International Journal of Remote Sensing* **4**, 235–45.

JOHNSON, D.W. 1921: *Battlefields of the World War*. New York: American Geographical Society Research Series No. 3.

_____ 1931: *Stream sculpture on the Atlantic Slope*. New York: Columbia University Press.

JOHNSON, W.M. 1963: The pedon and the polypedon. *Proceedings of the Soil Science Society of America* **27**, 212–15.

JOHNSTON, R.J. 1979: *Geography and geographers. Anglo American human geography since 1945*. London: Arnold.

_____ 1983a: *Geography and geographers. Anglo American human geography since 1945* (2nd edn). London: Arnold.

_____ 1983b: *Philosophy and human geography*. London: Arnold.

_____ 1983c: Resource analysis, resource management, and the integration of physical and human geography. *Progress in Physical Geography* **7**, 127–46.

_____ 1984: The world is our oyster. *Transactions of the Institute of British Geographers* NS **9**, 443–59.

_____ 1986: Four fixations and the quest for unity in geography. *Transactions of the Institute of British Geographers* **11**, 449–53.

_____ 1989: The Institute study groups, and a discipline without a core. *Area* **21**, 407–14.

_____ (ed.) 1993: *The challenge for geography. A changing world: a changing discipline*. Oxford: Blackwell.

_____ 1997: *Geography and geographers. Anglo American human geography since 1945* (5th edn). London: Arnold.

JOHNSTON, R.J., TAYLOR, P.J. and WATTS, M. (eds) 1995: *Geographies of global change: remapping the world in the late twentieth century*. Oxford: Blackwell.

JONES, D.K.C. 1980: British applied geomorphology: an appraisal. *Zeitschrift für Geomorphologie* Supplementband **36**, 48–73.

_____ 1983: Environments of concern. *Transactions of the Institute of British Geographers* NS **8**, 429–57.

_____ 1995: Environmental change, geomorphological change and sustainability. In D.F.M. McGregor and D.A. Thompson (eds) *Geomorphology and land management in a changing environment.* Chichester: John Wiley, 11–34.

JONES, M.D.H. and HENDERSON-SELLERS, A. 1990: History of the greenhouse effect. *Progress in Physical Geography* **14**, 1–18.

KATES, R.W. 1987: The human environment: the road not taken, the road still beckoning. *Annals of the Association of American Geographers* **77**, 525–34.

KE CHUNG KIM and WEAVER, R.D. (eds) 1994: *Biodiversity and landscapes. A paradox of humanity.* Cambridge: Cambridge University Press.

KELLER, E.A. 1996: *Environmental geology* (7th edn). Upper Saddle River, NJ: Prentice Hall.

KELLER, E.A. and PINTER, N. 1996: *Active tectonics. Earthquakes, uplift and landscape.* Upper Saddle River, NJ: Prentice Hall.

KEMP, D.D. 1990: *Global environmental issues: a climatological approach.* London: Routledge.

KENNEDY, B.A. 1992: Hutton to Horton: views of sequence, progression and equilibrium in geomorphology. *Geomorphology* **5**, 231–50.

_____ 1993: '. . . no prospect of an end'. *Geography* **78**, 124.

_____ 1994: Requiem for a dead concept. *Annals of the Association of American Geographers* **84**, 702–5.

_____ 1997: Classics in physical geography revisited. S.A. Schumm and R.W. Lichty 1965. Time, space and causality in geomorphology. *Progress in Physical Geography* **21**, 419–23.

KENT, M., JONES, A. and WEAVER, R. (eds) 1993: *Geographical information systems and remote sensing in land use planning. An* Applied Geography *theme volume. Applied Geography* **13**, 1–95.

KENT, M., GILL, W.J., WEAVER, R.E. and ARMITAGE, R.P. 1997: Landscape and plant community boundaries in biogeography. *Progress in Physical Geography* **21**, 315–53.

KENZER, M.S. (ed.) 1989: *Applied geography: issues, questions and concerns.* Dordrecht: Kluwer Academic.

_____ 1992: Applied and academic geography and the remainder of the twentieth century. *Applied Geography* **12**, 207–10.

KIDSON, C. 1982: Sea level changes in the Holocene. *Quaternary Science Reviews* **1**, 121–51.

KIMBALL, D. 1948: *Denudation chronology. The dynamics of river action.* Occasional Paper No. 8, University of London: Institute of Archaeology.

KINCER, J.B., *et al.* 1941: *Climate and man.* Washington, DC: Department of Agriculture Yearbook.

KING, C.A.M. 1966: *Techniques in geomorphology.* London: Arnold.

_____ 1972: *Beaches and coasts* (2nd edn). London: Arnold.

_____ 1980: *Physical geography.* Oxford: Blackwell.

KING, L.C. 1950: A study of the world's plainlands: a new approach in geomorphology. *Quarterly Journal of the Geological Society London* **106**, 101–27.

_____ 1953: Canons of landscape evolution. *Bulletin of the Geological Society of America* **C4**, 721–52.

_____ 1962: *The morphology of the Earth*. Edinburgh: Oliver & Boyd.

KING, P.S. and SCHUMM, S.A. 1980: *The physical geography (geomorphology) of W.M. Davis*. Norwich: Geo Books.

KING, R.B. 1970: A parametric approach to land system classification. *Geoderma* **4**, 37–46.

KINGSLEY, C. 1872: *Town geology*. London: Strahan.

KIRKBY, M.J. 1989: The future of modelling in physical geography. In B. Macmillan (ed.) *Remodelling geography*. Oxford: Blackwell, 255–72.

_____ (ed.) 1994. *Process models and theoretical geomorphology*. Chichester: John Wiley.

KIRKBY, M.J., NADEN, P.S., BURT, T.P. and BUTCHER, D.P. 1987: *Computer simulation in physical geography*. Chichester: John Wiley.

KLISKEY, A.D. and KEARSLEY, G.W. 1993: Mapping multiple perception of wilderness in southern New Zealand. *Applied Geography* **13**, 203–23.

KNAPP, P.A. and SOULE, P.T. 1996: Vegetation change and the role of atmospheric CO_2 enrichment on a relict site in Central Oregon 1960–1994. *Annals of the Association of American Geographers* **86**, 387–411.

KNIGHT, P.G. 1993: Glaciers. *Progress in Physical Geography* **17**, 349–53.

KNIGHTON, A.D. 1984: *Fluvial forms and processes*. London: Arnold.

_____ 1998: *Fluvial forms and processes. A new perspective*. London: Arnold.

KNOWLES, R.L. 1974: *An ecological approach to urban growth*. Cambridge, MA: MIT Press.

KNOX, J.C. 1972: Valley alluviation in southwestern Wisconsin. *Annals of the Association of American Geographers* **62**, 401–10.

_____ 1995: Fluvial systems since 20,000 years BC. In K.J. Gregory, L. Starkel and V.R. Baker (eds) *Global continental palaeohydrology*. Chichester: John Wiley, 87–108.

_____ 1999: Long-term episodic changes in magnitudes and frequency of floods in the Upper Mississippi River Valley. In A.G. Brown and T. Quine (eds) *Fluvial processes and environmental change*. Chichester: John Wiley, 255–82.

KRCHO, J. 1978: The spatial organization of the physical–geographical sphere as a cybernetic system expressed by means of measures of entropy. *Acta Facultatis Rerum Naturalium Universitatis Comenianal, Geographica* **16**, 57–147.

KRINSLEY, D.H. and DOORNKAMP, J.C. 1973: *Atlas of quartz sand surface textures*. Cambridge: Cambridge University Press.

KROPOTKIN, P. 1893: The teaching of physiography. *Geographical Journal* **2**, 350–59.

KRUMBEIN, W.C. 1941: The effect of abrasion on the size and shape and roundness of rock fragments. *Journal of Geology* **49**, 482–520.

KRUMBEIN, W.C. and GRAYBILL, F.A. 1965: *An introduction to statistical models in geology*. New York: McGraw Hill.

KUHN, T.S. 1962: *The structure of scientific revolutions*. Chicago: University of Chicago Press.

_____ 1977: *The essential tension*. Chicago: University of Chicago Press.

KUKLA, G.J. 1975: Loess stratigraphy in central Europe. In K.W. Butzer and G.Ll. Isaac (eds) *After the Australopithecines*. The Hague: Mouton, 99–188.

KUPFER, J.A. 1995: Landscape ecology and biogeography. *Progress in Physical Geography* **19**, 18–34.

L'VOVICH, M.I., GANGARDT, G.G., SARUKHANOV, G.L. and BEREN-ZER, A.S. 1982: Territorial redistribution of streamflow within the European USSR. *Soviet Geography* **22**, 391–405.

LAKATOS, I. 1970: Falsification and the methodology of scientific research programmes. In I. Lakatos and A. Musgrave (eds) *Criticism and the growth of knowledge*. Cambridge: Cambridge University Press.

_____ 1978: *The methodology of scientific research programmes* Philosophical Papers Vol. 1, edited by J. Worrall and G. Currie. Cambridge: Cambridge University Press.

LAMB, H.H. 1950: Types and spells of weather in the British Isles. *Quarterly Journal of the Royal Meteorological Society* **76**, 393–429.

LANDSBERG, H. 1941: *Physical climatology*. Dubois, PA: Gray Printing Co.

LANE, S.N. and RICHARDS, K.S. 1997: Linking river channel form and process: time, space and causality revisited. *Earth Surface Processes and Landforms* **22**, 249–60.

LANGBEIN, W.B., *et al.* 1949: Annual runoff in the United States. *US Geological Survey Circular* **52**.

LAZLO, E. 1972: *Introduction to systems philosophy*. London: Gordon & Breach.

LEATHERMAN, S.P. 1990: Modelling shore response to sea-level rise on sedimentary coasts. *Progress in Physical Geography* **14**, 447–64.

LEMONS, J. 1999: Environmental ethics. In D.E. Alexander and R.W. Fairbridge (eds) *Encyclopedia of environmental science*. Dordrecht: Kluwer Academic, 204–6.

LEOPOLD, L.B. 1969: Landscape esthetics. *Natural History* **Oct.**, 37–44.

_____ 1973: River channel change with time: an example. *Bulletin of the Geological Society of America* **84**, 1845–60.

_____ 1977: A reverence for rivers. *Geology* **5**, 429–30.

LEOPOLD, B. and EMMETT, W.W. 1965: Vigil network sites: a sample of data for permanent filing. *Bulletin of the International Association of Scientific Hydrology* **10**, 12–21.

LEOPOLD, B. and LANGBEIN, W.B. 1962: *The concept of entropy in landscape evolution.* US Geological Survey Professional Paper 500-A.

_____ 1963: Association and indeterminacy in geomorphology. In C.C. Albritton (ed.) *The fabric of geology.* Reading, MA: Addison-Wesley, 184–92.

LEOPOLD, L.B. and MARCHAND, M.O. 1968: On the quantitative inventory of riverscape. *Water Resources Research* **4**, 709–17.

LEOPOLD, L.B., CLARKE, F.E., HANSHAW, B.B. and BALSLEY, J.R. 1971: A procedure for evaluating environmental impact. *US Geological Survey Circular* **645**.

LEOPOLD, L.B., WOLMAN, M.G. and MILLER, J.P. 1964: *Fluvial processes in geomorphology.* San Francisco: Freeman.

LEWIN, J. 1980: Available and appropriate time scales in geomorphology. In R.A. Cullingford, D.A. Davidson and J. Lewin (eds) *Timescales in geomorphology.* Chichester: John Wiley, 3–10.

LEWIS, W.V. 1949: The function of meltwater in cirque formation: a reply. *Geographical Review* **39**, 110–28.

LEWIS, W.V. and MILLER, M.M. 1955: Kaolin model glaciers. *Journal of Glaciology* **2**, 533–8.

LIER, J. 1989: A climatologist's personal perspective on applied geography. In M.S. Kenzer (ed.) *Applied geography: issues, questions and concerns.* Dordrecht: Kluwer Academic, 75–98.

LIKENS, G.E., BORMANN, F.H., PIERCE, R.S., EATON, J.S. and JOHNSTON, N.M. 1977: *Biogeochemistry of a forested ecosystem.* Berlin: Springer-Verlag.

LINDEMAN, R.L. 1942: The trophic–dynamic aspect of ecology. *Ecology* **23**, 399–418.

LINSLEY, R.K., KOHLER, M.A. and PAULHUS, J.L.H. 1949: *Applied hydrology.* New York: McGraw Hill.

LINTON, D.L. 1951: The delimitation of morphological regions. In L.D. Stamp and S.W. Wooldridge (eds) *London essays in geography.* London: LSE, 199–218.

_____ 1957: The everlasting hills. *Advancement of Science* **14**, 58–67.

_____ 1965: The geography of energy. *Geography* **50**, 197–228.

_____ 1968: The assessment of scenery as a natural resource. *Scottish Geographical Magazine* **84**, 219–38.

LIVERMAN, D.M. 1999: Geography and the global environment. *Annals of the Association of American Geographers* **89**, 107–20.

LIVINGSTONE, I. and WARREN, A. 1996: *Aeolian geomorphology: an introduction.* Harlow: Longman.

LOCKWOOD, J.G. 1979a: *Causes of climate.* London: Arnold.

_____ 1979b: Causative factors in climatic fluctuations. *Progress in Physical Geography* **3**, 111–18.

_____ 1980: Milankovitch theory and ice ages. *Progress in Physical Geography* **4**, 79–87.

_____ 1983a: Modelling climatic change. In K.J. Gregory (ed.) *Background to palaeohydrology*. Chichester: John Wiley, 25–50.

_____ 1983b: The influence of vegetation on the Earth's climate. *Progress in Physical Geography* 7, 81–9.

_____ 1984: The southern oscillation and El Niño. *Progress in Physical Geography* 8, 102–10.

_____ 1996: Classics in Physical Geography revisited. Budyko, M.I. 1956, translated 1958. The heat balance of the earth's surface. *Progress in Physical Geography* 20, 337–43.

LOVELOCK, J.E. 1979: *Gaia: a new look at life on Earth*. Oxford: Oxford University Press.

LOWE, J.J. 1991: *Radiocarbon dating: recent applications and future potential*. Cambridge: Quaternary Research Association.

LOWE, J.J. and WALKER, M.J.C. 1997: *Reconstructing Quaternary environments* (2nd edn). Harlow: Addison Wesley Longman Limited.

LOWENTHAL, D. 1961: Geography, experience and imagination: towards a geographical epistemology. *Annals of the Association of American Geographers* 51, 241–60.

_____ (ed.) 1965: *Man and nature by George Perkins Marsh*. Cambridge, MA: Harvard University Press.

LOWRY, W.P. 1998: Urban effects on precipitation. *Progress in Physical Geography* 22, 477–520.

LULLA, K. 1983: The Landsat satellites and selected aspects of physical geography. *Progress in Physical Geography* 7, 1–45.

LYELL, C.W. 1830: *Principles of geology*. London: Murray.

MABBUTT, J.A. 1976: Report on activities of the IGU Working Group on desertification in and around arid lands. *Geoforum* 7, 147–52.

MACDONALD, G. and EDWARDS, K.J. 1991: Holocene palynology: I. Principles, population and community ecology, palaeoclimatology. *Progress in Physical Geography* 15, 261–89.

MACMILLAN, B. (ed.) 1989a: *Remodelling geography*. Oxford: Blackwell.

_____ 1989b: Modelling through: an afterword to *Remodelling geography*. In B. Macmillan (ed.) *Remodelling geography*. Oxford: Blackwell, 291–313.

MAGUIRE, D.J. 1991: An overview and definition of GIS. In D.J. Maguire, M.F. Goodchild and D.W. Rhind (eds) *Geographical information systems. Principles and applications*. Harlow: Longman, 9–20.

MAGUIRE, D.J., GOODCHILD, M.F. and RHIND, D.W. (eds) 1991: *Geographical information systems. Principles and applications*. Harlow: Longman.

MAHANEY, W.C. 1984: *Quaternary dating methods, developments in palaeontology and stratigraphy*. Rotterdam: Elsevier.

_____ 1990: Dating methods. *Progress in Physical Geography* 14, 389–94. See also 1991: 15, 304–9; 1993: 17, 76–80; 1994: 18, 136–42; 1995: 19, 130–7, 399–406.

MALONE, T.F. 1994: Geographers explore the road into the twenty-first century. *Annals of the Association of American Geographers* **84**, 725–8.

MALTBY, E. 1975: Numbers of soil micro-organisms as ecological indicators of changes resulting from moorland and reclamation on Exmoor. *Journal of Biogeography* **2**, 117–36.

MANLEY, G. 1952: *Climate and the British scene*. London: Collins.

MANNERS, I.R. and MIKESELL, M.W. 1974: *Perspectives on environment*. Association of American Geographers Publication No. 13.

MANNION, A.M. 1989a: Palaeoecological evidence for environmental change during the last 200 years. I. Biological data. *Progress in Physical Geography* **13**, 23–46.

_____ 1989b: Palaeoecological evidence for environmental change during the last 200 years. II. Chemical data. *Progress in Physical Geography* **13**, 192–215.

_____ 1991: *Global environmental change. A natural and cultural history*. Harlow: Longman Scientific and Technical.

_____ 1995: Biotechnology and environmental quality. *Progress in Physical Geography* **19**, 192–215.

_____ 1999: *Natural environmental change*. London: Routledge.

MARBUT, C.F. 1935: Translation of K.D. Glinka, *The great soil groups of the world and their development*. Ann Arbor: Edward Bros.

MARCUS, M.G. 1979: Coming full circle: physical geography in the twentieth century. *Annals of the Association of American Geographers* **69**, 521–32.

MAROTZ, G.A. 1989: Current status, trends, and problem-solving in applied climatology. In M.S. Kenzer (ed.) *Applied geography: issues, questions and concerns*. Dordrecht: Kluwer Academic, 99–114.

MARSH, G.P. 1864: *Man and Nature or physical geography as modified by human action*. New York: Charles Scribner.

MATHER, J.R., FIELD, R.T., KALKSTEIN, L.S. and WILLMOTT, C.J. 1980: Climatology: the challenge for the eighties. *Professional Geographer* **32**, 285–92.

MATHER, P.M. 1976: *Computational methods of multivariate analysis in physical geography*. London: John Wiley.

MATHEWS, J.A. 1992: *The ecology of recently deglaciated terrain*. Cambridge: Cambridge University Press.

MAUNDER, W.H. 1970: *The value of the weather*. London: Methuen.

_____ 1986: *The uncertainty business*. London: Methuen.

MAY, R.M. 1977: Thresholds and breakpoints in ecosystems with a multiplicity of stable states. *Nature* **269**, 471–7.

McDOWELL, L. 1994: The transformation of cultural geography. In D. Gregory, R. Martin and R. Smith (eds) *Human geography: society, space and social science*. London: Macmillan, 146–73.

McGREGOR, D.M. and THOMPSON, D.A. (eds) 1995: *Geomorphology and land management in a changing environment*. Chichester: John Wiley.

McHARG, I.L. 1969: Design with nature. New York: Natural History Press.

McHARG, I.L. 1992: *Design with nature* (2nd edn). Chichester: John Wiley.

MEIGS, P. III 1954: The geographic study of water on the land. In P.E. James and C.F. Jones (eds) *American geography. Inventory and prospect.* Syracuse University Press, Association of American Geographers, 396–409.

MEINZER, O.K. (ed.) 1942: *Hydrology.* New York: McGraw Hill.

MERRIAM, C.H. 1894: Laws of temperature control of the geographic distributions of terrestrial animals and plants. *National Geographic Magazine* **6**, 229–38.

MEYER, W.B., GREGORY, D., TURNER, B.L. and McDOWELL, P.F. 1992: The local–global continuum. In R.F. Abler, M.G. Marcus and J.M. Olson (eds) *Geography's inner worlds. Pervasive themes in contemporary American geography.* New Brunswick, NJ: Rutgers University Press, 255–79.

MEYERS, C.D. 1977: *Energetics: systems and analysis with application to water resources planning and decision-making.* Fort Belvoir, VA: US Army Engineering, Institute for Water Resources Report 77–6.

MICHENER, W., BRUNT, J. and STAFFORD, S. (eds) 1994: *Environmental information measurment and analysis: ecosystem to global scales.* London: Taylor and Francis.

MIDDLETON, N. 1996: *The global casino: an introduction to global issues.* London: Arnold.

_____ 1999: *The global casino: an introduction to global issues* (2nd edn). London: Arnold.

MIDDLETON, N. and THOMAS, D.G. 1992: *World atlas of desertification.* London: Arnold.

_____ (eds) 1997: *World atlas of desertification* (2nd edn). London: Arnold.

MILLER, R.L. and KAHN, J.S. 1962: *Statistical analysis in the geological sciences.* New York: John Wiley.

MILNE, G. 1935: Some suggested units of classification and mapping, particularly for East African soils. *Soil Research* **4**, 183–98.

MITCHELL, C.W. 1973: *Terrain evaluation.* London: Longman.

_____ 1991: *Terrain evaluation. An introductory handbook to the history, principles and methods of practical terrain assessment.* Harlow: Longman Scientific and Technical.

MITCHELL, T.D. and HULME, M. 1999: Predicting regional climate change: living with uncertainty. *Progress in Physical Geography* **23**, 57–78.

MOORE, J.J., FITZSIMMONS, P., LAMBE, E. and WHITE, J. 1970: A comparison and evaluation of some phytosociological techniques. *Vegetatio* **20**, 1–20.

MOORE, P.D., CHALONER, B. and STOTT, P. 1996: *Global environmental change*. Oxford: Blackwell.

MORE, R.J. 1967: Hydrological models and geography. In R.J. Chorley and P. Haggett (eds) *Models in geography*. London: Methuen, 145–85.

MORGAN, R.P.C. 1979: *Soil erosion*. London: Longman.

MORISAWA, M.E. 1968: *Streams: their dynamics and morphology*. New York: McGraw Hill.

MOSER, K.A., MacDONALD, G.M. and SMOL, J.P. 1996: Applications of freshwater diatoms to geographical research. *Progress in Physical Geography* **20**, 21–52.

MOSLEY, M.P. 1989. Perceptions of New Zealand river scenery. *New Zealand Geographer* **45**, 2–13.

MOSLEY, M.P. and ZIMPFER, G.L. 1978: Hardware models in geomorphology. *Progress in Physical Geography* **2**, 438–61.

MOSS, M.R. 1999: Environmental stability. In D.E. Alexander and R.W. Fairbridge (eds) *Encyclopedia of environmental science*. Dordrecht: Kluwer Academic, 227–9.

MOSS, R.P. 1968: Land use, vegetation and soil factors in south west Nigeria: a new approach. *Pacific Viewpoint* **9**, 107–27.

———— 1969a: The appraisal of land resources in tropical Africa. *Pacific Viewpoint* **10**, 18–27.

———— 1969b: The ecological background to land-use studies in tropical Africa, with special reference to the west. In M.F. Thomas and G.W. Whittington (eds) *Environment and landuse in Africa*. London: Methuen, 193–240.

———— 1970: Authority and charisma: criteria of validity in geographical method. *South African Geographical Journal* **52**, 13–37.

MUMFORD, L. 1931: *The Brown decades: a study of the arts in America 1865–1895*. New York: Dover.

MURRAY GRAY, J. 1997: Planning and landform: geomorphological authenticity or incongruity in the countryside? *Area* **29**, 312–24.

MURTON, B.J. 1968: Mapping the immediate pre-European vegetation on the east coast of the North Island of New Zealand. *Professional Geographer* **20**, 262–4.

NASH, R.F. 1982: *Wilderness and the American mind* (3rd edn). New Haven: Yale University Press.

———— 1989: *The rights of nature: a history of environmental ethics*. Madison: University of Wisconsin Press.

National Research Council 1992: *Restoration of aquatic ecosystems – science, technology and public policy*. Washington, DC: National Academy Press.

———— 1997: *Rediscovering geography. New relevance for science and society*. Washington, DC: National Academy Press.

National Research Council Committee on Watershed Management 1999:

New strategies for America's watersheds. Washington, DC: National Academy Press.

NELSON, G. and PLATNICK, N.I. 1981: *Systematics and biogeography: cladistics and vicariance.* New York: Columbia University Press.

NELSON, G. and ROSEN, D.E. (eds) 1981: *Vicariance biogeography: a critique.* New York: Columbia University Press.

NEMEC, J. 1995: General circulation models (GCMS), climatic change, scaling and hydrology. In G.W. Kite (ed.) *Time and the river.* Highland Ranch, CO: Water Resources Publications, 317–56.

NEWSON, M.D. (ed.) 1992a: *Managing the human impact on the natural environment: patterns and processes.* Chichester: John Wiley.

_____ 1992b: Twenty years of systematic physical geography: issues for a 'New Environmental Age'. *Progress in Physical Geography* **16**, 209–21.

_____ 1992c: *Land, water and development. River basin systems and their sustainable management.* London: Routledge.

_____ 1995: Fluvial geomorphology and environmental design. In A. Gurnell and G. Petts (eds) *Changing river channels.* Chichester: John Wiley, 413–32.

NEWSON, M.D. and LEWIN, J. 1991: Climatic change, riverflow extremes and fluvial erosion – scenarios for England and Wales. *Progress in Physical Geography* **15**, 1–17.

NGOWI, J. and STOCKING, M. 1989: Assessing land suitability and yield potential. *Applied Geography* **9**, 21–33.

NICKLING, W.G. (ed.) 1986: *Aeolian geomorphology.* Boston: Allen and Unwin.

NIKIFOROFF, C.C. 1942: Fundamental formula of soil formation. *American Journal of Science* **240**, 847–66.

_____ 1949: Weathering and soil evolution. *Soil Science* **67**, 219–30.

_____ 1959: Reappraisal of the soil: pedogenesis consists of transactions in matter and energy between the soil and its surroundings. *Science* **129**, 186–96.

NORTCLIFF, S.M. 1984: Spatial analysis of soil. *Progress in Physical Geography* **8**, 261–9.

NYE, J.F. 1952: The mechanics of glacier flow. *Journal of Glaciology* **2**, 82–93.

ODUM, H.T. and ODUM, E.C. 1976: *Energy basis for man and nature.* New York: John Wiley.

OKE, T.R. 1978: *Boundary layer climates.* London: Methuen.

_____ 1987: *Boundary layer climates* (2nd edn). London: Routledge.

OLDFIELD, F. 1977: Lakes and their drainage basins as units of sediment-based ecological study. *Progress in Physical Geography* **1**, 460–504.

_____ 1983a: The role of magnetic studies in palaeohydrology. In K.J. Gregory (ed.) *Background to palaeohydrology.* Chichester: John Wiley, 141–65.

_____ 1983b: Man's impact on the environment: some recent perspectives. *Geography* **68**, 245–56.

OLDFIELD, F., BATTARBEE, R.W. and DEARING, J.A. 1983: New approaches to recent environmental change. *Geographical Journal* **149**, 167–81.

OLIVER, J.E. 1973: *Climate and man's environment: an introduction to applied climatology.* New York: John Wiley.

_____ 1991: The history, status and future of climatic classification. *Physical Geography* **12**, 231–51.

OLLIER, C.D. 1977: Terrain classification: methods, applications and principles. In J.R. Hails (ed.) *Applied geomorphology.* Amsterdam: Elsevier, 277–316.

_____ 1979: Evolutionary geomorphology of Australia and Papua New Guinea. *Transactions of the Institute of British Geographers* NS **4**, 516–39.

_____ 1981: *Tectonics and landforms.* Edinburgh: Oliver & Boyd.

_____ 1988: *Volcanoes.* Oxford: Blackwell.

_____ 1995: Classics in physical geography revisited: King, L.C. 1953, Canons of Landscape Evolution. *Progress in Physical Geography* **19**, 371–7.

O'LOUGHLIN, C.L. and OWENS, I.F. 1987: Our dynamic environment. In P.G. Holland and W.B. Johnston (eds) *Southern approaches: geography in New Zealand.* Christchurch: New Zealand Geographical Society, 59–90.

O'RIORDAN, T. 1994: Civic science and global environmental change. *Scottish Geographical Magazine* **110**, 4–12.

OLSON, C.G. 1989: Soil geomorphic research and the importance of palaeosol stratigraphy to Quaternary investigations, Midwestern USA. *Catena Supplement* **16**, 129–142.

OSTERKAMP, W.R. and HUPP, C.R. 1996: The evolution of geomorphology, ecology and other composite sciences. In B.L. Rhoads and C.E. Thorn (eds) *The scientific nature of geomorphology.* Chichester: John Wiley, 415–41.

O'SULLIVAN, P.E. 1979: The ecosystem watershed concept in the environmental sciences – a review. *International Journal of Environmental Sciences* **13**, 273–81.

OWENS, S., RICHARDS, K. and SPENCER, T. 1997: Managing the earth's surface: science and policy. *Transactions of the Institute of British Geographers* NS **22**, 3–5.

PAIN, C.F. 1978: Landform inheritance in the central highlands of Papua New Guinea. In G.L. Davies and M.A.J. Williams (eds) *Landform evolution in Australia.* Canberra: Australian National University Press, 5–47.

PAINE, A.D.M. 1985: Ergodic reasoning in geomorphology – time for review of the term? *Progress in Physical Geography* **9**, 1–15.

PALM, R.I. 1990: *Natural hazards: an integrative framework for research and planning*. Baltimore: John Hopkins University Press.

PALM, R.I. and BRAZEL, A.J. 1992: Applications of geographic concepts and methods. In R.F. Abler, M.G. Marcus and J.M. Olson (eds) *Geography's inner worlds*. New Brunswick, NJ: Rutgers University Press, 342–62.

PARK, C.C. 1981: Man, river systems and environmental impacts. *Progress in Physical Geography* 5, 1–13.

_____ 1997: *The environment: principles and applications*. London: Routledge.

PARKER, D.J. and HARDING, D.M. 1979: Natural hazard evaluation, perception and adjustment. *Geography* 64, 307–16.

PARRY, M.L. 1993: Geography and the impact of climate change. In R.J. Johnston (ed.) *The challenge for geography*. Oxford: Blackwell, 138–47.

PATON, T.R., HUMPHREYS, G.S. and MITCHELL, P.B. 1995: *Soils: a new global view*. London: UCL Press.

PATTON, P.C. and SCHUMM, S.A. 1975: Gully erosion, northern Colorado: a threshold phenomenon. *Geology* 3, 88–90.

PEARSON, M.G. 1978: Snowstorms in Scotland 1831 to 1861. *Weather* 33, 392–9.

PELTIER, L.C. 1950: The geographic cycle in periglacial regions as it is related to climatic geomorphology. *Annals of the Association of American Geographers* 40, 214–36.

_____ 1954: Geomorphology. In P.E. James and C.F. Jones (eds) *American geography: inventory and prospect*. Syracuse University Press, Association of American Geographers, 362–81.

_____ 1975: The concept of climatic geomorphology. In W.N. Melhorn and R.C. Flemal (eds) *Theories of landform development*. Binghamton: State University of New York.

PENCK, A. and BRUCKNER, E. 1901–9: *Die Alpen im Eiszeitalter*. Leipzig: Tauchnitz.

PENCK, W. 1924: *Die Morphologische Analyse*. Stuttgart: Engelhom.

PENMAN, H.C. 1948: Natural evaporation from open water, bare soil and grass. *Proceedings of the Royal Society of London A* 193, 120–45.

_____ 1950: Evaporation over the British Isles. *Quarterly Journal of the Royal Meteorological Society* 76, 372–83.

PENNING-ROWSELL, E.C. 1981a: Consultancy and contract research. *Area* 13, 9–12.

_____ 1981b: Fluctuating fortunes in gauging landscape value. *Progress in Human Geography* 5, 25–41.

PENNING-ROWSELL, E.C. and HARDY, D.I. 1973: Landscape evaluation and planning policy: a comparative survey in the Wye valley Area of Outstanding Natural Beauty. *Regional Studies* 7, 153–60.

PERRY, A.H. 1981: *Environmental hazards in the British Isles*. London: Allen & Unwin.

_____ 1992: The economic impacts, costs and opportunities of global warming. *Progress in Physical Geography* **16**, 97–100.

PERRY, A.W. 1995: New climatologists for a new climatology. *Progress in Physical Geography* **19**, 280–5.

_____ 1998: Netting research data for the climatologist. *Progress in Physical Geography* **22**, 121–6.

PERRY, G.L.W. 1998: Current approaches to modelling the spread of wildfire: a review. *Progress in Physical Geography* **22**, 222–45.

PETTS, G.E. 1984: *Impounded rivers: perspectives for ecological management.* Chichester: John Wiley.

PETTS, G.E. and AMOROS, C. (eds) 1996: *Fluvial hydrosystems.* London: Chapman and Hall.

PEWE, T.L. (ed.) 1969: *The periglacial environment.* Montreal: McGill-Queens University Press.

PHILLIPS, J.D. 1988: Nonpoint source pollution and spatial aspects of risk assessment. *Annals of the Association of American Geographers* **78**, 611–23.

_____ 1989: An evaluation of the state factor model of soil ecosystems. *Ecological Modelling* **45**, 165–77.

_____ 1995: Self-organization and landscape evolution. *Progress in Physical Geography* **19**, 309–21.

_____ 1999a: Divergence, convergence and self-organization in landscapes. *Annals of the Association of American Geographers* **89**, 466–88.

_____ 1999b: *Earth surface systems. Complexity, order, and scale.* Oxford: Blackwell.

PIERCE, C. and VAN DE VEER, D. (eds) 1995: *People, penguins and plastic trees: basic issues in environmental ethics.* Belmont, CA: Wadsworth.

PINNOCK 1823: *Catechism of geography, being an easy introduction to a knowledge of the world and its inhabitants.* London: G.B.Whittaker.

PITTY, A.F. 1982: *The nature of geomorphology.* London: Methuen.

PIWOWAR, J.M. and LeDREW, E.F. 1995: Hypertemporal analysis of remotely sensed sea-ice data for climate change studies. *Progress in Physical Geography* **19**, 216–42.

PLAYFAIR, J. 1802: *Illustrations of the Huttonian theory of the Earth.* Reprinted 1964. New York: Dover.

PORTER, P.W. 1978: Geography as human ecology. *American Behavioural Science* **22**, 15–39.

POSER, H. 1947: Dauerfrostboden und Temperaturverhältnisse während der Wurm-Eiszeit im nichtvereisten Mittel- und Westeuropa. *Naturwissenschaften* **34**.

PRENTICE, I.C. 1983: Postglacial climatic change: vegetation dynamics and the pollen record. *Progress in Physical Geography* **7**, 273–86.

PRICE, R.J. 1973: *Glacial and fluvioglacial landforms.* Edinburgh: Oliver & Boyd.

PRIOR, D.B. 1977: Coastal mudslide morphology and processes on Eocene clays in Denmark. *Geografisk Tidsshrift* **76**, 14–33.

PYE, K. 1987: *Aeolian dust and dust deposits*. London: Academic Press.

PYE, K. and TSOAR, H. 1990: *Aeolian sand and sand dunes*. London: Unwin-Hyman.

QUINE, T.A., GOVERS, G., WALLING, D.E., ZHANG, X., DERMET, P.J.J., ZHANG, Y. and VANDAELE, K. 1997: Erosion processes and landform evolution in agricultural land – new perspectives from Caesium-137 measurements and topographic based erosion modelling. *Earth Surface Processes and Landforms* **22**, 799–816.

RAPER, J. (ed.) 1989: *Three dimensional applications in geographical information systems*. London: Taylor and Francis.

_____ 1991: Geographical information systems. *Progress in Physical Geography* **15**, 438–44.

RAPP, A. 1960: Recent development of mountain slopes in Karkevagge and surroundings, northern Scandinavia. *Geografiska Annaler* **42**, 73–200.

RAYNER, J.N. and HOBGOOD, J.S. 1991: Dynamic climatology: its history and future. *Physical Geography* **12**, 207–19.

REDISCOVERING GEOGRAPHY COMMITTEE 1997: *Rediscovering geography. New relevance for science and society*. Washington, DC: National Academy Press.

RHIND, D. 1989: Computing, academic geography and the world outside. In B. Macmillan (ed.) *Remodelling geography*. Oxford: Blackwell, 177–90.

RHOADS, B.L. 1994: On being a 'real' geomorphologist. *Earth Surface Processes and Landforms* **19**, 269–72.

RHOADS, B.L. and THORN, C.E. 1994: Contemporary philosophical perspectives on physical geography with emphasis on geomorphology. *Geographical Review* **84**, 90–101.

_____ (eds) 1996: *The scientific nature of geomorphology*. Chichester: John Wiley.

RHOADS, B.L. and HERRICKS, E. 1996: Naturalization of headwater streams in Illinois: challenges and possibilities. In A. Brookes and F.D. Shields (eds) *River channel restoration: guiding principles for sustainable projects*. Chichester: John Wiley, 331–68.

RHOADS, B.L., WILSON, D., URBAN, M. and HERRICKS, E. 1999: Interaction between scientists and nonscientists in community-based watershed management: emergence of the concept of stream naturalization. *Environmental Management* **24**, 297–308.

RICHARDS, K.S. 1982: *Rivers: form and process in alluvial channels*. London: Methuen.

_____ 1990: 'Real' geomorphology. *Earth Surface Processes and Landforms* **15**, 195–7.

_____ 1994: 'Real' geomorphology revisited. *Earth Surface Processes and Landforms* **19**, 277–82.

_____ 1996: Samples and cases: generalisation and explanation in geomorphology. In B.L. Rhoads and C.E. Thorn (eds) *The scientific nature of geomorphology*. Chichester: John Wiley, 171–90.

RICHARDS, K.S. and WRIGLEY, N. 1996: Geography in the United Kingdom 1992–1996. *Geographical Journal* **162**, 41–62.

RICHARDS, K.S., BROOKS, S., CLIFFORD, N., HAMS, T. and LANE, S. 1997: Theory, measurement and testing in 'real' geomorphology and physical geography. In D.R. Stoddart (ed.) *Process and form in geomorphology*. London: Routledge, 265–92.

ROBERTS, C.R. 1989: Flood frequency and urban-induced channel change: some British examples. In K. Bevan and P. Carling (eds) *Floods. Hydrological, sedimentological and geomorphological implications*. Chichester: John Wiley, 57–82.

ROBERTS, N. 1989: *The Holocene*. Oxford: Blackwell.

_____ (ed.) 1994: *The changing global environment*. Oxford: Blackwell.

_____ 1996: Long-term environmental stability and instability in the tropics and subtropics. In T.S. Driver and G.P. Chapman (eds) *Time-scales and environmental change*. London: Routledge, 25–38.

ROBINSON, D.A. and WILLIAMS, R.B.G. (eds) 1994: *Rock weathering and landform evolution*. Chichester: John Wiley.

RODDA, J.C., DOWNING, R.A. and LAW, F.M. 1976: *Systematic hydrology*. London: Butterworth.

RODRIGUEZ ITURBE, I. and RINALDO, A. 1998: *Fractal river basins*. Cambridge: Cambridge University Press.

ROGERS, J.C. 1995: Applied climatology. *Progress in Physical Geography* **19**, 555–60.

ROHLI, R.V. and ROGERS, J.C. 1993: Atmospheric teleconnections and citrus freezes in the southern United States. *Physical Geography* **14**, 1–15.

ROSS, S. 1992: Gardens, earthworks and environmental art. In S. Kemal and I. Gaskell (eds) *Landscape, natural beauty and the arts*. Cambridge: Cambridge University Press, 158–82.

ROSTANKOWSKI, P. 1982: Transformation of nature in the Soviet Union: proposal, plans and reality. *Soviet Geography* **22**, 381–90.

RUDEFORTH, C.C. 1982: Handling soil survey data. In E.M. Bridges and D.A. Davidson (eds) *Principles and applications of soil geography*. London: Longman, 97–131.

RUELLAN, A. 1971: The history of soils: some problems of definition and interpretation. In D.H. Yaalon (ed.) *Palaeopedology*. Jerusalem: International Society of Soil Science and Israel Universities Press, 3–13.

RUMNEY, G.R. 1970: *The geosystem dynamic integration of land, sea and air*. Dubuque, Iowa: Wm. C. Brown Company.

RUNGE, E.C.A. 1973: Soil development sequence and energy models. *Soil Science* **115**, 183–93.

RUSSELL, J.E. 1957: *The world of the soil*. London: Collins.

RUSSELL, R. 1949: Geographical geomorphology. *Annals of the Association of American Geographers* 39, 1–11.

SAARINEN, T.F. 1966: *Perception of the drought hazard on the Great Plains.* Chicago: University of Chicago Press.

SACK, D. 1991: The trouble with antitheses: the case of G.K. Gilbert, geographer and educator. *Professional Geographer* 43, 28–37.

_____ 1992: New wine in old bottles: the historiography of a paradigm change. *Geomorphology* 5, 251–63.

SANT, M. 1992: Comment: Applied geography and a place for passion. *Applied Geography* 12, 295–8.

SAVIGEAR, R.A.G. 1952: Some observations on slope development in South Wales. *Transactions of the Institute of British Geographers* 18, 31–52.

_____ 1965: A technique of morphological mapping. *Annals of the Association of American Geographers* 55, 514–38.

SCHAEFER, F.K. 1953: Exceptionalism in geography: a methodological examination. *Annals of the Association of American Geographers* 43, 226–49.

SCHEIDEGGER, A.E. 1961: *Theoretical geomorphology.* Berlin: Springer-Verlag.

_____ 1970: *Theoretical geomorphology* (2nd edn). Berlin: Springer-Verlag.

_____ 1990: *Theoretical geomorphology* (3rd edn). Berlin: Springer-Verlag.

SCHEIDEGGER, A.E. and LANGBEIN, W.B. 1966: Probability concepts in geomorphology. *US Geological Survey Professional Paper* 500C.

SCHUMM, S.A. 1956: Evolution of drainage systems and slopes in badlands at Perth Amboy, New Jersey. *Bulletin of the Geological Society of America* 67, 597–46.

_____ 1963a: The disparity between present rates of denudation and orogeny. *US Geological Survey Professional Paper* 454H.

_____ 1963b: A tentative classification of river channels. *US Geological Survey Circular* 477.

_____ 1965: Quaternary palaeohydrology. In H.E. Wright and D.G. Frey (eds) *The Quaternary of the United States.* Princeton: Princeton University Press, 783–94.

_____ 1968: Speculations concerning palaeohydrologic controls of terrestrial sedimentation. *Bulletin of the Geological Society of America* 79, 1573–88.

_____ 1969: River metamorphosis. *Proceedings of the American Society of Civil Engineers, Journal of the Hydraulics Division* 95, 255–73.

_____ 1977: *The fluvial system.* New York: John Wiley.

_____ 1979: Geomorphic thresholds: the concept and its applications. *Transactions of the Institute of British Geographers* NS 4, 485–515.

_____ 1985: Explanation and extrapolation in geomorphology: seven reasons for geologic uncertainty. *Transactions of the Japanese Geomorpholgical Union* **6**, 1–18.

_____ 1991: *To interpret the Earth: ten ways to be wrong.* Cambridge: Cambridge University Press.

_____ 1994: Erroneous perceptions of fluvial hazards. *Geomorphology* **10**, 129–38.

SCHUMM, S.A. and LICHTY, R.W. 1965: Time, space and causality in geomorphology. *American Journal of Science* **263**, 110–19.

SCHUMM, S.A., MOSLEY, M.P. and WEAVER, W.E. 1987: *Experimental fluvial geomorphology.* Chichester: John Wiley Interscience.

SEAR, D.J. 1994: River restoration and geomorphology. *Aquatic Conservation: Marine and Freshwater Ecosystems* **4**, 169–77.

SELBY, M.J. 1993: *Hillslope materials and processes* (2nd edn). Oxford: Oxford University Press.

SHACKLETON, N.J. and HALL, M.A. 1983: Stable isotope record of hole 504 sediments: high resolution record of the Pleistocene. *Initial Reports of the Deep Sea Drilling Project 69.* Washington: US Government Printing Office.

SHAPERE, D. 1987: Method in the philosophy of science and epistomology. In N.J. Nersessian (ed.) *The process of science.* Dordrecht: Martinus Nijhoff, 1–39.

SHARMA, S. 1995: *Landscape and memory.* London: Harper Collins.

SHAW, J. 1994: A qualitative view of sub ice-sheet landscape evolution. *Progress in Physical Geography* **18**, 159–84.

SHERMAN, D.J. 1989: Geomorphology: praxis and theory. In M.S. Kenzer (ed.) *Applied geography: issues, questions and concerns.* Dordrecht: Kluwer Academic, 115–34.

_____ 1996: Fashion in geomorphology. In B.L. Rhoads and C.E. Thorn (eds) *The scientific nature of geomorphology.* Chichester: John Wiley, 87–114.

SHERLOCK, R.L. 1922: *Man as a geological agent.* London: Witherby.

_____ 1923: The influence of man as an agent in geographical change. *Geographical Journal* **61**, 258–73.

SHERMAN, D.J. and BAUER, B.O. 1993: Coastal geomorphology through the looking glass. *Geomorphology* **7**, 225–49.

SHREVE, F.S. 1936: Plant life in the Sonoran desert. *Scientific Monthly* **42**, 213.

SIMBERLOFF, D. 1972: Models in biogeography. In T.J.M. Schopf (ed.) *Models in palaeobiology.* San Francisco: Freeman, Cooper & Co., 160–91.

SIMMONS, I.G. 1974: *The ecology of natural resources.* London: Arnold.

_____ 1978: Physical geography in environmental science. *Geography* **63**, 314–23.

_____ 1979a: *Biogeography: natural and cultural.* London: Arnold.

_____ 1979b: Conservation of plants, animals and ecosystems. In K.J.

Gregory and D.E. Walling (eds) *Man and environmental processes.* Folkestone: Dawson, 241–58.

———— 1980: Biogeography. In E.H. Brown (ed.) *Geography yesterday and tomorrow.* Oxford: Oxford University Press, 146–66.

———— 1987: Lilies and peacocks: a view of biogeography. In M.J. Clark, K.J. Gregory and A.M. Gurnell (eds) *Horizons in physical geography.* Basingstoke: Macmillan, 165–80.

———— 1989: *Changing the face of the Earth. Culture, environment, history.* Oxford: Blackwell.

———— 1990a: No rush to grow green. *Area* **22**, 384–7.

———— 1990b: Ingredients of a green geography. *Geography* **75**, 98–105.

———— 1991: *Earth, air and water. Resources and environment in the late 20th century.* London: Arnold.

———— 1993a: *Interpreting nature: cultural constructions of the environment.* London: Routledge.

———— 1993b: *Environmental history: a concise introduction.* Oxford: Blackwell.

———— 1997: *Humanity and environment: a cultural ecology.* Harlow: Addison Wesley Longman.

SIMMONS, I.G. and TOOLEY, M.J. (eds) 1981: *The environment in British prehistory.* London: Duckworth.

SIMONS, M. 1962: The morphological analysis of landforms: a new review of the work of Walther Penck. *Transactions of the Institute of British Geographers* **31**, 1–14.

SIMONSON, R.W. 1959: Outline of the generalized theory of soil genesis. *Proceedings of the Soil Science Society of America* **23**, 152–61.

———— 1986: Historical aspects of soil survey and soil classification. Part 1. 1899–1910. *Soil Survey Horizons* **27**, 3–11.

SIMPSON, R.H. and RIEHL, H. 1981: *The hurricane and its impact.* Oxford: Blackwell.

SISSONS, J.B. 1958: Supposed ice-dammed lakes in Britain with particular reference to the Eddleston valley, southern Scotland. *Geografiska Annaler* **40**, 159–87.

———— 1960: Some aspects of glacial drainage channels in Britain. Part 1. *Scottish Geographical Magazine* **79**, 131–46.

———— 1961: Some aspects of glacial drainage channels in Britain Part II. *Scottish Geographical Magazine* **77**, 15–36.

———— 1967: *The evolution of Scotland's scenery.* Edinburgh: Oliver & Boyd.

———— 1976: *Scotland.* London: Methuen.

———— 1977: Former ice dammed lakes in Glen Moriston, Ivernesshire and their significance in upland Britain. *Transactions of the Institute of British Geographers* NS **2**, 224–42.

SLAYMAKER, H.O. (ed.) 1991: *Field experiments and measurement programs in geomorphology.* Rotterdam: Balkema.

_____ (ed.) 1994: *Geomorphic hazards*. Chichester: John Wiley.

_____ 1997: A pluralist problem-focused geomorphology. In D.R. Stoddart (ed.) *Process and form in geomorphology*. London: Routledge, 328–39.

SLAYMAKER, H.O., DUNNE, T. and RAPP, A. 1980: Geomorphic experiments on hillslopes. *Zeitschrift für Geomorphologie* Supplementband 35, v–vii.

SLAYMAKER, H.O. and SPENCER, T. 1998: *Physical geography and global environmental change*. Harlow: Longman.

SMALLEY, I.J. and VITA-FINZI, C. 1969: The concept of 'system' in the earth sciences. *Bulletin of the Geological Society of America* 80, 1591.

SMART, J.S. 1978: The analysis of drainage network composition. *Earth Surface Processes* 3, 129–70.

SMART, P.L. and FRANCIS, P.D. (eds) 1991: *Quaternary dating methods – a user's guide*. Quaternary Research Association Technical Guide No. 4.

SMIL, V. 1979: Controlling the Yellow River. *Geographical Review* 69, 251–72.

_____ 1991: *General energetics: energy in the biosphere and civilisation*. Chichester: John Wiley.

SMITH, D.E., MORRISON, J., JONES, R.L. and CULLINGFORD, R.A. 1980: Dating the main postglacial shoreline in the Montrose area, Scotland. In R.A. Cullingford, D.A. Davidson and J. Lewin (eds) *Timescales in geomorphology*. Chichester: John Wiley, 225–45.

SMITH, D.I. and NEWSON, M.D. 1974: The dynamics of solutional and mechanical erosion in limestone catchments on the Mendip Hills, Somerset. In K.J. Gregory and D.E. Walling (eds) *Fluvial processes in instrumented watersheds*. Institute of British Geographers Special Publication 6, 155–68.

SMITH, K. 1975: *Principles of applied climatology*. London: McGraw Hill.

_____ 1992: *Environmental hazards: assessing risk and reducing disaster*. London: Routledge.

SMITH, K. and WARD, R.C. 1998: *Floods – physical processes and human impacts*. Chichester: John Wiley.

SMITH, T.R. 1984: Artificial intelligence and its applicability to geographical problem solving. *Professional Geographer* 36, 147–58.

SMYTH, C. and DEARDEN, P. 1998: Attitudes of environmental management personnel involved in surface coal mine reclamation in Alberta and British Columbia, Canada. *Applied Geography* 18, 275–95.

SNYTKO, V.A., SEMENOV, YU. M. and DAVYDOVA, N.D. 1981: A landscape-geochemical evaluation of geosystems for purposes of rational nature management. *Soviet Geography* 22, 569–78.

SPEDDING, N. 1997: On growth and form in geomorphology. *Earth Surface Processes and Landforms* 22, 261–5.

SPENCER, T. 1995: Potentialities, uncertainties and complexities in the response of coral reefs to future sea level rise. *Earth Surface Processes and Landforms* **20**, 49–64.

SPENCER, T., FRENCH, J.R. and REED, D.J. 1995: Editorial – Geomorphic response to sea level rise: existing evidence and future impacts. *Earth Surface Processes and Landforms* **20**, 1–6.

SPERBER, I. 1990: *Fashions in science: opinion leaders and collective behaviour in the social sciences.* Minneapolis: University of Minnesota Press.

ST ONGE, D.A. 1981: Presidential address. Theories, paradigms, mapping and geomorphology. *Canadian Geographer* **25**, 307–15.

STARKEL, L. 1976: The role of extreme (catastrophic) meteorological events in contemporary evolution of slope. In E. Derbyshire (ed.) *Geomorphology and climate.* London: John Wiley, 203–46.

_____ (ed.) 1981: *The evolution of the Wisloka valley near Debica during the late Glacial and Holocene.* Krakow: *Folia Quaternaria.*

_____ 1983: The reflection of hydrologic changes in the fluvial environment on the temperate zone during the last 15 000 years. In K.J. Gregory (ed.) *Background to palaeohydrology.* Chichester: John Wiley, 213–36.

_____ 1987: Man as a cause of sedimentological changes in the Holocene. Anthropogenic sedimentological changes in the Holocene. *Striae* **26**, 5–12.

STARKEL, L., GREGORY, K.J. and THORNES, J.B. (eds) 1991: *Temperate palaeohydrology: fluvial processes in the temperate zone during the last 15,000 years.* Chichester: John Wiley.

STATHAM, I. 1977: *Earth surface sediment transport.* Oxford: Clarendon Press.

STEERS, J.A. 1948: *The coastline of England and Wales.* Cambridge: Cambridge University Press.

STEPHENS, N. 1980: *Geomorphology in the service of man.* Inaugural lecture, University College of Swansea.

STEWART, J.B., ENGMAN, E.T., FEDDES, R.A. and KERR, Y. (eds) 1996: *Scaling up in hydrology and remote sensing.* Chichester: John Wiley.

STOCKING, M.A. 1977: Rainfall energy in erosion: some problems and applications. *University of Edinburgh Department of Geography, Research Paper* **13**.

_____ 1980: Soil loss estimation for rural development: a position for geomorphology. *Zeitschrift für Geomorphologie* Supplementband **36**, 264–73.

STODDART, D.R. 1962: Catastrophic storm effects on the British Honduras reefs and cays. *Nature* **196**, 512–15.

_____ 1966: Darwin's impact on geography. *Annals of the Association of American Geographers* **56**, 683–98

_____ 1967: Organism and ecosystem as geographical models. In R.J. Chorley and P. Haggett (eds) *Models in geography.* London: Methuen, 511–48.

_____ 1968: Climatic geomorphology: review and assessment. *Progress in Geography* **1**, 160–222.

_____ 1975: 'That Victorian Science': Huxley's physiography and its impact on geography. *Transactions of the Institute of British Geographers* **66**, 17–40.

_____ 1978: Progress report: biogeography. *Progress in Physical Geography* **2**, 514–28.

_____ (ed.) 1981: *Geography, ideology and social concern.* Oxford: Basil Blackwell.

_____ 1983: Biogeography: Darwin devalued or Darwin revalued? *Progress in Physical Geography* **7**, 256–64.

_____ 1986: *On geography and its history.* Oxford: Blackwell.

_____ 1987: To claim the high ground: geography for the end of the century. *Transactions of the Institute of British Geographers* NS **12**, 327–36.

_____ 1996: Correspondence. *Geographical Journal* **162**, 354–5.

_____ 1997a: Carl Sauer: geomorphologist. In D.R. Stoddart (ed.) *Process and form in geomorphology.* London: Routledge, 340–79.

_____ 1997b: Richard J. Chorley and modern geomorphology. In D.R. Stoddart (ed.) *Process and form in geomorphology.* London: Routledge, 383–99.

STOTT, P. 1981: *Historical plant geography: an introduction.* London: Allen and Unwin.

_____ 1998: Biogeography and ecology in crisis: the urgent need for a new metalanguage. *Journal of Biogeography* **25**, 1–2.

STRAHLER, A.N. 1950a: Davis's concept of slope development viewed in the light of recent quantitative investigations. *Annals of the Association of American Geographers* **40**, 209–13.

_____ 1950b: Equilibrium theory of erosional slopes approached by frequency distribution analysis. *American Journal of Science* **248**, 673–96 and 800–14.

_____ 1952: Dynamic basis of geomorphology. *Bulletin of the Geological Society of America* **63**, 923–37.

_____ 1956: The nature of induced erosion and aggradation. In W.L. Thomas (ed.) *Man's role in changing the face of the Earth.* Chicago: University of Chicago Press, 621–38.

_____ 1964: Quantitative geomorphology of drainage basins and channel networks. In V.T. Chow (ed.) *Handbook of applied hydrology.* New York: McGraw Hill, 4-39–4-76.

_____ 1966: Tidal cycle of changes in an equilibrium beach, Sandy Hook, New Jersey. *Journal of Geology* **74**, 247–68.

_____ 1980: Systems theory in physical geography. *Physical Geography* **1**, 1–27.

_____ 1992: Quantitative/dynamic geomorphology at Columbia 1945–60: a retrospective. *Progress in Physical Geography* **16**, 65–84.

STRAHLER, A.N. and STRAHLER, A.H. 1976: *Elements of physical geography*. London: John Wiley.

STUART, L.C. 1954: Animal geography. In P.E. James and C.F. Jones (eds) *American geography: inventory and prospect*. Syracuse University Press, Association of American Geographers, 442–51.

SUGDEN, D.E. 1978: Glacial erosion by the Laurentide ice sheet. *Journal of Glaciology* 20, 367–91.

_____ 1982: *Arctic and Antarctic: a modern geographical synthesis*. Oxford: Blackwell.

_____ 1996: The East Antarctic Ice Sheet: unstable ice or unstable ideas? *Transactions of the Institute of British Geographers* NS 21, 443–54.

SUGDEN, D.E. and JOHN, B.S. 1976: *Glaciers and landscape. A geomorphological approach*. London: Arnold.

SUGDEN, D.E., SUMMERFIELD, M.A. and BURT, T.P. 1997: Editorial: Linking short-term geomorphic processes to landscape evolution. *Earth Surface Processes and Landforms* 22, 193–4.

SUMMERFIELD, M.A. 1981: Macroscale geomorphology. *Area* 13, 3–8.

_____ 1991: *Global geomorphology*. Harlow: Longman.

SUMNER, G.N. 1978: *Mathematics for physical geographers*. London: Arnold.

SUNDBORG, A. 1956: The river Klaralven, a study of fluvial processes. *Geografiska Annaler* 38, 127–316.

SUSLOV, R.P. 1961: *Physical geography of Asiatic Russia*, translated by N.D. Gershevsky, edited by J.E. Williams. San Francisco: W.H. Freeman.

SVERDRUP, H.U., JOHNSON, M.W. and FEMING, R.H. 1942: *The oceans*. New York: Prentice Hall.

SVOBODA, J. 1999: Homosphere. In D.E. Alexander and R.W. Fairbridge (eds) *Encyclopedia of environmental science*. Dordrecht: Kluwer Academic Publishers, 324–5.

TALLING, P.J. and SOWTER, M.J. 1999: Drainage density on progressively tilted surfaces with different gradients, Wheeler Ridge, California. *Earth Surface Processes and Landforms* 24, 809–24.

TANDARICH, J.P., DARMODY, R.G. and FOLLMER, L.R. 1988: The development of pedological thought: some people involved. *Physical Geography* 9, 162–74.

TANSLEY, A.G. 1935: The use and abuse of vegetational concepts and terms. *Ecology* 16, 284–307.

_____ 1946: *Introduction to plant ecology*. London: Allen and Unwin.

TAPSELL, S.M. 1995: River restoration: what are we restoring to? A case study of the Ravensbourne River in London. *Landscape Research* 20, 98–111.

TATHAM, G. 1951: Geography in the nineteenth century. In G. Taylor (ed.) *Geography in the twentieth century*. New York: Philosophical Library, 28–69.

TERJUNG, W. 1976: Climatology for geographers. *Annals of the Association of American Geographers* **66**, 199–222.

THOM, B.G. 1988: Australian physical geography: an introduction. *Progress in Physical Geography* **12**, 157–8.

THOM, R. 1975: *Structural stability and morphogenesis: an outline of a general theory of model*. Translated by D.H. Fowler. Reading, MA: Benjamin.

THOMAS, D.G. 1993: Sandstorm in a teacup? Understanding desertification. *Geographical Journal* **159**, 318–31.

———— (ed.) 1997: *Arid zone geomorphology – process, form and change in drylands* (2nd edn). Chichester: John Wiley.

THOMAS, M.F. 1978: Denudation in the tropics and the interpretation of the tropical legacy in higher latitudes – a view of the British experience. In C. Embleton, D. Brunsden and D.K.C. Jones (eds) *Geomorphology. Present problems and future prospects*. Oxford: Oxford University Press, 185–202.

———— 1980: Preface to Hagedorn, H. and Thomas, M. Perspectives in Geomorphology. *Zeitschrift für Geomorphologie* Supplementband **36**, v–vi.

———— 1994: *Geomorphology in the tropics: a study of weathering and denudation in low latitudes*. Chichester: John Wiley.

THOMAS, W.L. (ed.) 1956: *Man's role in changing the face of the Earth*. Chicago: University of Chicago Press.

THOMPSON, R.D., MANNION, A.M., MITCHELL, C.W., PARRY, M. and TOWNSHEND, J.R.G. 1986: *Processes in physical geography*. London: Longman.

THOMPSON, R.D. and PERRY, A. (eds) 1997: *Applied climatology: principles and practice*. London: Routledge.

THORN, C.E. 1988: *Introduction to theoretical geomorphology*. Boston: Unwin Hyman.

THORN, C.E. and WELFORD, C.M.R. 1994: Equilibrium concepts in geomorphology. *Annals of the Association of American Geographers* **84**, 666–96.

THORNBURY, W.D. 1954: *Principles of geomorphology*. New York: John Wiley.

THORNE, C.R., BATHURST, J.C. and HEY, R.D. (eds) 1987: *Sediment transport in gravel-bed rivers*. Chichester: John Wiley.

THORNE, C.R., ABT, S.R., BARENDS, F.B.J., MAYNORD, S.T. and PILARCZYK, K.W. (eds) 1995: *River, coastal and shoreline protection: erosion control using riprap and armourstone*. Chichester: John Wiley.

THORNES, J.B. 1979: Processes and interrelationships, rates and changes. In C. Embleton and J.B. Thornes (eds) *Process in geomorphology*. London: Arnold, 378–87.

———— 1983a: Geomorphology, archaeology and recursive ignorance. *Geographical Journal* **149**, 326–33.

_____ 1983b: Evolutionary geomorphology. *Geography* **68**, 225–35.

_____ 1987: The palaeoecology of erosion. In J.M. Wagstaff (ed.) *Landscape and culture – geographical and archaeological perspectives.* Oxford: Blackwell, 37–55.

_____ 1989: Geomorphology and grass roots models. In B. Macmillan (ed.) *Remodelling geography.* Oxford: Blackwell, 3–21.

_____ 1990: The interaction of erosional and vegetational dynamics in land degradation: spatial outcomes. In J.B. Thornes (ed.) *Vegetation and erosion: processes and environments.* Chichester: John Wiley, 41–53.

_____ 1995: Global environmental change and regional response: the European Mediterranean. *Transactions of the Institute of British Geographers* NS **20**, 357–67.

THORNES, J.B. and BRUNSDEN, D. 1977: *Geomorphology and time.* London: Methuen.

THORNES, J.B. and FERGUSON, R.I. 1981: Geomorphology. In N. Wrigley and R.J. Bennett (eds) *Quantitative geography: a British view.* London: Routledge & Kegan Paul, 284–93.

THORNES, J.B. and GREGORY, K.J. 1991: Unfinished business: a continuing agenda. In L. Starkel, K.J. Gregory and J.B. Thornes (eds) *Temperate palaeohydrology: fluvial processes in the temperate zone during the last 15,000 years.* Chichester: John Wiley, 521–36.

THORNES, J.E. 1982: Atmospheric management. *Progress in Physical Geography* **6**, 561–78.

_____ 1992: The impact of weather and climate on transport in the UK. *Progress in Physical Geography* **16**, 187–208.

_____ 1999: *John Constable's skies.* Birmingham: University of Birmingham Press.

THORNTHWAITE, C.W. 1948: An approach towards a rational classification of climate. *Geographical Review* **38**, 55–94.

THRIFT, N. and WALLING, D.E. forthcoming: Geography in the United Kingdom 1996–2000. *Geographical Journal* **166**, in press.

TICKELL, C. 1992: The Presidential Address. *Geographical Journal* **158**, 322–5.

TINKLER, K.J. 1985: *A short history of geomorphology.* London: Croom Helm.

_____ 1989: *History of geomorphology from Hutton to Hack.* London: Unwin Hyman.

TIVY, J. 1971: *Biogeography. A study of plants in the ecosphere.* Edinburgh: Oliver & Boyd.

TOOLEY, M.J. 1978: *Sea level changes: North-West England during the Flandrian Stage.* Oxford: Clarendon Press.

_____ 1994: Sea level response to climate. In N. Roberts (ed.) *The changing global environment.* Oxford: Blackwell, 173–89.

TOOLEY, M.J. and JELGERSMA, S. (eds) 1992: *Impacts of sea level rise on European coastal lowlands.* Institute of British Geographers Special Publication 27. Oxford: Blackwell.

TOOTH, S. and NANSON, G.E. 1995: The geomorphology of Australia's fluvial systems: retrospect, perspect and prospect. *Progress in Physical Geography* **19**, 35–60.

TOWNSHEND, J.R.G. (ed.) 1981a: *Terrain analysis and remote sensing.* London: Allen & Unwin.

——— 1981b: The spatial resolving power of earth resources satellites. *Progress in Physical Geography* **5**, 32–55.

——— 1981c: Prospect: a comment on the future role of remote sensing in integrated terrain analysis. In J.R.G. Townshend (ed.) *Terrain analysis and remote sensing.* London: Allen & Unwin, 219–23.

——— 1987: Remote sensing – global and local views. In M.J. Clark, K.J. Gregory and A.M. Gurnell (eds) *Horizons in physical geography.* Basingstoke: Macmillan, 62–85.

——— 1991: Environmental data bases and GIS. In D.J. Maguire, M.F. Goodchild and D.W. Rhind (eds) *Geographical information systems. Principles and applications.* Harlow: Longman, 201–16.

TOWNSHEND, J.R.G. and HANCOCK, P.J. 1981: The role of remote sensing in mapping surficial deposits. In J.R.G. Townshend (ed.) *Terrain analysis and remote sensing.* London: Allen Unwin, 204–18.

TOY, T.J. 1982: Accelerated erosion: process, problems and prognosis. *Geology* **10**, 524–9.

TRICART, J. 1957: Application du concept de zonalité à la géomorphologie. *Tijdschrift van het Koninklijk Nederlandsch Aardrijikskundig Geomootschap,* 422–34.

TRICART, J. and SHAEFFER, R. 1950: L indice d'émoussé des galets. Moyen d'étude des systèmes d'erosion. *Revue de Géomorphologie Dynamique* **1**, 151–79.

TRICART, J. and CAILLEUX, A. 1965: *Introduction à la géomorphologie climatique.* Paris: Sedes.

——— 1972: *Introduction to climatic geomorphology,* translated by C.J.K. De Jonge. London: Longman.

TRIMBLE, S. 1983: A sediment budget for Coon Creek basin in the driftless area, Wisconsin 1853–1977. *American Journal of Science* **283**, 454–74.

TRIMBLE, S.W. and COOKE, R.U. 1991: Historical sources for geomorphological research in the United States. *Professional Geographer* **43**, 212–28.

TRIMBLE, S.W. and MENDEL, A.C. 1995: The cow as a geomorphic agent – a critical review. *Geomorphology* **13**, 233–53.

TRUDGILL, S.T. 1977: *Soil and vegetation systems.* London: Oxford University Press.

——— 1983: Soil geography: spatial techniques and geomorphic relationships. *Progress in Physical Geography* **7**, 345–60.

——— (ed.) 1986: *Solute processes.* Chichester: John Wiley.

_____ 1987: Soil processes and their significance. In M.J. Clark, K.J. Gregory and A.M. Gurnell (eds) *Horizons in physical geography*. Basingstoke: Macmillan,181–95.

_____ 1990: *Barriers to a better environment*. London: Belhaven Press.

_____ 1991: Environmental issues. *Progress in Physical Geography* **15**, 84–90.

_____ (ed.) 1995: *Solute modelling in catchment systems*. Chichester: John Wiley.

TRUDGILL, S.T. and RICHARDS, K.S. 1997: Environmental science and policy: generalizations and context sensitivity. *Transactions of the Institute of British Geographers* NS **22**, 5–12.

TRUSOV, Y. 1969: The concept of the noosphere. *Soviet Geography* **10**, 220–36.

TUAN, Yi Fu. 1974: *Topophilia. A study of environmental perception, attitudes and values*. New Jersey: Prentice Hall.

_____ 1989: Surface phenomena and aesthetic experience. *Annals of the Association of American Geographers* **79**, 233–41.

_____ 1992: Desert and ice: ambivalent aesthetics. In S. Kemal and I. Gaskell (eds) *Landscape, natural beauty and the arts*. Cambridge: Cambridge University Press, 136–57.

TUNSTALL, S. and PENNING-ROWSELL, E.C. 1998: The English beach: experiences and values. *Geographical Journal* **164**, 319–32.

TURNER, B.L. 1989: The specialist-synthesis approach to the revival of geography: the case of cultural ecology. *Annals of the Association of American Geographers* **79**, 88–100.

TURNER, B.L., CLARK, W.C., KATES, R.W., RICHARDS, J.F., MATHEWS, J.T. and MEYER, W.B. (eds) 1990: *The Earth as transformed by human action*. Cambridge: Cambridge University Press.

TWIDALE, C.R. and LAGEAT, Y. 1994: Climatic geomorphology: a critique. *Progress in Physical Geography* **18**, 319–34.

UDVARDY, M.D.F. 1981: The riddle of dispersal: dispersal theories and how they affect vicariance biogeography. In G. Nelson and D.E. Rosen (eds) *Vicariance biogeography: a critique*. New York: Columbia University Press, 6–29.

UNESCO 1977: *Hydrological maps*. Paris: Unesco, WMO.

United Nations 1999: *Global environmental outlook 2000*. Geneva: United Nations.

UNWIN, D.J. 1977: Statistical methods in physical geography. *Progress in Physical Geography* **1**, 185–221.

_____ 1981: Climatology. In N. Wrigley and R.J. Bennett (eds) *Quantitative geography: a British view*. London: Routledge & Kegan Paul, 261–72.

_____ 1989: Three questions about modelling in physical geography. In B. Macmillan (ed.) *Remodelling geography*. Oxford: Blackwell, 53–7.

VAN GARDINGEN, P.R., FOODY, G.M. and CURRAN, P.J. 1997:

Scaling-up: from cell to landscape. Cambridge: Cambridge University Press.

VAN VALKENBURG, S. 1951: The German school of geography. In G. Taylor (ed.) *Geography in the twentieth century.* New York: Philosophical Library, 91–115.

VERSTAPPEN, H.Th. 1983: *Applied geomorphology: geomorphological surveys for environmental development.* Amsterdam: Elsevier.

VILES, H. (ed.) 1988: *Biogeomorphology.* Chichester: John Wiley.

VILES, H. and SPENCER, T. 1996: *Coastal problems: geomorphology, ecology and society at the coast.* London: Arnold.

VINK, A.P.A. 1968: The role of physical geography in integrated surveys in developing countries. *Tijdschrift voor Economische Sociale Geografie* **294**, 5–68.

_____ 1983: *Landscape ecology and land use.* Translated from Dutch and edited by D.A. Davidson. London: Longman.

VITA-FINZI, C. 1969: *The Mediterranean valleys.* Cambridge: Cambridge University Press.

_____ 1999: Geoarchaeology and ancient environments. In D.E. Alexander and R.W. Fairbridge (eds) *Encyclopedia of environmental science.* Dordrecht: Kluwer Academic Publishers, 274–5.

VITEK, J.D. and GIARDINO, J.R. 1992: Preface: A perpective on getting to the frontier. In J.D. Vitek and J.R. Giardino (eds) *Geomorphology: the research frontier and beyond.* Amsterdam: Elsevier, i–xii.

VON BERTALANFFY, L. 1962: General systems theory – a critical review. *General Systems Yearbook* **1**, 1–20.

_____ 1972: *General systems theory.* New York: Braziller.

VON ENGELN, O.D. 1942: *Geomorphology: systematic and regional.* New York: Macmillan.

VREEKEN, W.J. 1973: Soil variability in small loess watersheds: clay and organic matter content. *Catena* **1**, 181–96.

_____ 1975: Principal kinds of chronosequences and their significance in soil history. *Journal of Soil Science* **26**, 378–94.

WAGSTAFF, J.M. (ed.) 1987: *Landscape and culture – geographical and archaeological perspectives.* Oxford: Blackwell.

WALKER, B. and STEFFEN, W. (eds) 1996: *Global change and terrestrial ecosystems.* Cambridge: Cambridge University Press.

WALKER, D. and GUPPY, J.C. (ed.) 1978: *Biology and quaternary environments.* Canberra: Australian Academy of Science.

WALKER, J. and GRABAU, W. (ed.) 1993: *The evolution of geomorphology. A nation-by-nation summary of development.* Chichester: John Wiley and Sons.

WALLEN, R.N. 1992: *Introduction to physical geography.* Dubuque: Wm. C. Brown.

WALLING, D.E. 1974: Suspended sediment and solute yields from a small catchment prior to urbanisation. In K.J. Gregory and D.E. Walling (eds)

Fluvial processes in instrumented watersheds. Institute of British Geographers, Special Publication 6, 169–91.

———— 1979a: The hydrological impact of building activity: a study near Exeter. In G.E. Hollis (ed.) *Man's impact on the hydrological cycle in the United Kingdom.* Norwich: GeoBooks, 135–52.

———— 1979b: Hydrological processes. In K.J. Gregory and D.E. Walling (eds) *Man and environmental processes.* London: Dawson, 57–81.

———— 1983a: The sediment delivery problem. *Journal of Hydrology* **65**, 209–37.

———— 1983b: Physical hydrology. *Progress in Physical Geography* **7**, 97–112.

———— 1987: Hydrological processes. In K.J. Gregory and D.E. Walling (eds) *Human activity and environmental processes.* Chichester: John Wiley, 53–85.

———— 1996a: Erosion and sediment yield in a changing environment. In J. Branson, A.G. Brown and K.J. Gregory (eds) *Global continental changes: the context of palaeohydrology.* Geological Society Special Publication **115**, 43–56.

———— 1996b: Suspended sediment transport by rivers: a geomorphological and hydrological perspective. *Advances in limnology* **47**, *Suspended particulate matter in rivers and estuaries* (Special Issue, Archiv für Hydrobiologie), 1–27.

WALLING, D.E. and GREGORY, K.J. 1970: The measurement of the effects of building construction on drainage basin dynamics. *Journal of Hydrology* **11**, 129–44.

WALLING, D.E. and WEBB, B.W. 1983: Patterns of sediment yield. In K.J. Gregory (ed.) *Background to palaeohydrology.* Chichester: John Wiley, 69–100.

WALLING, D.E., OWENS, P.N. and LEEKS, G.J.L. 1998: The role of channel and floodplain storage in the suspended sediment budget of the River Ouse, Yorkshire, UK. *Geomorphology* **22**, 225–42.

WARBURTON, J. 1993: Energetics of alpine proglacial geomorphic processes. *Transactions of the Institute of British Geographers* NS **18**, 197–206.

WARD, R.C. 1967: *Principles of hydrology.* London: McGraw Hill.

———— 1971: *Small watershed experiments. An appraisal of concepts and research developments.* University of Hull Occasional Papers in Geography No. 18, 254.

———— 1978: *Floods. A geographical perspective.* London: Macmillan.

———— 1979: The changing scope of geographical hydrology in Great Britain. *Progress in Physical Geography* **3**, 392–412.

WARD, R.G. 1960: Captain Alexander Maconochie RNKH 1787–1860. *Geographical Journal* **126**, 459–68.

WARRICK, R. and FARMER, G. 1990: The greenhouse effect, climatic change and rising sea level: implications for development. *Transactions of the Institute of British Geographers* NS **15**, 5–20.

WASHBURN, A.L. 1973: *Periglacial processes and environments*. London: Arnold.

WATERS, R.S. 1958: Morphological mapping. *Geography* **43**, 10–17.

WATKINS, J.W.N. 1970: Against 'normal science'. In I. Lakatos and A. Musgrave (eds) *Criticism and the growth of knowledge*. London: Cambridge University Press.

WATTS, D.R. 1978: The new biogeography and its niche in physical geography. *Geography* **63**, 324–37.

WAYLEN, P. 1995: Global hydrology in relation to palaeohydrological change. In K.J. Gregory, L. Starkel and V.R. Baker (eds) *Global continental palaeohydrology*. Chichester: John Wiley, 61–86.

WEAVER, W. 1958: A quarter century in the natural sciences. *Annual Report, The Rockefeller Foundation*, New York: 7–122.

WEBSTER, R., LESSELLS, C.M. and HODGSON, J.M. 1976: 'DECODE' – computer program for translating soil profile descriptions into text. *Journal of Soil Science* **27**, 218–26.

WEHMILLER, J.F. 1982: A review of amino acid racemization studies in Quaternary mollusks: stratigraphic and chronologic applications in coastal and interglacial sites, Pacific and Atlantic coasts, United States, United Kingdom, Baffin Island, and tropical islands. *Quaternary Science Reviews* **1**, 83–120.

WEICHEL, B.J. and ARCHIBOLD, O.W. 1989: An evaluation of habitat potential for wild rice (*Zizania palustris* L.) in northern Saskatchewan. *Applied Geography* **9**, 161–75.

WERRITTY, A. 1972: The topology of stream networks. In R.J. Chorley (ed.) *Spatial analysis in geomorphology*. London: Methuen, 167–96.

_____ 1997: Chance and necessity in geomorphology. In D.R. Stoddart (ed.) *Process and form in geomorphology*. London: Routledge, 312–27.

WHALLEY, W.B. (ed.) 1978: *Scanning electron microscopy in the study of sediments*. Norwich: Geo Abstracts.

WHITE, G.F. 1973: Natural hazards research. In R.J. Chorley (ed.) *Directions in geography*. London: Methuen, 193–216.

_____ (ed.) 1974a: *Natural hazards: local, national, global*. New York: Oxford University Press.

_____ 1974b: Natural hazards research: concepts, methods, and policy implications. In G.F. White (ed.) *Natural hazards: local, national, global*. New York: Oxford University Press, 3–16.

WHITE, G.F., CALEF, W.C., HUDSON, J.W., MAYER, H.M., SHEAFFER, J.R. and YOLK, D.J. 1958: *Changes in urban occupance of floodplains in the United States*. University of Chicago, Department of Geography Research Paper No. 57.

WHITE, I.D., MOTTERSHEAD, D.N. and HARRISON, S.J. 1984: *Environmental systems: an introductory text*. London: George Allen & Unwin.

WHITE, K. 1998: Remote sensing. *Progress in Physical Geography* **22**, 95–102.

WHITMORE, T.C., FLENLEY, J.R. and HARRIS, D.R. 1982: The tropics as the norm in biogeography. *Geographical Journal* **148**, 8–21.

WHITTAKER, R.H. 1953: A consideration of climax theory – the climax as a population and pattern. *Ecological Monographs* **23**, 41–78.

WHITTOW, J.B. 1980: *Disasters: the anatomy of environmental hazards.* London: Allen Lane.

WILBY, R.A. (ed.) 1997: *Contemporary hydrology.* Chichester: John Wiley.

WILBY, R.A. and WIGLEY, T.M.L. 1997: Downscaling general circulation model output: a review of methods and limitations. *Progress in Physical Geography* **21**, 530–48.

WILKINSON, H. 1963: Man and the natural environment. Department of Geography, University of Hull Occasional Papers in Geography No. 1.

WILLIAMS, G.P. 1978: Hydraulic geometry of river cross sections – theory of minimum variance. *US Geological Survey Professional Paper* **1029**.

—— 1983: Improper use of regression equations in the earth sciences. *Geology* **11**, 195–7.

WILLIAMS, M. 1990: Review of H.H. Birks, H.J.B. Birks, P.E. Kalard and D. Moe (eds) 1989 *The cultural landscape – past, present and future. Progress in Physical Geography* **14**, 270–2.

WILLIAMS, M.A.J., DUNKERLEY, D.L., DE DECKER, P., KERSHAW, A.P. and STOKES, T.J. 1993: *Quaternary environments.* London: Arnold.

WILLIAMS, M.A.J. and BALLING, R.C. 1996: *Interactions of desertification and climate.* London: Arnold.

WILLIAMS, P.J. 1982: *The surface of the earth: an introduction to geotechnical science.* London: Longman.

WILLIAMS, P.W. 1991: Tectonic geomorphology, uplift rates and geomorphic response in New Zealand. *Catena* **18**, 439–52.

WILLMOTT, C.J. and GAILE, G.L. 1992: Modeling. In R.F. Abler, M.G. Marcus and J.M. Olson (eds) *Geography's inner worlds.* New Brunswick, NJ: Rutgers University Press, 163–86.

WILSON, A.G. 1981: *Geography and the environment: systems analytical methods.* Chichester: John Wiley.

WILSON, A.G. and KIRKBY, M.J. 1974: *Mathematics for geographers and planners.* Oxford: Clarendon Press.

WINDLEY, B.F. 1977: *The evolving continents.* London: John Wiley.

WINKLEY, B.R. 1972: River regulation with the aid of Nature. *International Commission Irrigation and Drainage, Eighth Congress,* 433–57.

WISCHMEIER, W.H. 1976: Use and misuse of the Universal Soil Loss Equation. *Journal Soil and Water Conservation* **31**, 5–9.

WISE, M.J. 1983: Three founder members of the IBG: R. Ogilvie Buchanan,

Sir Dudley Stamp, S.W. Wooldridge. A personal tribute. *Transactions of the Institute of British Geographers* NS 8, 41–54.

WOLMAN, M.G. 1967a: A cycle of sedimentation and erosion in urban river channels. *Geografiska Annaler* 49A, 385–95.

_____ 1967b: Two problems involving river channel changes and background observations. In *Quantitative geography Part II: Physical and cartographic topics*. Northwestern Studies in Geography 14, 67–107.

WOLMAN, M.G. and MILLER, J.P. 1960: Magnitude and frequency of forces in geomorphic processes. *Journal of Geology* 68, 54–74.

WOLMAN, M.G. and GERSON, R.A. 1978: Relative scales of time and effectiveness of climate in watershed geomorphology. *Earth Surface Processes* 3, 189–208.

WOOD, A. 1942: The development of hillside slopes. *Proceedings of the Geologists Association* 53, 128–40.

WOOLDRIDGE, S.W. 1932: The cycle of erosion and the representation of relief. *Scottish Geographical Magazine* 48, 30–36.

_____ 1949: On taking the ge- out of geography. *Geography* 34, 9–18.

_____ 1958: The trend of geomorphology. *Transactions of the Institute of British Geographers* 25, 29–36.

WOOLDRIDGE, S.W. and MORGAN, R.S. 1937: *The physical basis of geography*. London: Longman.

WOOLDRIDGE, S.W. and LINTON, D.L. 1939: *Structure, surface and drainage in South East England*. London: Philip.

WOOLDRIDGE, S.W. and EAST, W.G. 1951: *The spirit and purpose of geography*. London: Hutchinson.

WRIGHT, R.L. 1972: Some perspectives on environmental research for agricultural land-use planning in developing countries. *Geoforum* 10, 15–33.

WRIGHT, W.B. 1937: *The Quaternary ice age*. London: Macmillan.

WRIGLEY, N. (ed.) 1979: *Statistical applications in the spatial sciences*. London: Pion.

WRIGLEY, N. and BENNETT, R.J. (ed.) 1981: *Quantitative geography: a British view*. London: Routledge & Kegan Paul.

WRIGLEY, N. and MATHEWS, S. 1986: Citation classics and citation levels in geography. *Area* 18, 185–94.

XU, CHONG-YU. 1999: From GCMs to river flow: a review of downscaling methods and hydrologic modelling approaches. *Progress in Physical Geography* 23, 229–49.

YAALON, D.M. 1975: Conceptual models in pedogenesis: can soil-functions be solved? *Geoderma* 14, 189–205.

YATSU, E. 1966: *Rock control in geomorphology*. Tokyo: Sozosha.

_____ 1988: *The nature of weathering: an introduction*. Tokyo: Sozosha.

_____ 1992: To make geomorphology more scientific. *Transactions of the Japanese Geomorphological Union* 13, 87–124.

YE GRISHANKOV, G. 1973: The landscape levels of continents and geographic zonality. *Soviet Geography* 14, 61–77.

YOUNG, A. 1960: Soil movement by denudational processes on slopes. *Nature* **188**, 120–22.

_____ 1963: Deductive models of slope evolution. *Nachrichten der Akademie der Wissenschaften in Göttingen, II: Mathematisch-Physikalische Klasse* 5, 45–66.

_____ 1972: *Slopes*. Edinburgh: Oliver & Boyd.

_____ 1974: The rate of slope retreat. In E.H. Brown and R.S. Waters (eds) *Progress in geomorphology*, Institute of British Geographers Special Publication No. 7, 65–78.

_____ 1998: *Land resources: now and for the future*. Cambridge: Cambridge University Press.

YOUNG, A. and GOLDSMITH, P.F. 1977: Soil survey and land evaluation in developing countries. A case study in Malawi. *Geographical Journal* **143**, 407–38.

ZAKRZEWSKA, B. 1967: Trends and methods in landform geography. *Annals of the Association of American Geographers* 57, 128–65.

ZEUNER, F.E. 1945: *The Pleistocene Period: its climate, chronology and faunal successions*. London: Hutchinson.

_____ 1958: *Dating the past: An introduction to geochronology*. London: Methuen.

Index of Authors

General Index

advisory unit 220
air masses 35
alluvial chronology 182
animal geography 43
anthropogeomorphology 174
applied climatology 217
applied geomorphology 45, 64, 216
applied physical geography 198–224
archaeology 147, 179, 182, 259
arid areas 162
artificial intelligence 76
atmosphere 35, 61, 284
atmospheric physiography 33, 248, 287
Australian geomorphology 77
autoecology 38

Baconian route to scientific explanation 57–8
bioenergetics 95
biogeochemical fluxes 100, 253
biogeography 6, 8–11, 28, 33, 36–8, 42–4, 66–7, 83–4, 91, 94–5, 117, 125–6, 165–7, 171, 173, 179–80, 193, 201, 203, 208, 218, 230, 260, 279, 282–6
biogeomorphology 166
biogeochemical cycles and cycling 103, 117, 202, 232, 258, 276
biotechnology 269
boundary layer climates 62, 181
branches of physical geography 5, 32–8, 42, 61–8, 83, 115–24, 171, 173, 199, 201, 207–19, 257, 279, 282–3, 285

canons of landscape evolution 142
cascading systems 87
catastrophe theory 132, 163
catastrophic explanations 132
catastrophists 260
catechism 20, 289
catena 46, 85, 116
chairs of geography 26
channelization 215, 264
chaos theory 15, 57, 281
chronofunctions 164
civic science 250
chronozones 151
cladistics 165
climate 18, 33, 63, 166
climatic change 19, 62, 168, 217, 245, 249–50
climatic classifications 35, 42, 229
climatic geomorphology 45, 127–30, 142–3
climatogenetic geomorphology 129, 143
climatology 6, 8–10, 15, 28, 35–6, 42, 45, 56, 60–63, 83, 85, 91–4, 117–18, 124–5, 171, 173, 181–2, 203, 208, 217, 256, 258, 279, 282, 285
climatomorphogenetic zones 129
climax vegetation 37
coastal geomorphology 44, 120, 126
Columbia school 45, 64, 119, 123
complexity 68–9, 141
complex response 161
composite sciences 30, 49
conceptual approach 13, 152–67